Salters Horners Advanced Physics
for Edexcel AS Physics

STUDENT BOOK

A PEARSON COMPANY

Pearson Education
Edinburgh Gate
Harlow
Essex
CM20 2JE
United Kingdom

and Associated Companies throughout the world

www.pearson.com

First published 2000
This edition published 2008
Second impression 2008

ISBN 978-1-40589-6023

Designed and illustrated by Pantek Arts, Maidstone, Kent
Indexer John Holmes
Printed and bound by Graficas Estella, Bilboa, Spain

The publisher's policy is to use paper manufactured from sustainable forests.

Please cite as: Salters Horners Advanced Physics Project, AS Student Book, Edexcel Pearson, London, 2008.

Picture Credits

The publisher would like to thank the following for their kind permission to reproduce their photographs:

(Key: b-bottom; c-centre; l-left; r-right; t-top)

2 Reuters: Aly Song (b); Benoit Tessier (t). 3 Reuters: Sukree Sukplang. 4 Colorsport. 13 No Trace: (t) (b). 17 John Cleare Mountain Photography: (t) (c) (b). 23 John Cleare Mountain Photography. 29 Colorsport: (l). Roger Scruton: (r). 30 Colorsport: (tl) (tc) (tr). No Trace: G. Lewis (b). 35 Colorsport. 40 No Trace: (l) (r). 41 Corbis: Albrecht G. Schaefer. 44 Colorsport. 48 No Trace. 66 No Trace. 67 Alamy Images: David O'Shea. 70 No Trace: (t) (c) (b). 71 Chris Butlin: (b). CSC Scientific Company, Fairfax, Virginia: (t). 76 No Trace: (t) (b). 77 No Trace. 79 Science Photo Library Ltd: ADAM HART-DAVIS. 82 Science Photo Library Ltd: FOOD & DRUG ADMINISTRATION. 84 No Trace: (t). Science Photo Library Ltd: VAUGHAN FLEMING (br); KAJ R. SVENSSON (bl). 85 No Trace. 86 Stable Micro Systems: (l) (r). 87 No Trace: (all). 88 No Trace: (t). Science & Society Picture Library: (b). 89 Instron Corporation: (t). No Trace: (b). 92 NASA. 93 No Trace. 94 Food Features. 102 Science Photo Library Ltd. 103 Alamy Images: Mira (l). Getty Images: Jonathan Kirn (r). 108 Colorific: Alfred Wolf (t) (b). 114 Otto Bock HealthCare GmbH: (t). tbc: (b). 115 Smith & Nephew plc: (l) (r). 116 No Trace. 117 Perplas Medical Ltd. 118 Chris Butlin: (l) (r). 130 Alamy Images: JUPITERIMAGES/ Brand X. 131 Science Photo Library Ltd: Tim Malyon. 136 Science Photo Library Ltd: EADWEARD MUYBRIDGE COLLECTION/ KINGSTON MUSEUM. 145 Picture Viewer: (l) (c) (r). 147 Picture Viewer. 149 David J. Rudio. 151 Picture Viewer. 159 Jupiter Unlimited. 160 Science Photo Library Ltd: DR JEREMY BURGESS (l) (r). 165 University of York Science Education Group. 171 Science Photo Library Ltd: JEROME WEXLER (t) (b). 173 Chris Butlin. 175 Science Photo Library Ltd: DEPT. OF PHYSICS, IMPERIAL COLLEGE. 192 NASA. 193 NASA: Marshall Space Flight Center. 194 Rutherford Appleton Laboratory: (t) (b). 195 Rutherford Appleton Laboratory: (t). Surrey Satellite Technology Ltd, Centre for Satellite Engineering Research, Univ. of Surrey: (b). 196 NASA. 199 Chris Butlin. 202 Science Photo Library Ltd: NASA. 214 NASA: Johnson Space Center (b). Science Photo Library Ltd: NASA (t). 215 NASA: (t); Marshall Space Flight Center (b). 216 NASA: (b). Science Photo Library Ltd: Lockheed Martin Corporation/Nasa (t). 222 Science Photo Library Ltd. 225 Daimler-Benz Stuttgart. 227 Ciel et Espace magazine June 1996 , p23 AFA Paris. 236 Science & Society Picture Library. 237 Science Photo Library Ltd: NASA. 238 NASA. 242 NASA. 243 Science Photo Library Ltd. 244 Science Photo Library Ltd: NASA/JPL/University of Colorado. 247 Science Photo Library Ltd: NASA. 248 Getty Images: Time & Life Pictures (b). NASA. 260 Ancient Art & Architecture: (l) (r). 261 Ancient Art & Architecture. 262 Archaeological Prospection Services of Southampton. 263 No Trace. 273 Alamy Images: Andrew Holt (b). PA Photos: Associated Press (t). Science Photo Library Ltd: EMMANUEL LAURENT / EURELIOS (c). 276 No Trace: (t). Prof. Ian Isherwood, University of Manchester: (br). Science Photo Library Ltd: ALEXANDER TSIARAS (bl). Roger Scruton: (bc). 277 English Heritage Photo Library: (t). York Archaeological Trust: (b). 278 York Archaeological Trust. 279 York Archaeological Trust: (t) (b). 280 akg-images Ltd. 283 tbc. 285 Cambridge Science Media: (tr) (tc) (tl). Dr. Runying Chen, Associate Professor, East Carolina University: (r/A-D). 286 Science Photo Library Ltd: SUSUMU NISHINAGA (r); STEVE GSCHMEISSNER (l) (c). 288 Science Photo Library Ltd: Andrew Lambert Photography (l); DR DAVID WEXLER, COLOURED BY DR JEREMY BURGESS (r). 289 Andrew Wilson, University of Bradford. Science Photo Library Ltd: Astrid & Hanns-Frieder Michler (l); Maximilian Stock Ltd (r). 292 Werner Forman Archive Ltd: (t) (b). 293 Alamy Images: POPPERFOTO (c). Natural History Museum Picture Library: (r). Science Photo Library Ltd: John Reader (l). 294 reproduced with kind permission of IOP Publishing Ltd: (c) 2008

All other images © Pearson Education

Picture Research by: Kay Altwegg

Every effort has been made to trace the copyright holders and we apologise in advance for any unintentional omissions. We would be pleased to insert the appropriate acknowledgement in any subsequent edition of this publication.

Contributors

Many people from schools, colleges, universities, industries and the professions have contributed to the Salters Horners Advanced Physics (SHAP) project and the preparation of SHAP course materials.

Authors of this AS edition

Steven Chapman (Institute of Education, University of London)
Frances Green (Watford Girls' Grammar School)
Greg Hughes (de Ferrers College, Staffs)
Paul Lee (Franklin College, Grimsby)
Chris Pambou (City & Islington College, London)
Sandy Stephens (formerly of The Lady Eleanor Holles School, Middlesex)
Wendy Swarbrick (Lancing College, Sussex)
Elizabeth Swinbank (University of York)
David Swinscoe (City & Islington College, London)
Carol Tear (York)
Clare Thomson (Institute of Physics)

Project director and general editor

Elizabeth Swinbank (University of York)

Acknowledgements

We would like to thank the following for their advice and contributions to this edition.

Cameron Clapp   Robin Millar (University of York)   Sandra Wilmott (University of York)
Graham Meredith (Gloucester)   Kenny Webster (Thinktank, Birmingham)

Authors of previous edition

This edition is based on the original SHAP course materials and incorporates the work of the following authors:

Jonathan Allday   Alasdair Kennedy   David Sang
Chris Butlin   Bob Kibble   Tony Sherborne
Steve Cobb   Maureen Maybank   Richard Skelding
Tony Connell   Averil Macdonald   Elizabeth Swinbank
Howard Darwin   David Neal   Carol Tear
Nick Fisher   Kerry Parker   Nigel Wallis

Sponsors

We are grateful to the following for sponsorship that has continued to support the Salters Horners Advanced Physics project after its initial development and has enabled the production of this edition.

The Worshipful Company of Horners
The Worshipful Company of Salters
Corus UK Ltd

Advisory Committee for the initial development

Prof. Frank Close   Prof. Robin Millar
Prof. Cyril Hilsum FRS   Prof. Sir Derek Roberts FRS

Contents

How to use this book

Context-led study

Welcome to the AS part of the Salters Horners Advanced Physics course.

Each teaching unit in the course starts by looking at particular situations in which physics is used or studied, and then develops the physics you need to learn to explore this 'context'.

We have tried to select contexts to give you some idea of how physics can help improve people's lives, how physics is used in engineering and technology, and how physics research extends our understanding of the physical world at a fundamental level. These will show you just some of the many physics-related careers and further study that might be open to you in the future.

Within each chapter, you will develop your knowledge and understanding on one or more areas of physics. In later chapters, you will meet many of these ideas again – in a completely different context – and develop them further. In this way, you will gradually build up your knowledge and understanding of physics and learn to apply key principles of physics to a variety of contexts.

About this book

Each chapter includes the following features:

Main text

This presents the context of each teaching unit and explains the relevant physics as you need it.

Within the main text, some words are printed in **bold**. These are key terms relating to the physics. We suggest that you make your own summary of the these terms (and others if you wish) as you go along. Then you can refer back to it when you revisit a similar area of physics later in the course and when you revise for exams.

Activities

The text refers to many Activities. These include practical work, the use of information technology (e.g. CD-ROMs and the Internet), reading, writing, data handling and discussion. Some activities are best carried out with one or more other students, others are intended for you to do on your own. For some activities, there are handout sheets giving further information, details about apparatus and so on.

Activity 4 Non-uniform motion

Use ticker-tape, a stop-frame video or a camcorder to record your own motion when sprinting from a crouched start and plot a graph of your displacement against time.

Calculate your velocity in each small interval between dots or between frames, and hence also calculate your acceleration in each small time interval.

Questions

There are two types of Questions in this book. Some are to do as you go along, as a self-check; the answers to these questions appear at the back of each chapter.

Questions on the whole chapter are intended as a summary to what you have learned in each chapter: your teacher will provide the answers to these questions.

Once you have had a go at a question, check your answer.

If you have gone wrong, use the answer (and the relevant part of the book chapter) to help you sort out your ideas. Working in this way is not cheating! Rather, it helps you to learn.

Maths notes

Maths references in the main text will direct you to the Maths notes, which are to help you with the maths needed in physics. This may involve calculations, rearranging equations, plotting graphs, and so on. You will probably have covered most of what's needed at GCSE, but you may not be used to using it in physics.

The Maths notes at the end of the book summarise the key maths ideas that you need in the AS course, and show how to apply them to situations in physics.

> **Maths reference**
>
> Manipulating powers on a calculator
> See Maths note 1.4

Study notes

These notes in the margin are intended to help you to get to grips with the physics – for example, they indicate links with other parts of the course.

> **Study note**
>
> When π appears in a description of phase the units of radians are taken for granted and are sometimes omitted.

Further investigations

If you continue into the second year of this course, you will be asked to plan and carry out an extended experiment for your coursework and you might be asked to suggest your own topic for this. Alternatively, you might have the opportunity to carry out a major piece of experimental investigative work for a separate Extended Project qualification.

As you proceed through the course, keep a note of any areas you might like to pursue; we have included some suggestions under the heading Further Investigations, but any unanswered question that intrigues you could form the basis of an investigation.

> **Further investigations**
>
> Explore the behaviour of various materials that appear 'bone-like', such as cuttle-fish bone (available from pet-shops), seaside rock and 'oasis' (used in flower-arranging). Try to measure the Young modulus and ultimate compressive strength. Observe the way they fracture. Comment on their resemblance (or otherwise) to real bone.

Achievements

At the end of each chapter you will find a list of Achievements. This is a summary of the key points that you have covered in that chapter, and shows what you can expect to be tested on in the exams. (It is copied from the Exam Specification.) Look through the Achievements when you check back over your work after finishing a chapter. If there is anything that looks unfamiliar, or that you think you have not properly understood, consult your teacher and the explanations in this book.

CHAPTER 1

Higher, Faster, Stronger

Why a chapter called Higher, Faster, Stronger?

'Unthinkable' they said. 'Surely no-one will ever run a mile in under four minutes.' Not only were the commentators proved wrong, but in the fifty years since then, 15 seconds have been lopped off the world record mile. More impressive still, a full six minutes have been hacked off the women's 5000 metre record. Off the track, the story is the same. Before the authorities stepped in and changed the design of the javelin, javelin throws were beginning to endanger the crowd at the other end of the stadium.

What's going on? Are we becoming a more powerful species? The place to search for an explanation of this record-breaking frenzy is in the laboratory. With physiology, psychology and physics, science has revolutionised sport, giving us a better understanding of bodily and mental processes with which to train winners (Figure 1.1a, b, c).

Figure 1.1a Highest

Figure 1.1b Fastest

The new scientific discipline of sports science has emerged to help build tomorrow's champions. Using tools such as computer-linked video, researchers can now analyse the movements involved in sporting activity in minute detail. This allows trainers to correct even tiny errors in an athlete's performance. Advances in the science of materials have brought equally dramatic changes, with almost every conceivable property of an athlete's clothing and footwear optimised for performance. And with so much resting on winning, measurement technology now enables races to be decided on differences of just a few milliseconds.

Higher, Faster, Stronger is (loosely) the Olympic motto. In this chapter, you will see how many basic physics concepts can be applied to sports, ranging from sprinting to bungee jumping, to help athletes go 'higher, faster, stronger'.

Figure 1.1c Strongest

Overview of physics principles and techniques

In this chapter, you will study the physics of motion, force and energy. You may be familiar with some of the ideas from GCSE. In this chapter you will deepen your understanding of these concepts by applying them to solve real problems. Many of the concepts, like force and energy, are fundamental to the whole of physics, and you will meet them in almost every part of the course.

You will have many opportunities for practical work, investigations and computer data-logging. You will learn how to extract additional information from a set of measurements, and you will also use computer software to analyse various sporting activities, just as sports science researchers do.

In later chapters you will do further work on:

- vectors in *Probing the Heart of Matter*
- graphs in *Digging Up the Past*, *The Medium is the Message* and *Probing the Heart of Matter*
- kinematics and dynamics in *Good Enough to Eat* and *Transport on Track*
- kinetic energy and work in *Transport on Track*, *Probing the Heart of Matter* and *Reach for the Stars*
- properties of materials in *Good Enough to Eat*, *Spare Part Surgery* and *Build or Bust?*

1 Running

1.1 Biomechanics

By studying physics, you are following in the footsteps of one of the world's best athletes. Durham physics graduate Jonathan Edwards (Figure 1.2) became world record holder in the triple jump in the summer of 1995 at the World Athletics Championships in Gothenburg, Sweden. He leapt his way to into the record books by jumping in excess of 18 m. In the space of a couple of months he became an MBE and BBC Sports Personality for 1995. He won the silver medal in the Olympics in Atlanta the following year.

To be a good jumper, Jonathan Edwards had to train to be a very fast sprinter. He also had to develop very careful timing in his jumping techniques. The science of biomechanics is devoted to trying to help sports people get the best possible results by helping them to improve their technique. Biomechanics concerns itself with the forces on a human body and sports equipment, and the effects of these forces. Much work has been conducted at Loughborough University (where former world record holder Sebastian Coe studied and trained) to help training methods and improve performance.

Figure 1.2 Jonathan Edwards jumping the triple jump

External forces (such as those exerted by a cyclist on a pedal) are measured using transducers to convert the force into an electrical signal. Internal forces exerted by the muscles and applied to the bones and joints in the human body are harder to measure directly (transducers would have to be implanted), but they can be calculated indirectly by a technique known as inverse dynamics. If the mass of each part of the human body is known and the accelerated motion measured, then forces can be calculated. Position–time data from video can be used to calculate accelerations of each segment of the body. We will look at some examples of inverse dynamics later on in this part of the chapter. First we will see how motion can be recorded and analysed.

1.2 Describing motion

The performance of an athlete in a running event is usually recorded simply as the time taken to complete the distance. Table 1.1 lists some times for the men's and women's 10 000 metre world record. It has been suggested that women will soon be running faster than men.

Year	Time /minutes	Time/minutes
	women	men
1970	35.50	27.65
1975	34.02	27.50
1980	31.75	27.37
1985	30.98	27.22
1990	30.22	27.13
1995	29.52	26.72
2000	29.52	26.38
2005	29.52	26.30

Table 1.1 World record times for the 10 000 m event

Activity 1 Record times

Plot the data from Table 1.1 on a single set of axes. Choose a sensible scale for your graph and label it clearly.

Continue (extrapolate) each graph forward in time and discuss (with reasons) whether you think women will one day beat the men in the 10 000 m event.

Speed

Data such as those in Table 1.1 allow us to calculate the average speed of an athlete:

$$\text{average speed} = \text{distance travelled} \div \text{time taken} \quad (1)$$

Expressed in the symbols that are normally used, Equation 1 becomes

$$v = \frac{\Delta s}{\Delta t} \qquad (1a)$$

The symbol Δ is the Greek capital letter delta, and is used to mean 'change in' or 'difference in', so if s represents the athlete's position, then Δs means 'change in position' i.e. the distance travelled. Δt represents the time interval. Notice that Δ is *not* a number multiplying s or t.

> **Maths reference**
>
> The symbol Δ
> See Maths note 0.2

In the right direction

An orienteer runs from post A 300 m, then 500 m, then 400 m, finally reaching post B. How far is she from the first post A? Figure 1.3 shows some possibilities. You do not know how far she is from A, because the directions were not specified. An orienteer would use a compass to get her bearings and find out which direction to travel in. She would need to specify the distance to be travelled in a given direction.

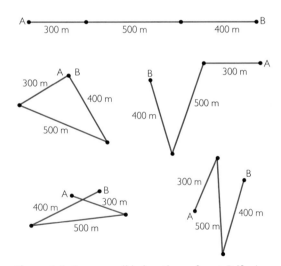

Figure 1.3 Some possible locations of post B if directions are not specified

When we mean 'distance in a specified direction' we use the word **displacement**. Similarly, speed in a specified direction is called **velocity**. Physical quantities where *direction is as important as size* are called **vector quantities**. (The size of a vector is often called its **magnitude**.) A quantity where there is no question of a direction being involved is called a **scalar**. A good example of a scalar quantity is *temperature*: 5 °C south-east hardly makes sense. When measuring vector quantities we need to pay attention to how the measuring instrument is directed: as a rather trivial example you can hardly measure an object's weight with a horizontal spring balance. But if a patient's temperature is being taken, it does not matter in which direction the thermometer is pointing out of their mouth!

Activity 2 Vectors and scalars

Write down a list of as many physical quantities as you can (not just mechanical – you could try electrical as well). Also write the SI unit by the side of each.

Draw up two columns headed 'Vector' and 'Scalar' and allocate each quantity to its correct column. You can start with *displacement/metre* and *temperature/kelvin*. Remember the question you have to ask of each is 'does direction matter?' (It seems quite straightforward but there are one or two deceptive ones – work? energy?)

In Part 1 of this chapter we will be concerned only with motion in one dimension (back and forth along a straight line), and we will use positive and negative signs to denote the direction.

Study note

In Part 2 of this chapter you will see how to deal with vectors in two dimensions.

Acceleration

When we say that something is moving, we are not giving much information away. Is an athlete travelling at constant speed? Is she changing direction? Is she speeding up or slowing down? In other words, we are interested in how the velocity changes:

> change in velocity = final velocity – initial velocity   (2)

Conventionally, u represents the initial velocity and v the final velocity, and Δv the change in velocity. We can write Equation 2 as:

$$\Delta v = v - u \qquad (2a)$$

This leads to a definition of **acceleration**:

> acceleration = change in velocity ÷ time taken   (3)

In symbols:

$$a = \frac{\Delta v}{\Delta t} \quad \text{or} \quad a = \frac{(v - u)}{\Delta t} \qquad (3a)$$

Another useful version of Equation 3 is:

$$v = u + a\Delta t \qquad (3b)$$

Notice that change of velocity is used and not change in speed. Acceleration is a vector of quantity, having both magnitude and direction. Equation 3 can only be used in situations with **uniform acceleration** as in Table 1.2, which shows motion with an acceleration of 2 m s^{-2}, ie the velocity is increasing by 2 m s^{-1} every second.

An important example of uniform acceleration is that of an object in **free fall**, i.e. moving only under the influence of gravity. In the initial part of a parachute jump or a bungee jump, you are in free fall, accelerating vertically downwards with the **acceleration due to gravity**, symbolised g, which (close to the Earth's surface) is always 9.8 m s^{-2}.

Time	Velocity
0 s	1.5 m s^{-1}
1 s	1.5 m s^{-1} + 2 m s^{-1} = 3.5 m s^{-1}
2 s	3.5 m s^{-1} + 2 m s^{-1} = 5.5 m s^{-1}
3 s	5.5 m s^{-1} + 2 m s^{-1} = 7.5 m s^{-1}

Table 1.2 Changing velocity

Activity 3 Free fall

Carry out some simple activities to illustrate the constant acceleration of objects falling under gravity.

Questions

1 A jogger who is initially running at 2.0 m s^{-1} accelerates uniformly at a rate of 1.5 m s^{-2} for 3.0 s. Calculate the final velocity.

2 How fast will a bungee jumper be moving after he has been in free fall for 2.5 s?

3 A squash ball travelling at 9.0 m s^{-1} horizontally to the right is hit by a racket which stops it in 0.003 s. Calculate the acceleration.

4 A tennis ball moving to the right at a velocity of 5.0 m s^{-1} is struck by a tennis racket and accelerated to the left, leaving the tennis racket at a speed of 25 m s^{-1} to the left. If the contact time is 0.012 s, calculate the average acceleration.

> **Maths reference**
>
> Index notation and powers of 10
> See Maths note 1.1
>
> Index notation and units
> See Maths note 2.2

1.3 Motion graphs

Graphs are often used to represent motion. Not only do they give a visual record but they also enable us to extract additional information about the motion. Figure 1.4 shows two **displacement–time graphs**. In Figure 1.4(a), the velocity is constant: the graph is a straight line, and dividing any given displacement Δt gives the same answer for the velocity – the velocity is equal to the **gradient** of the displacement–time graph.

> **Maths reference**
>
> Gradient of a linear graph
> See Maths note 5.3

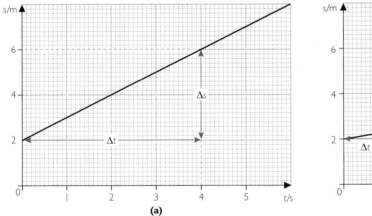

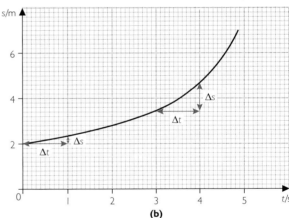

Figure 1.4 Displacement–time graphs for motion with (a) uniform velocity (b) non-uniform velocity

In Figure 1.4(b) you can tell that the velocity is non-uniform, because the graph does not show equal displacements in equal time intervals – the graph is not straight. A calculation of $\dfrac{\Delta s}{\Delta t}$ gives the average velocity in a given time interval Δt.

Figure 1.5 shows three **velocity–time graph**s. In Figure 1.5(a) the velocity does not change – the acceleration is zero. Figure 1.5(b) is a plot of the data in Table 1.2: the acceleration is uniform (the velocity changes by equal amounts in equal time intervals) so the graph is a straight line, and dividing any given change in velocity Δv by the corresponding time interval Δt gives the same answer – the acceleration is the gradient of the velocity–time graph. In Figure 1.5(c) the graph is not straight, because the acceleration is not uniform – the velocity does not change by equal amounts in equal time intervals.

In everyday language, 'accelerate' just means 'get faster', and 'decelerate' means 'get slower'. In physics we tend always to use the term 'accelerate' and use appropriate signs to indicate its direction. A negative acceleration is *not* necessarily a slowing down – if something that is already moving in the negative direction experiences acceleration in the same direction it will get faster!

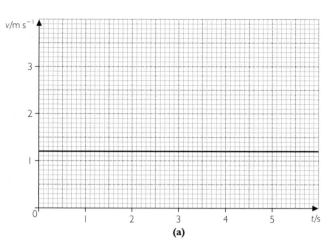

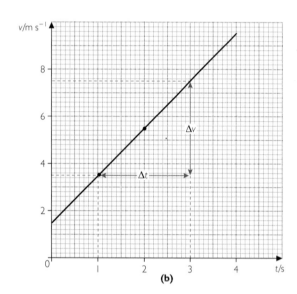

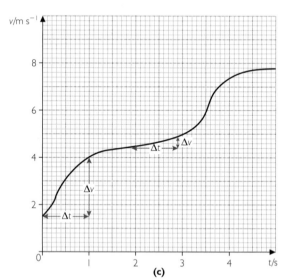

Figure 1.5 Velocity–time graphs for motion with (a) zero acceleration, (b) uniform (non-zero) acceleration and (c) non-uniform acceleration

Questions

5 Find the velocity of the motion shown in Figure 1.4(a) by (a) using the values of Δs and Δt shown and (b) drawing another triangle on a copy of Figure 1.4(a) that has a different Δs and Δt.

6 How would you interpret (a) a displacement–time graph that sloped more steeply than Figure 1.4(a) and (b) a displacement–time graph that sloped down from left to right?

7 Calculate the acceleration of the motion shown in Figure 1.5(b).

8 How would you show a negative acceleration on a velocity–time graph?

Small changes

Athletes such as sprinters, hurdlers, triple jumpers and so on improve their technique by observing their motion on a video playback frame by frame, or in a strobe photograph.

This allows them to study their motion in detail and then work with a coach to develop ways of improving their technique. In Activities 4 and 5 you will do something similar, analysing motion in some detail.

In most real-life examples of motion, the velocity is not uniform. Usually there is some acceleration from rest, and often the velocity changes before the final deceleration to rest. Acceleration, too, is not usually uniform. It can change because of wind resistance or changing muscle effort, or running on a slope. It is still possible to use Equation 2, however, if you split the motion into time intervals where the velocity in *nearly* uniform (as in Figure 1.4(b)). Likewise, acceleration can be calculated using Equation 3 for a small time interval where it is *nearly* uniform as in Figure 1.5(c).

Activity 4 Non-uniform motion

Use ticker-tape, a stop-frame video or a camcorder to record your own motion when sprinting from a crouched start and plot a graph of your displacement against time.

Calculate your velocity in each small interval between dots or between frames, and hence also calculate your acceleration in each small time interval.

If you have many very small time intervals, you need to perform lots of calculations. This can be very boring and so a computer program is usually used. The CD-ROM package *Multimedia Motion* contains a catalogue of video clips that can be played back frame by frame, and the motion can be analysed to find displacement, velocity and acceleration.

Activity 5 Producing graphs of motion

Use *Multimedia Motion* (version I or II) to produce graphs showing the displacement, velocity and acceleration for one or more of the following (see Figure 1.6): sprint start, squash, soccer, tennis. Alternatively use a motion sensor to generate graphs of your own motion. With *Logger Pro* you can synchronise the motion sensor data to your video clip to get a closer analysis of how your body moves. Keep a copy of your graphs for use in later activities.

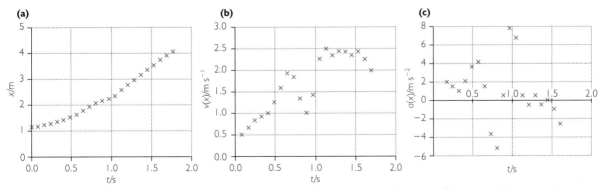

Figure 1.6 The motion of a sprinter shown as graphs of (a) displacement–time, (b) velocity–time and (c) acceleration–time

In Activity 4 you found the **instantaneous velocity** by dividing a small change in displacement by a small change in time. If you are working directly form a curved displacement–time graph, it is difficult to read values that are very close together. It is better to draw a **tangent** (a straight line touching the curve) at the required point as shown in Figure 1.7 and then work out its gradient. Likewise, you can find the **instantaneous acceleration** by drawing a tangent to a velocity–time graph and working out its gradient.

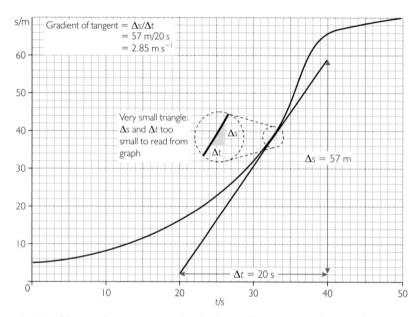

Figure 1.7 Working out instantaneous velocity from a displacement–time graph

Activity 6 Tangents and gradients

By drawing tangents on a displacement–time graph from Activity 5, find the velocity at two different times. Compare your answers with the velocities at those times calculated by the *Multimedia Motion* or similar software (if used). Similarly, find the acceleration at two times by drawing tangents on a velocity–time graph.

Going the distance

You have seen how a record of displacement can be used to deduce an athlete's velocity, and how velocity data can, in turn, be used to find acceleration. But can the same thing be done in reverse? Can a record of velocity be used to deduce displacement?

In Figure 1.8(a) the displacement in the first 4 s is:

$$1.5 \text{ m s}^{-1} \times 4.0 \text{ s} = 6.0 \text{ m}$$

This displacement is equal to the area of the shaded portion of the graph. In Figure 1.8(b) the velocity is not uniform, but if we choose a time interval small enough that the velocity v is *nearly* uniform, then the displacement in that small time interval Δt is given by:

$$\Delta s = v\Delta t$$

which is equal to the area of the narrow shaded strip. The total displacement in a longer time interval can be found by adding up all the areas of the narrow strips (each with a different height). In other words, displacement can be found from the **area under a velocity–time graph**.

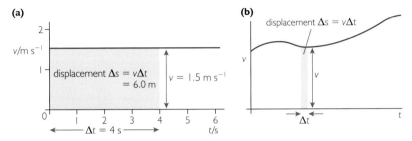

Figure 1.8 Working out displacement from a velocity–time graph (a) with **uniform velocity** and (b) with non-uniform velocity

Uniform acceleration

If the acceleration is uniform, then the velocity–time graph is a straight line, as in Figure 1.9, and the displacement can be found by adding together the areas of the rectangle and the triangle as shown.

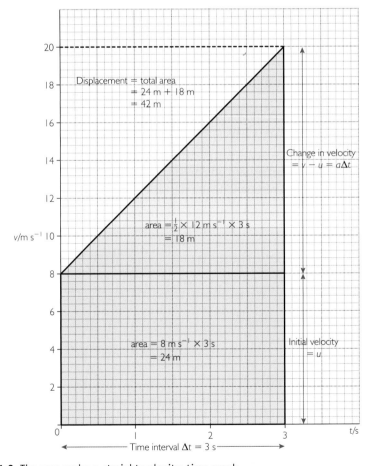

Figure 1.9 The area under a straight velocity–time graph

Expressing the areas of the rectangle and triangle in symbols leads to another useful equation for uniformly accelerated motion, which lets us calculate displacement directly without going via a graph:

$$\Delta s = u\Delta t + \frac{1}{2}a(\Delta t)^2 \qquad\qquad (4)$$

Usually you will see Equation 4 written using just t (not Δt) to represent the overall time taken and s to represent the overall displacement:

$$s = ut + \frac{1}{2}at^2 \qquad\qquad (4a)$$

Equations 3b and 4a can be combined to produce another useful relationship that relates change of velocity directly to the displacement:

$$v^2 = u^2 + 2as \qquad\qquad (5)$$

You can derive Equation 5 by squaring Equation 3b (and dropping the Δ):

$$v^2 = (u + at)^2 = u^2 + 2aut + a^2t^2$$

and multiplying Equation 4 by $2a$:

$$2as = 2aut + a^2t^2$$

Comparing the right-hand sides leads to Equation 5.

> **Maths reference**
>
> Algebra and elimination
> See Maths note 3.4

Activity 7 Free fall again

Carry out some explorations of freely falling objects that show how the time of fall is related to the distance fallen. Use your results to find the acceleration due to gravity.

Non-uniform acceleration

If the velocity–time graph is curved then the area can be found by using a computer program to work out and add together the areas of many very narrow strips, or by counting squares on the graph paper.

When you are counting squares, the vertical axis must start at zero otherwise the height of each strip does not represent the velocity. Also, be careful to use the scales of the graph and not the actual sizes of the squares.

Questions

9 A sprinter is running at a uniform speed of 8.00 m s^{-1} as she approached the finish, then she puts on a spurt to overtake her rival and accelerates at 0.70 m s^{-2} for 3 s before crossing the line. What is her final velocity, and what distance does she cover in the final three seconds of her sprint?

10 Figure 1.10 shows velocity–time graphs for two athletes in a race. After 30 seconds, who is ahead and by approximately how much?

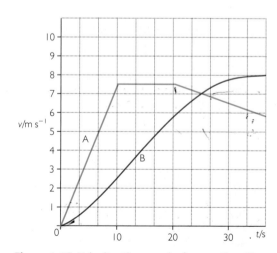

Figure 1.10 Velocity–time graphs for question 10

1.4 Force and acceleration

You have seen how acceleration can be deduced from measurements of velocity or displacement. Sports scientists often find it useful to go one stage further and use so-called inverse dynamics to work out the forces that provide the accelerations, using the relationship

$$F = ma \qquad (6)$$

where m is the mass of the accelerated object and F the net force acting on it. Like acceleration, force is a vector, and when dealing with one-dimensional motion we can use positive and negative signs to indicate the direction of a force. It is sometimes useful to combine Equations 3a and 6 to give:

$$F = ma = \frac{m\Delta v}{\Delta t} \qquad (6a)$$

Equation 6 expresses **Newton's second law of motion**.

Newton's first law of motion states that an object moves at constant velocity or remains at rest unless an unbalanced force acts on it, while the second law relates to the size of the unbalanced force to the change that it causes.

Questions

11 When starting a race, a sprinter of mass 65 kg accelerates forwards at 2.0 m s^{-2} (Figure 1.11). What must be the net forward force acting on his body?

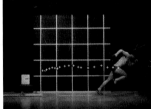

Figure 1.11 A sprint start

12 A tennis ball of mass 120 g approaches a racket at 5.0 m s^{-1} and is hit back in the opposite direction at 25 m s^{-1} (Figure 1.12). If the contact time with the racket is 0.015 s, what is the average force exerted on the ball by the racket?

Figure 1.12 A tennis ball in contact with a racket

Activity 8 Inverse dynamics

Using your results from Activity 5, estimate the net force acting to accelerate the person or object you studied. You might need to estimate the mass of the sprinter, or the squash, tennis or soccer ball.

Activity 9 Measuring forces directly

Use a force sensor with graphing software, or bathroom scales calibrated in newtons, to measure the forces involved in various activities such as jumping, throwing and catching.

In Activity 9 you probably took it for granted that the bathroom scales or force sensor register a non-zero vertical force even when you are standing still. But how does this tie in with Newton's first and second laws of motion, and Equation 6? The explanation involves some careful thinking about forces – and another of Newton's laws of motion.

Weight and gravitational field

The relationship between force, mass and acceleration (Equation 6) gives us another way of looking at free fall. Close to the Earth's surface, *any* object in free fall has an acceleration of 9.8 m s^{-2}. The gravitational force responsible for the acceleration must, therefore, depend on the object's mass – an object of mass 1 kg must experience a force of 9.8 N, a 2 kg object must experience a force of 2 × 9.8 N, and so on. We can express this by saying that, close to its surface, the Earth's **gravitational field strength**, symbolised g, is 9.8 N kg^{-1}.

The gravitational force acting on an object is called its **weight**. This can be confusing, because in everyday language the word 'weight' is used to mean the same thing as mass. Weight in measured in newtons and is a vector (it acts downwards). The weight W of an object of mass m is given by a special case of Equation 6:

$$W = mg \tag{7}$$

Question

13 Estimate your own weight.

If gravity is the only force acting, a downward acceleration will result, so, if you are standing still, your weight must be balanced by an upward force. How does this come about?

Pairs of forces

The sprinter in Question 11 exerts a 'backwards' force on the starting block, but he accelerates because the block (and the Earth to which it is attached) exerts a 'forwards' force on him. This is an example of **Newton's third law of motion**, which can be stated as 'all forces involve the interaction between two objects' or 'all forces come in pairs'. The two forces are always:

- between two different objects
- equal in size
- opposite in direction
- of the same type (e.g. both electrostatic or both gravitational).

In the example of the sprinter, he pushes backwards on the starting-block-plus-Earth with a force of 130 N, and the block pushes him forwards also with a force of 130 N (Figure 1.13). A force of 130 N on a 65 kg person gives him a significant acceleration, but the same size force acting on the block-plus-Earth produces such a minute acceleration that it is not detectable.

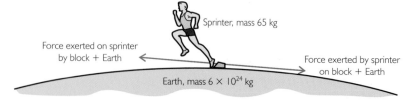

Sprinter, mass 65 kg

Force exerted on sprinter by block + Earth

Force exerted by sprinter on block + Earth

Earth, mass 6 × 10^{24} kg

Figure 1.13 Forces involved in the interaction between sprinter and starting block

Consider another example: when a bungee jumper is in free fall, there is a pair of gravitational forces acting between him and the Earth. The downward force on the bungee jumper (his weight) produces an acceleration of 9.8 m s^{-2}, but an upward force of the same size acting on the Earth produces no noticeable effect (Figure 1.14).

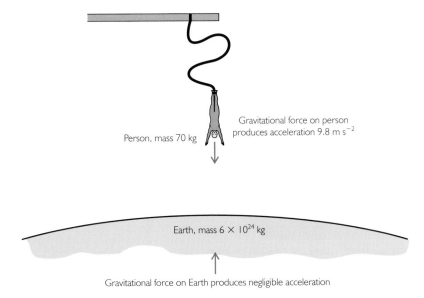

Person, mass 70 kg

Gravitational force on person produces acceleration 9.8 m s^{-2}

Earth, mass 6 × 10^{24} kg

Gravitational force on Earth produces negligible acceleration

Figure 1.14 Forces involved in the interaction between bungee jumper and Earth

If you stand on the floor, there is still a pair of gravitational forces acting between you and the Earth. But now you exert a downward force on the floor (equal to your weight) and the compressed material in the floor exerts an equal force upwards on you – you and the floor are interacting according to Newton's third law. In Activity 9, you recorded the force exerted by the scales (or a force sensor): you and the scales interact via a pair of forces and, in turn, the scales and the floor also interact. If you are at rest, the force registered is equal to your own weight; but if you jump off the scales, you do so by pushing downwards with additional force and the scales in turn exert an upward force upon you to produce an upward acceleration.

The forces involved in the interactions between athletes and their surroundings are not only important in producing the required accelerations. They are also responsible for injuries – if you hit something with a force, you also experience a force of the same size exerted by the object.

Questions

14 Identify the third-law pairs of forces that involve a javelin thrower and say what type of force they are.

15 A squash player hits the ball of mass 0.024 kg with her racket. The ball is decelerated at 12 200 m s^{-2}. Calculate the size and direction of the force exerted on the ball, and write down the size and direction of the force exerted by the ball on the racket.

16 (a) What are the pairs of 'Newton's third law' forces involved when a (not very good) diver (i) is in free fall and then (ii) splashes into the water?

(b) Explain why splashing awkwardly into the water is painful, whereas a smooth dive is not.

(c) In Figure 1.14, what is the magnitude of the acceleration of the Earth?

1.5 Summing up Part 1

In this part of the chapter you have some key ideas about forces and motion and seen how graphs can be used to display and analyse motion. You will use all these ideas again later in this chapter and elsewhere in your study of physics. Activity 10 is intended to help you review your work so far, and Question 17 shows that ideas about motion are relevant to situations other than sport!

Activity 10 Summing up Part 1

Spend a few minutes checking through Part 1, making sure you understand the meanings of all the key terms printed in bold. Then use at least five of those terms to describe the forces involved in sprinting and hence to explain how the 'cushion' in the sole of a running shoe helps prevent damage to the sprinter's feet.

Further investigations

In practice, many falling objects do not fall freely – they are affected by air resistance as well as by gravity. If you have an opportunity, you could investigate some of the records of falling objects (and people) included in *Multimedia Motion* to see what extent their motion is affected by air resistance.

A shuttlecock is designed to be affected by air resistance. You could investigate how the shape and weight distribution of a cone-shaped object affect its motion in the air.

Question

17 An instrument called a dynamometer is used to test the performance of trains. It can measure and record, amongst other things, a train's speed, acceleration and distance travelled, together with the time at which the measurements were taken. Figure 1.15 shows a record for a train of mass 200 000 kg until it reached its maximum speed of 35 m s^{-1}.

(a) Between what times was the train's acceleration uniform?

(b) What was the magnitude of this uniform acceleration?

(c) What was the magnitude of the instantaneous acceleration at $t = 80$ s?

(d) Calculate the size of the net force that produced the uniform acceleration.

(e) How far had the train travelled during the interval from $t = 0$ s to $t = 40$ s?

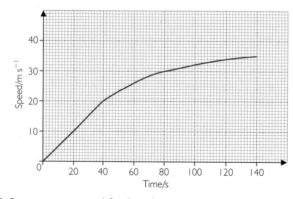

Figure 1.15 Dynamometer record for Question 17

2 Rock climbing

Groping with my other gloved hand in the crack, I found a solid fist-jam, my axe hanging from its strap around my wrist. Pulling into the rock and releasing my other axe, I lifted it high above my head and hooked it over a chockstone in the crack. I straightened my arm, and relaxed the muscles, so that I could feel the strain tugging from within my shoulders. I placed the front points of my boot on the original flat ledge, which caused my left hand to rip from the crack. I palmed it against the rock on the left. I was totally absorbed. I stood up on the right foot, first transferring my weight from the left and shivering it from the crack. With my left axe I reached as high as I could into the back of the groove where it opened out like the base of a peapod. I hit only rock beneath powdery snow. 'Shit!' My right calf muscles began to complain. Rather desperately I kicked my left foot back into the crack and pushed up again, leaning into the base of the groove, scratching both my axes across the snow. I found a patch of névé and sank my teeth into it. I exhaled and pulled up into a resting position. ... I gazed up the groove towards the sun and the deep blue sky. 'Brilliant' I said jubilantly. 'This climbing is brilliant!' The drag of fatigue was forgotten.

Fanshawe, Andy. *Coming Through* (p160)
Hodder and Stoughton, 1990.

In the passage above, climber Andy Fanshawe is describing a climb in the Himalayas. The sport of rock climbing is quite a contrast to athletics. However complicated or contorted the move in rock climbing, all the forces must somehow combine to produce **equilibrium** (ie a net force of zero) almost all of the time, unlike athletics where forces and accelerations can be deliberately large. The only occasion when a rock climber is not in equilibrium is when there might be an acceleration in transferring position – or of course when falling!

In this part of the chapter we are going to explore the equilibrium of forces and also look at some of the physical properties of materials that help to make the sport of rock climbing safe.

2.1 Hanging on

The climber in Figure 1.16(a) is clearly in equilibrium – there are no horizontal forces acting and the upward vertical force exerted by the rock face balances his weight vertically downwards. But so also is the climber in Figure 1.16(b). Here, though, the situation is a little more complicated – there are three points of contact with the rock which together with his weight (the rope is more or less slack) produce four forces all acting in different directions. Figure 1.16(c) shows yet another situation (a so-called *Tyrolean traverse*) where forces are acting in different directions.

Figure 1.16 Rock climbers in equilibrium

Activity 11 Forces in different directions

Can you see any similarities between the equilibrium situations of Figures 1.16(b) and (c)? (Try to wipe the 'physical context' from the pictures and think just about the directions of forces acting on the climbers.)

Use the arrangement shown in Figure 1.17 to explore the effect of pulling ropes in different directions but still trying to maintain equilibrium.

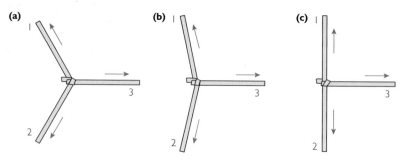

Figure 1.17 Diagram for Activity 11

It should be clear from Activity 11 that, when forces combine with each other, we don't just have to think about how large each one is, but also in which direction it is acting. In Activity 13 you will extend the qualitative ideas of Activity 11 by taking measurements of tension in, and angle of, a rope. Before you do that, what sort of behaviour might you expect? We can get a clue from considering what is probably the simplest vector quantity of all – displacement.

Combining displacement vectors

You will be familiar with displacement vectors if you have ever done orienteering or any navigation across open country. It is simply an instruction to move from one point to another, and it must contain the *two* pieces of information: how far? which way? (unlike the situation in Figure 1.3).

Figure 1.18(a) shows two displacement vectors: d_1 is 5 km north (call this direction 0° and d_2 is 3 km in a direction 40° clockwise from north. Let us start from point O and carry out the displacement d_1 first, then d_2. This is shown in Figure 1.18(b) and takes us to point P. Alternatively we could do d_2 first, then d_1, as in Figure 1.18(c).

Study note

In printed texts vectors are usually in **bold** or ***bold italic*** type. When writing them yourself it is usual to underline them with a wavy line. The magnitude (size) of a vector is a scalar quantity, so is shown in the same way as any other scalar, i.e. in *italic*, non-bold type.

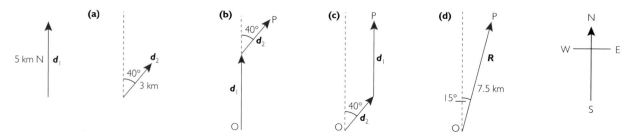

Figure 1.18 Combining displacement vectors (1 cm = 2.5 km)

Whichever way we combine the vectors, the net effect is exactly the same. The vector from O to P is the same in both diagrams and is the *single* vector that replaces the two separate ones. Careful measurement on Figure 1.18(d) shows that it is 7.5 km in a direction 15° from N. This single vector is called the **resultant** vector of d_1 and d_2: if we denote it by R we write

$$R = d_1 + d_2 \qquad (8)$$

If we need to combine more than two vectors then we just continue the process of joining them 'head-to-tail', ending up with a **vector polygon** as in Figure 1.19. The resultant is always found by joining the starting point O to finishing point P.

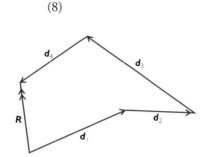

Figure 1.19 A vector polygon

Zero resultant

We can rewrite Equation 8 as:

$$d_1 + d_2 - R = 0 \qquad (9)$$

which we can interpret as the combination of the vector $-R$ with the sum $(d_1 + d_2)$, giving a resultant of zero. If in Figure 1.18(b) or (c) we added the vector $-R$ at point P (the minus sign simply means reverse the direction), then it is clear that we end up back at O and the resultant is zero.

> **Study note**
>
> Note that the '+' sign in Equation 8 is not the same as an ordinary arithmetic (scalar) addition, but we borrow the sign to show that we are combining the vectors and we describe the procedure as vector addition. It is obviously not true that $R = d_1 + d_2$ (scalars, magnitude only).

Activity 12 Vector polygon

Draw on graph paper a vector polygon similar to Figure 1.19. It need not be an exact match. Draw up a table of two column headed 'Magnitude' and 'Direction' and with the help of a ruler and protractor enter the information for each of the vectors.

Now combine the vectors in a variety of different orders. You will get quite different polygons, but you ought to find the resultant is the same in each case.

Combining force vectors by drawing

Figure 1.20 shows three forces W, T_1 and T_2 that are in equilibrium and so their resultant is zero. Do they combine like displacement vectors? Activity 13 provides the answer.

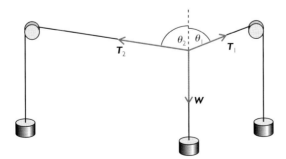

Figure 1.20 Three force vectors in equilibrium

Activity 13 Forces in equilibrium

Use the apparatus shown in Figure 1.21 to investigate whether force vectors combine like displacement vectors. For each equilibrium arrangement, construct a vector addition diagram and see if it is closed – do we end up back where we started? The critical step is to choose a scale so that each force is represented by a line of length proportional to its magnitude, e.g. 2 cm to 1 N.

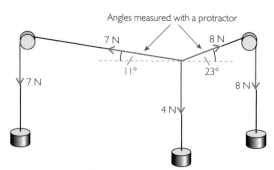

Maths reference

The symbol Σ
See Maths note 0.3

Figure 1.21 Apparatus for Activity 13

Careful measurements show that forces do indeed behave as displacement vectors when they combine. This means that when any number of forces are in equilibrium, the vectors form a closed polygon. This can be expressed in symbols:

$$\Sigma F = 0 \qquad\qquad (10)$$

The symbol Σ (Greek sigma) means 'the sum of all'. When applied to vectors, the sum must take account of their directions as well as the magnitudes.

A common special case is when there are just three forces – here the polygon is a triangle usually called the **triangle of forces**.

Question

18 Figure 1.22 shows a simplified end on view of a cable car (looking along the length of the cable) in operation. It is being blown by high wind, which can produce a sideways force of up to 5000 N. The total weight of the car is 2.5×10^4 N. Assume that it is in equilibrium (ie. not swinging). T is the force exerted on the car by the support arm, which is fixed rigidly to the car but can rotate on the cable.

Draw a triangle of forces for this arrangement. Choose a suitable scale, working on graph paper for convenience, and start with a vector whose details you know. (Does it matter which?) As accurately as you can, measure the values of T and θ.

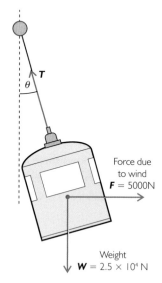

Figure 1.22 Forces acting on a cable car

Activity 14 What stops a rock climber falling?

By tracing the directions of the forces in Figure 1.16(b), you can deduce an important result about forces in equilibrium.

It is always true that if three **coplanar** non-parallel forces (coplanar means acting in the same plane) are in equilibrium, their lines of action pass through the same point. With a bit of imagination many rather complicated real-life situations can be reduced to three forces in equilibrium, as in Activity 15.

Activity 15 Forces acting on a power cable

Study (from a safe distance!) an overhead power cable. Make a scale drawing of the cable as accurately as you can and use it to estimate the force that the cable is exerting on the pylon if the mass per unit of length of the cable is 2 kg m^{-1}.

Free-body diagram

What you have just been drawing in Activities 14 and 15 are **free-body diagrams**.

We often need to know whether an object is in equilibrium, or whether there is a resultant force acting on it. To help analyse the situation it is often useful to draw a diagram of the object by itself, representing it by a dot, and then mark on the size and direction of all of the forces acting on it. The object can be represented by a dot because this marks its **centre of gravity** – the point through which all of its weight may be considered to act (for a person the centre of gravity is approximately at the belly button). Figure 1.23 is a free-body diagram for the cable car shown in Figure 1.22.

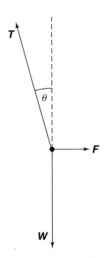

Figure 1.23 Free-body diagram for a cable car

Question

19 Draw free-body diagrams for

(a) a plane flying at constant velocity in still air

(b) a skier being dragged at constant velocity uphill on a ski-tow

(c) a friction-free puck sliding on ice at constant velocity

(d) a ball that has just been thrown in the air (neglecting the very small amount of air resistance).

Combining force vectors by calculation

Since most force situations of interest will involve different directions, it is obviously important, not least from considerations of strength and safety, to be able to deal accurately with them (just think about buildings, bridges or aircraft). Is there any way of doing it apart from scale drawing?

As an introduction, consider the case when two force vectors are acting at a point at right angles to each other (Figure 1.24(a)). The vector addition diagram, including the resultant **R**, is obviously a right-angled triangle as in Figure 1.24(b). (Remember that **R** *replaces* F_1 and F_2 and is equivalent to the other two acting together – we are *not* introducing a third extra force.) From Pythagoras we can say:

$$R^2 = F_1^2 + F_2^2$$
$$\text{or } R = \sqrt{(F_1^2 + F_2^2)} \qquad\qquad (11)$$

and also:

$$\tan\theta = \frac{F_2}{F_1}$$
$$\text{or } \theta = \tan^{-1}\left(\frac{F_2}{F_1}\right) \qquad\qquad (12)$$

so in this special case we can find the magnitude and direction of **R** by calculation alone.

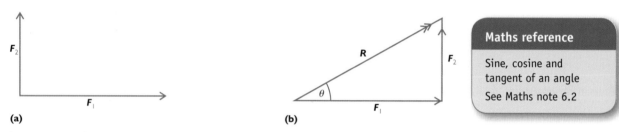

(a)   (b)

Maths reference

Sine, cosine and tangent of an angle

See Maths note 6.2

Figure 1.24 Two forces at right angles

Question

20 Figure 1.25 shows a climber supported by a rope and 'walking' down a vertical rock face. Sketch the triangle of forces acting on the climber when she is momentarily at rest and hence calculate the magnitudes of the horizontal force exerted by the rock and the tension in the rope.

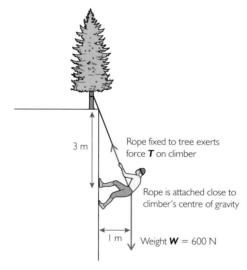

3 m

Rope fixed to tree exerts force **T** on climber

Rope is attached close to climber's centre of gravity

1 m   Weight **W** = 600 N

Figure 1.25 Diagram for question 20

Resolving a force into perpendicular components

But what about two vectors that are not at right angles, as in Figure 1.18 and Activity 14? The trick is to take each force in turn and resolve it (split it up) into two perpendicular **components** (parts). For example, the vector d_2 in Figure 1.18 can be thought of as the resultant of two displacements – one due north (d_N) and one due east (d_E) (Figure 1.26). These two components form a right angled triangle with d_2 as the hypotenuse, from which we can see that the magnitudes of the components are given by:

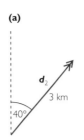
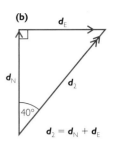

Figure 1.26 Resolving a displacement vector into perpendicular components

$$d_N = d_2\cos40°$$

$$\text{and } d_E = d_2\sin40°$$

The overall displacement $d_1 + d_2$ can then be treated as the sum of two northerly displacements plus an easterly displacement, which together make two sides of a right-angled triangle, so the magnitude and direction of the resultant can be calculated.

This example illustrates a general rule. Any vector P making an angle θ with a particular direction can be resolved into two components, one parallel to the chosen direction and one perpendicular to it:

$$\text{magnitude of parallel component} = P\cos\theta \qquad (13a)$$

$$\text{magnitude of perpendicular component} = P\sin\theta \qquad (13b)$$

We now have another way of looking at equilibrium of forces. In any direction you choose, all the components of forces in that direction must combine to produce a resultant of zero. Activity 16 illustrates this.

Study note

Note that there is nothing special about the directions of the components – provided they are at 90° to each other. In practice they will often be horizontal and vertical, or parallel and perpendicular to a surface.

Think about the two special cases $\theta = 0$ and $\theta = 90°$. What would be the northerly and easterly components of the displacement vector 5 km north?

Activity 16 Components of force vectors

Return to your data for Activity 13. Using your values of tensions, weight and directions, draw up a table of the horizontal and vertical components of each force.

By adding separately the horizontal and vertical components, show to what extent the condition for equilibrium is satisfied experimentally.

Tyrolean traverse

The Tyrolean traverse (Figure 1.16(c)) is a technique for crossing a deep chasm suspended on a rope. In the pioneering days of mountaineering the rope was thrown across a chasm and lassoed onto a suitable spike. Figure 1.27 shows a close-up of how it might work in practice. In Question 21 and Activity 17 you are going to use what you have learned about forces to explore the Tyrolean traverse in a bit more detail, and Question 22 applies the same ideas to a more complicated situation where the two ends are not on a level.

Figure 1.27 Tyrolean traverse

Suppose for convenience the two parts of the rope each make an angle of 15° with the horizontal and the climber has a weight of 600 N. How can we find the tension in the rope? A free-body diagram (Figure 1.28) helps. From the symmetry, the tension in the two halves must be the same. (Warning! – only of course true when the two angles are the same.) If we resolve vertically:

$$T\cos75° + T\cos75° = 600 \text{ N}$$

$$(\text{or } T\sin15° + T\sin15° = 600 \text{ N})$$

$$2T\cos75° = 600 \text{ N}$$

$$0.52T = 600 \text{ N}$$

$$T = 1160 \text{ N}$$

i.e. the tension is nearly twice the climber's weight.

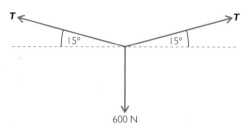

Figure 1.28 Free-body diagram for a person on a Tyrolean traverse

Question

21 (a) Repeat the calculation in the text for progressively smaller angles with the horizontal. (Hint: call the angle with the horizontal θ and produce a formula for T in terms of θ.) What is your general conclusion?

(b) If the rope is designed to take a maximum tension of 15 kN, what is the smallest angle the rope must be allowed to make with the horizontal for this particular weight of climber?

Activity 17 Model of Tyrolean traverse

Set up the apparatus as in Figure 1.29 so that the tensions in both sides are *equal* and much bigger that *W*. Gently pull the load across the traverse from one side to the other by lowering one of the side masses, and investigate how the sag *y* varies with distance *x* from the centre (Figure 1.30).

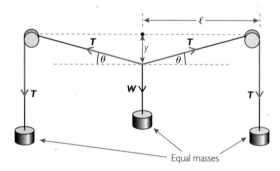

Figure 1.29 The starting arrangement for Activity 17

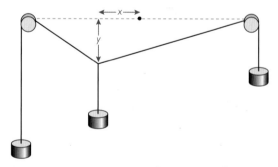

Figure 1.30 Moving the load

Question

22 A cable car is being pulled up a mountain in calm conditions. The cable above it is at an angle of 24° with the horizontal and the cable down to the valley station is at 23°. The weight of the car is 1.0×10^5 N.

(a) Draw a free-body diagram and a triangle of forces for the cable-car.

(b) By resolving horizontally and vertically, form a pair of simultaneous equations containing the tension in each part of the cable. Solve these to find the tensions.

(c) Why is it impossible for the two angles to be the same?
(Note: this question becomes much more manageable if you first draw a diagram and then write down your working carefully step by step.)

Out of equilibrium

So far, we have dealt only with situations where forces combine to produce a zero resultant. But if a climber is pushing on a foothold to give himself an upward acceleration, or if a rope snaps and allows him to fall, then the resultant force is not zero. The resultant force produces an acceleration as described by Equation 6, which can be written as:

$$\Sigma \boldsymbol{F} = m\boldsymbol{a} \qquad\qquad\qquad (6b)$$

The direction of the acceleration vector is the same as the direction of the resultant force.

Worked example

Q A rock climber of mass 71 kg (including kit) pushes down against a foothold with a force of 900 N at 10° to the vertical. Draw a free-body diagram showing forces on the climber and use a vector addition diagram to find the resultant force acting on her. (There is no tension in her climbing rope.) What is her acceleration?

A She experiences a force in the opposite direction to the one she exerts on the foothold. Figure 1.31 shows the two forces acting on her.

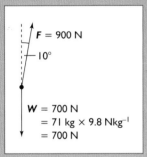

$F = 900$ N
$\quad$ 10°

$W = 700$ N
$\quad = 71$ kg $\times 9.8$ Nkg^{-1}
$\quad = 700$ N

Figure 1.31 Free-body diagram for a climber

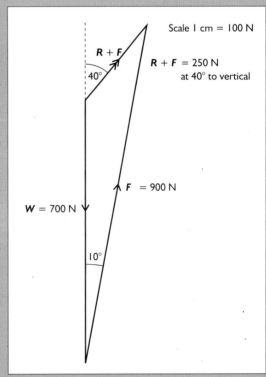

Scale 1 cm = 100 N

$R + F$
40°

$R + F = 250$ N
at 40° to vertical

$F = 900$ N

$W = 700$ N

10°

Figure 1.32 Force vector diagram for the climber in Figure 1.31

From the vector diagram in Figure 1.32, The resultant force is approximately 250 N at 40° to the vertical.

The magnitude of the acceleration is $a = 250$ N/71 kg = 3.5 m s^{-2}. She accelerates in the same direction as the resultant force i.e. 40° to the vertical.

Question

23 A sky-diver falling vertically experiences a net vertical force of 500 N. A sideways gust of wind exerts a force of 100 N.

(a) What are the magnitude and direction of the resultant force on the sky-diver?

(b) If the diver plus kit has a total mass of 100 kg, what is the acceleration?

2.2 On the ropes

Climbing ropes obviously have to be **strong** (a large tension is needed to break them). Hanging from a horizontally supported rope can produce tensions several times the hanging weight, and if you fall while climbing, the safety rope must be able to exert large decelerating forces on your body.

Climbing ropes are also **elastic** – they stretch when put under tension, and return (not always completely) to the original length when the load is removed. A useful term to compare different samples is **stiffness**: one rope is stiffer than another if the extension is smaller for the same force. (The opposite of stiffness is **compliance**.)

The stiffness k of a sample is defined as:

$$k = \frac{F}{x} \tag{14}$$

where F is the net applied force and x is the resulting extension. The SI units of stiffness are N m^{-1}.

Activity 18 Tension and extension

Use a variety of different 'ropes' (at least, fibres that could be made up into ropes) with the arrangement shown in Figure 1.33 to produce graphs showing how the extension varies with tension and to find the breaking strength of your sample.

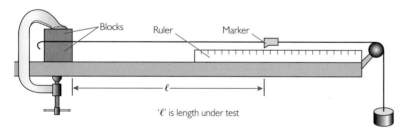

Blocks   Ruler   Marker

ℓ

'ℓ' is length under test

Figure 1.33 Diagram for Activity 18

Hooke's law

In Activity 18 you might have found that some of your samples had a constant stiffness for a range of loads. Samples that behave in this way are said to obey **Hooke's law**. For such a sample:

tension ∝ extension

or $F \propto x$ (15)

and the stiffness k is constant. In graphical terms, for Hooke's law to hold, the force–extension graph has to be linear *and pass through the origin*. You will probably find

Maths reference

Graphs and proportionality
See Maths note 5.1

that your results show that Hooke's law is not very closely followed. A point where the graph starts to deviate significantly from Hooke's law is called the **limit of proportionality**.

Questions

24 A certain rope extends by 0.020 m when supporting a load of 800 N and by 0.012 m when supporting a load of 600 N. Does the rope obey Hooke's law? Calculate the stiffness of the rope under each of the loads.

25 Suppose a 2 m length of climbing rope obeys Hooke's law and has a stiffness of 60 kN m^{-1}.

(a) If the rope supports the weight of a 650 N climber, by how much does the rope extend?

(b) If the same climber was supported by 4 m of the same rope, what would be the extension?

Drop tests

The skill in designing climbing ropes for different tasks is to get the right combination of stiffness and strength. Climbing ropes are subjected to a standard drop test for safety certification: a mass of 80 kg attached to 5 m of the rope is dropped freely and the force exerted on the rope is measured as it brings the mass to rest.

2.3 Summing up Part 2

In this part of the chapter you have studied two areas of physics, both of which you will revisit shortly. In Section 2.1 you have seen how to resolve and combine displacement and force vectors in two dimensions – you will meet these ideas again in Parts 3 and 5 of this chapter. You have also begun to study elastic properties of materials, which you will meet again in Part 4 of this chapter and later in this course.

Activity 19 Summing up Part 2

Check through Part 2 and make sure you know the meaning of all the terms printed in bold. Then discuss the following questions in a small group.

- What can you say (qualitatively) about the stiffness of a rope suitable for a Tyrolean traverse?
- When setting up a Tyrolean traverse, should you aim to get the rope nearly horizontal, or to let it sag?
- Thinking about the design of a rope for rock climbing, what are the consequences for the climber for making it either very stiff or very compliant?
- A climbing rope catalogue states that the standard drop test produces an extension of 7.5% of the original length. What extension(s) do(es) this correspond to for your samples in Activity 18? Would any sample have broken already in trying to reach this extension?
- What can you say about the motion of a falling climber from the instant the rope starts to go under tension?
- Suppose you had rope samples of widely varying stiffness and you subjected them to the standard drop test. Sketch a sequence of graphs (on the same axes of force against time) to show the effect of decreasing stiffness. (Qualitative only – no calculation.)

Maths reference

Fractions and percentages

See Maths notes 3.1

3 Working out work

3.1 Energy return shoes

Walk in to a sports shop, looking for a new pair of trainers, and you're confronted with an incredible choice. You can attach space-age technology to your feet, packed with gels, fluids and air bags. Companies spend millions on research each year, with engineers putting new designs and materials through their paces with the help of computer-linked sensors in mechanical testing laboratories. The most significant innovation has been the advances in cushioning. A runner hits the ground with a force up to three times his or her body weight, so athletes have welcomed materials that lessen the impact and the chances of injury.

Cushioning aside, how do you choose between all the varieties? Sales people will often try to blind you with science and technological jargon. In this section we will examine one particular claim – that some shoes can supply you with extra energy, so you can jump higher and run further. These 'energy return' shoes are supposed to store energy like a spring, in the cushioning material of the sole, when your foot hits the ground, and return it as you move off.

Activity 20 Selling science

Read the article in Figure 1.34 about the 'Recoil' trainer, one of the original energy return shoes. Discuss in a group the following questions.

• How does the article use science to try to persuade the reader?

• What scientific claims do the manufacturers make?

• Are the claims supported by any evidence?

How well does science imitate nature? Do 'energy return' shoes justify the hype? We will try to find out if there is any truth in the claims, using ideas about energy.

THE NEW RECOIL TRAINER FROM Z-TECH

The Z-Tech Recoil Trainer is the most exciting innovation in running shoes since rubber soles. Run in them once, and you'll never want to run in anything else.

THE Z-TECH SPRING...THE ULTIMATE SHOCK ABSORBER

When you run, you hit the ground with a force between 2-4 times your body weight. The majority of this force is transferred into your body immediately upon impact with the ground. This quick rate of impact is one of the main reasons so many runners suffer injuries. That's why we developed the Z-Tech Spring. The Spring offers the greatest impact absorption of any running shoe, significantly reducing impact related injuries.

The graph below illustrates the more gradual rate of impact of the Recoil running shoes vs. the competition. The reduction in impact translates into more effective and safer training.

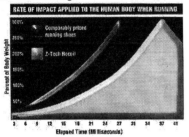

CUSTOMIZED SHOCK ABSORPTION... THREE SPRING TENSIONS

Since no two runners are alike, each Recoil Spring is available in three tensions based on your weight, training pace, and foot strike specifications. The Recoil is the only running shoe giving you customized shock absorption based on your body and running style. Each Spring is easily interchangeable and replaceable.

Research was conducted by the Los Alamos National Laboratory under the auspices of the U.S. Department of Energy. Testing performed compared shock absorption and energy return of the Recoil vs comparably priced running shoes. Testing procedures may not accurately reproduce the biomechanical forces and effects that would be experienced by an actual runner.

THE Z-PLATE... MAXIMIZING STABILITY

To ensure absolute stability on any surface, we've created the revolutionary Z-Plate. Located above the Spring midsole, it stabilizes and supports the heel, arch and center foot, while the Spring works independently.

ENERGY RETURN...

Other running shoes dissipate your precious energy, thus wasting it. The Recoil running shoe returns up to 49% of the impact energy with each step forward. Once compressed, hundreds of pounds of tension are built up and stored in the Z-Tech Spring. Upon release, this built up energy is catapulted back into your legs causing you to be propelled forward with a burst of power.

LEAVING THE COMPETITION IN THE DUST...

They may stare, some may even laugh, but not for long. Not after they realize that the Recoil makes you run more comfortably with less fear of injury. Not after you pass them on the track, or keep going long after they've tired. Especially not after they too, try a pair on. Because it only takes one try to know, right down to your bones, that the Recoil is everything you've ever wanted in a running shoe.

©1997 Z-tech Inc., Inc. All Rights Reserved
info@Ztech-inc.com

Figure 1.34 The Recoil trainer

Energy

In your GCSE work you probably learned that there is gravitational potential energy, electrical energy, chemical energy – you may know several others. They are all measured in the same units, joules (J), but do you know what makes, say, gravitational energy different from chemical energy? When energy seems to come in so many varieties, things can start to become confusing. You may be relieved to know that, basically there are only two 'types' of energy – **kinetic energy** and **potential energy**. All the others are really either kinetic or potential or a combination of the two.

Kinetic energy is the energy an object has because of its movement. Figure 1.35 shows two examples of kinetic energy: the sprinter clearly has kinetic energy but, at a molecular level, so does a hot drink because its molecules are in rapid motion.

Figure 1.35 Two examples illustrating kinetic energy

Potential energy is the energy a body has due to its position or the arrangement of its parts. Look at the examples of potential energy in Figure 1.36.

Figure 1.36 Three examples illustrating potential energy

Gravitational energy, due to being raised up, is such a common example of potential energy, you can be forgiven for thinking it is the only one. But 'chemical energy' in the drink is really potential energy, due to the arrangement of atoms in its molecules giving them the 'potential' (the possibility) of taking part in a chemical reaction. The 'elastic energy' stored in the running-shoe sole is also potential energy because it is due to the rearrangement of molecules as the material is compressed.

You will also have met the idea of **energy conservation**: in any process, energy can be transferred but cannot be created or destroyed. While bearing in mind that all energy is kinetic or potential, it is still sometimes useful to use other terms to 'label' the stages in an energy transfer process in order to keep track of what is happening.

Activity 21 Talking energy

Describe the energy transfers that are shown in Figure 1.37.

Can you think of a way of communicating with a friend that does not involve an energy transfer?

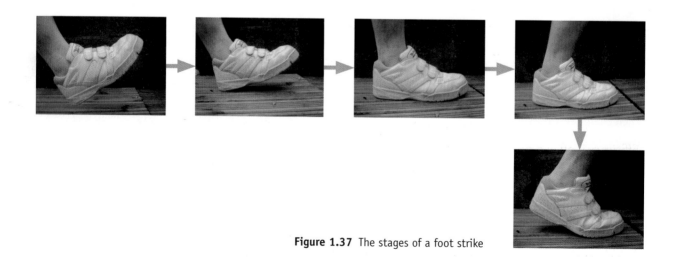

Figure 1.37 The stages of a foot strike

The claim for energy return shoes is that they store significant amounts of the runner's kinetic energy and return it as kinetic energy perhaps, to make you run faster, or jump higher. To investigate this further, we need to be able to measure energy.

Work

We measure the energy an object has by the **work** it can do. 'Work' in physics has a more precise meaning than in everyday life: work is done when something is moved by a force. So a body has energy if it can do work, i.e. move something else with a force. Look back at Figures 1.35 and 1.36 and convince yourself that all these examples show situations where work can be done. Work is defined as follows:

$$\text{work} = \text{force} \times \text{displacement } \textit{in direction of force} \qquad (16)$$

When work is done, energy is transferred, which gives us a useful way of measuring energy:

$$\text{work done} = \text{energy transferred} \qquad (17)$$

In symbols:

$$\Delta E = \Delta W = F \Delta s \qquad (17a)$$

For example, at the start of a race, a rower pulls her oars backwards a distance of 0.8 m during a stroke, exerting a constant force of 250 N (Figure 1.38):

$$\text{work done} = 250 \text{ N} \times 0.8 \text{ m} = 200 \text{ J}$$

Study note

Force and displacement are both vectors but Equation 17(a) involves their magnitudes. Work and energy are scalars.

Figure 1.38 A rower at the start of a race

There has been an energy transfer: 200 J of potential energy (stored in her muscles) has been transferred so that she now has 200 J of kinetic energy.

Efficiency

In addition to the 200 J of 'useful' energy transfer, some energy will be 'wasted' in heating – rowing makes you hot! – so rather more than 200 J will have been transferred from her muscles altogether. The **efficiency** of any energy-transfer process is defined as:

$$\text{efficiency} = \frac{\text{energy usefully transferred}}{\text{total energy transferred}} \qquad (18)$$

Maths reference

Fractions and percentages
See Maths note 3.1

Efficiency is often expressed as a percentage. For example, if the rower's muscles are 20% efficient, 20% (one-fifth) of the energy transferred does work on the oars, while the remaining 80% is wasted in making her hot.

Calculating energy return

Using the concept of work to measure energy, it is possible to test the claims about energy return shoes. For an adult male runner, Figure 1.39 shows the pattern of compressions in the sole of an 'energy return' shoe. As a rough approximation, the average force on the shoe during a running step is about 2.5 times body weight (about 2000 N for a person of mass 80 kg). From such pictures, we can estimate the average compression to be about 5 mm (0.005 m). So:

work done = 2000 N × 0.005 m = 10 J

This figure agrees well with what researchers have found. They have also found that only about 6 J of this may be recovered (the rest is 'wasted' in heating the shoe and the surroundings). Could 6 J enable a basketball player to jump much higher? Can it give you extra speed while you're running? To answer these questions, we need some more formulae.

Kinetic and potential energy

As an object falls, it loses gravitational potential energy. If an object of mass m falls through a height Δh then the force acting is just the object's weight, and so:

loss of gravitational potential energy = weight × loss of height

(19)

If the object is moved upwards, then the process is reversed and it regains its gravitational energy. Using ΔE_{grav} to represent change in gravitational potential energy, and Equation 7 to express the object's weight, we get:

$$\Delta E_{grav} = mg\Delta h \qquad (19a)$$

What about running? How can we calculate kinetic energy? From

Activity 22 Gravitational energy in a jump

Measure the height through which you can jump and hence calculate the change in your gravitational potential energy.

What difference would an additional 6 J from 'energy return' trainers make to the height of your jump?

your earlier work you are probably familiar with the formula for **kinetic energy** E_k of an object of mass m at speed v:

$$E_k = \tfrac{1}{2}mv^2 \qquad (20)$$

This expression can be derived using others that you have used in Part 1 of this unit. Suppose a force of magnitude F accelerates an object mass m from rest so that it reaches a speed v after moving through a distance Δs. Putting $u = 0$ in Equation 5 we can write:

$$v^2 = 2a\Delta s$$

so $a\Delta s = \tfrac{1}{2}v^2$

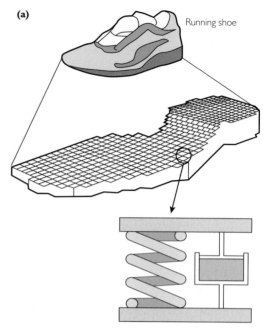

(a)   Running shoe

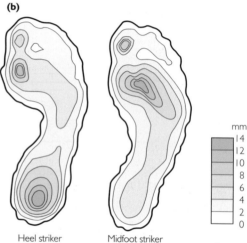

(b)

Heel striker   Midfoot striker

mm
14
12
10
8
6
4
2
0

Figure 1.39 Computer-generated pictures showing (a) a model of how the sole of a shoe is compressed during impact and (b) peak deflections of a sole caused by two types of runner

Study note

Equation 19(a) is often stated as $E_{grav} = mgh$, where h is the height above some zero reference point, e.g. the bench, or floor. But since there is no absolute reference point, you can really only calculate a *change* in potential energy.

We can also use Equations 6 and 17a to write:

$$\Delta E_k = F\Delta s = ma\Delta s = \tfrac{1}{2}mv^2$$

Maths reference

Algebra and elimination
See Maths note 3.4

 Activity 23 Kinetic energy in running

Measure your kinetic energy when running. Try to devise a way to measure your speed as accurately as possible.
What difference would 6 J from 'energy return' trainers make to your speed?

It is likely that you found that the energy returned was too small to give you much extra speed or height, so it looks as if the effect of energy return shoes is minimal. However, published research with real athletes has found that running in a shoe with a gas-inflated cushioning system reduces the oxygen consumption by 2%, compared to a regular foam-cushioned shoe (although the researchers suggested that this could have been due to factors other than energy return).

Nature has already endowed us with an energy return mechanism, one which, it turns out, is rather more efficient. When your heel strikes the ground, about 70 J of energy are stored and returned by the quadriceps muscle in the lower leg, and another 70 J by the Achilles tendon. Energy return training shoes owe a lot more to marketing than to science.

Questions

26 Suppose the rower discussed above has an efficiency of 20%. How much energy is transferred from her muscles when she does 200 J of work?

27 Calculate the kinetic energy of a jet boat, mass 200 kg, that can tear across the water at 160 km h^{-1} (44 m s^{-1}).

28 Which do you think has more kinetic energy (don't calculate it), a speed skier at 150 mph or a small car travelling at 50 mph? Use the following typical figures to check whether your guess was right:

   mass of car: 600 kg   mass of skier: 80 kg

   speed of car: 22 m s^{-1} speed of skier: 66 m s^{-1}

29 Acapulco, Mexico, is the home of high diving championships, where divers jump from a ravine 35 m above the sea.

   (a)   Calculate the loss in gravitational potential energy, and hence the gain in kinetic energy, of a diver of mass 65 kg, and so calculate the speed on hitting the sea.

   (b)   Explain whether or not a more massive diver would hit the sea at the same speed. (Assume the diver falls freely, i.e. gravity is the only force acting, and use g = 9.8 N kg^{-1} = 9.8 m s^{-2}.)

3.2 Speed skiing

Speed skiing (Figure 1.40) is the fastest non-motorised sport on Earth. The world record is close to 156 miles per hour – about the top speed of a sports car. Wearing only a rubber suit, and with feet strapped into two 2.4 m long boards, the speed skier hurtles down a steep 'waterfall' of ice.

The speed skier has a huge amount of kinetic energy, which in a collision could easily break every bone in the body, and prove fatal. Because of this the ski authorities limit the speed that the skiers can reach in competitions by specifying a maximum length of the acceleration zone (see Figure 1.41). They can calculate the maximum allowable distance using Equations 17 and 19. But can you see a problem? Equation 17 says that we must use the displacement in the direction of the force, but the skier's motion is not parallel to the vertical force of gravity.

Figure 1.40 Speed skier

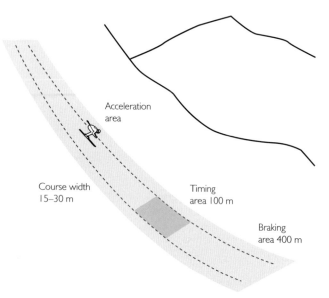

Figure 1.41 The RTS speed skiing course at Willamette Pass, Oregon, is one of three sanctioned courses in the USA and ten in the world. One section has a 52° slope.

From Figure 1.42 you can see that, while the skier travels a distance Δs along the slope, the displacement in the direction of the force is the change in vertical height Δh. Using the trigonometric rules for a right-angled triangle:

$$\Delta h = \Delta s \, \cos\theta$$

In other words, Δh is the vertical component of the skier's displacement vector. It is the component of the displacement parallel to the direction of the force, as described by Equation 12 in Part 2 of this unit. We can modify Equation 17 to take account of the angle θ between force and displacement:

$$\Delta E = \Delta W = F\Delta s \, \cos\theta \qquad (21)$$

> **Maths reference**
>
> Sine, cosine and tangent
> See Maths note 6.2

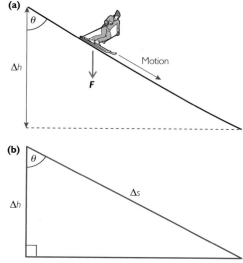

Figure 1.42 Skiing down a slope

You can also arrive at Equation 21 by considering the component of the force that acts along the slope, $F\cos\theta$, and by multiplying by the overall displacement Δs.

Question

30 Calculate the maximum length of an acceleration zone at 50° to the horizontal if skiers are not to exceed 66 m s^{-1} (about 150 mph). (Hint: look at your answers to Questions 28 and 29.)

3.3 Pumping iron

It's not just strong men and weight-lifters who pump iron. These days, gyms are full of people who build 'resistance training' into their workouts (Figure 1.43). Even if you don't play sport competitively, lifting weights can benefit your health. Stronger muscles are less prone to injury, and it is believed that building muscle can help you 'burn' your food more quickly.

Figure 1.43 Pumping iron

Power

Weight-training simply means increasing the resistance to a muscle's movement. Making the muscles work harder stimulates their growth. But for most athletes it's not strength itself that is the main objective, it's **power**. Loosely speaking, power is 'fast strength'. Whether it's in the legs, back, arms, or shoulders, power is important for almost every sport. If you look in a sports training manual, you might see power defined as:

$$\text{power} = \text{force} \times \text{velocity} \qquad (22)$$

But if you consult a physics text book you will find it is defined as 'the rate of doing work' or 'the rate of energy transfer':

$$\text{power} = \frac{\text{work done or energy transferred}}{\text{time taken}} \qquad (23)$$

$$P = \frac{\Delta W}{\Delta t}$$

$$\text{or } P = \qquad\qquad (23a)$$

The SI units of power are W or J s^{-1}; 1 W = 1 J s^{-1}.

Using Equations 17 and 23:

$$P = \frac{F\Delta s}{\Delta t} = F\frac{\Delta s}{\Delta t}$$

From Equation 1, $\frac{\Delta s}{\Delta t} = v$ and so we have:

$$P = Fv \qquad\qquad (22a)$$

So the two meanings are equivalent – they are both useful in different circumstances. (Remember, though: just as Δs is the displacement in the direction of the force, so v is the component of velocity in the direction of the force.)

Worked examples

Q A woman training on a stepping machine 'climbs' 150 m in 2 minutes. If her mass is 60 kg, calculate her power.

A Using Equations 7, 17 and 23:

work done against gravity = $F\Delta s = mg\Delta s$

$$\text{power} = \frac{mg\Delta s}{\Delta t}$$

$$= \frac{(60 \text{ kg} \times 9.8 \text{ N kg}^{-1} \times 150 \text{ m})}{120 \text{ s}}$$

$$= 736 \text{ W}$$

Q A formula 1 racing car travelling at its top speed of 95 m s^{-1} has an engine power of 15 kW. Calculate the thrust of the engine (i.e. the force it produces in the car).

A Rearranging Equation 22:

$$F = \frac{P}{v} = \frac{1.5 \times 10^4 \text{ W}}{95 \text{ m s}^{-1}} = 158 \text{ N}.$$

In the second example, you might wonder why the engine needs to provide a thrust if the car is not accelerating. The thrust from the engine is balanced by an equal, opposite force from air resistance and friction with the road, and the car moves at constant velocity.

Questions

31 An athlete is working out, doing 'bench presses'. Each lift raises 60 kg through a distance of 60 cm. If he wants to generate 150 W of power, how quick does each lift have to be?

32 A sprinter, mass 60 kg, accelerates to her top speed of 10 m s^{-1} in 3 s. Calculate her average power while accelerating.

Measuring power

Which type of sport are you better at: 'explosive' sports like sprinting, or activities that require more endurance? One reason why people often fall into one or other category is the composition of their muscles. There are two distinct types of muscle fibre (Figure 1.44), called slow twitch (ST) and fast twitch (FT). FT fibres produce more force and power but the can only operate over a short period. We all have different compositions of FT and ST muscle fibres. People who have a high percentage of ST muscle may have an advantage in prolonged endurance, whereas those with predominantly FT fibres are often better suited to short-term explosive activities. With world-class athletes the contrast is striking. Marathon champions, for instance, have over 90% ST fibres, whereas sprinters' muscles contain 75% FT fibres.

Every time you need a short burst of maximum power, you're relying on your FT fibres. Unlike ST fibres, which need oxygen to break down fuel, your FT fibres can work without it – anaerobically – but at a cost. If you continue to work at high power, a substance called lactic acid builds up in your muscles, making them hurt and eventually forcing you to stop or slow down.

In any physical activity, the output power that you measure (as in Activities 24 and 25) is less than the overall input power from your muscles, because the process is never 100% efficient and you have not measured the energy wasted in heating. Table 1.3 on page 38 lists some typical input powers for various activities. You might like to compare them with your output power measured in Activity 24.

Figure 1.44 A micrograph of human muscle

Activity 24 Anaerobic power

In a laboratory or a gym, devise experiments to measure your power while running up stairs, cycling, jumping or sprinting. Compare your results with the averages shown in Figure 1.45.

Activity 25 The power of an athlete

Use *Multimedia Motion* to estimate the power of a weight-lifter, or of the Space Shuttle as it takes off.

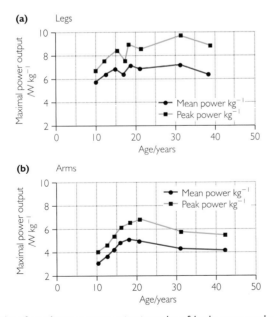

Figure 1.45 Graphs of maximum power output per kg of body mass against age

In any physical activity, the output power that you measure (as in Activities 24 and 25) is less than the overall input power from your muscles, because the process is never 100% efficient and you have not measured the energy wasted in heating. Table 1.3 (p38) lists some typical input powers fro various activities. you might like to compare then with your output power measured in Activity 24.

Activity	Input power/W kg^{-1}
Resting, lying down	1.2
Sitting	1.2
Standing	1.2
Eating	1.2
Dressing/undressing	2.3
Showering	4.1
Typing at a computer	2.3
Walking (5.5 km/h)	6.4
Cycling (15 km/h)	5.8
Jogging (8 km/h)	9.4
Fast running (6 min mile)	18
Swimming (fast crawl)	18
Playing musical instrument	2.9
Playing cricket	4.7
Playing table tennis	5.3
Playing tennis	7.1
Skiing	9.4
Dancing (energetically)	7.6
Playing football	11
Gymnastics	12

Table 1.3 Some activities and their input power demands

Activity 26 Aerobic power

Find out what athletes mean by 'VO$_2$ max', and how this relates to oxygen uptake and aerobic power. Find out how you can measure your own aerobic power in a 'shuttle run'. Compare your own measurements with the data in Table 1.4.

Activity 27 Energy demands

Use the data in Table 1.3 to estimate how much energy you need to take in (from food) in a typical day.

A Fuse bar supplies about 1000 kJ. Use Table 1.3 to estimate how many Fuse bars you would 'burn off' during a sporting activity.

In a discussion or by exchanging notes, compare your results with someone else's.

Event	Typical power
click beetle leaping into the air	0.5 W
TV when switched on	50 W
leaping red deer	2 kW
sports car	150 kW

Table 1.4 Power ratings. Where do you fit in?

Questions

33 A sports car can generate a maximum 60 kW of power. If it has a mass of 800 kg, what is the minimum time in which the car could accelerate from rest to 30 m s^{-1}?

34 Your 'basal metabolic rate', the amount of power you generate while resting (pumping blood, breathing and keeping warm), is about 90 W. Calculate the amount of energy you need per day for simply existing.

35 A 'kinetic' watch needs no battery. A tiny weight is set in motion by the slightest movements of your arm and, according to the manufacturer, enough energy to drive the watch for 2 weeks is stored electrically. If the power needed to drive the watch is 0.1 mW (1×10^{-4} W), estimate the amount of energy stored in the watch.

3.4 Summing up Part 3

In this part of the chapter you have seen how energy can be measured using the concept of work. You have applied the conservation of energy, and formulae for kinetic and gravitational potential energy, to model different sporting situations. Finally, you have explored the relationship between power and energy.

Activity 28 Summing up Part 3

A good way to reinforce the new ideas you have learned is to produce a 'concept map'. First make a list of all the terms printed in bold in this part of the chapter– plus any others that you think important – spread out on a large sheet of paper. Then draw lines between all related terms. Finally label each line with a phrase or equation describing the link.

Alternatively, download 'freemind' software which allows you to produce an interactive revision map. You can use visual aids and colour to help highlight areas, and paste in hyperlinks to websites and animations on the Internet.

4 Stretching and springing

In this part of the chapter, you are going to look at two sports that rely on elastic materials: bungee jumping (Figures 1.46 and 1.47) and pole vaulting. In doing so, you will revisit and use ideas from Parts 1 to 3 of this chapter.

Figure 1.46 The popular sport of bungee jumping

Figure 1.47 One of the authors doing research for this unit

4.1 Bungee jumping

Standing on a platform built from the side of a sheer cliff face, I looked down. Forty-five metres below me the blue waters lay in wait, glistening in the sun. I had never been so terrified in my life. There was no way my brain was going to let me jump – it was as though an invisible force held me back.

'Stretch your arms out wide' came the voice from behind me, 'and whatever you do, don't grab the rope as you go down.' I tightened my shoelaces once more. 'Are you sure the rope won't slip off my ankles?' I asked.

The two technicians went through the equipment checklist a second time.

'Harnesses,' 'Checked.'

'Static sleeve,' 'Checked.'

'Hey, Bob. This carabiner's a bit loose. Do you think that matters?'

'No. It's been like that for ages.'

Bungee technicians took great pleasure in scaring first-timers.

'Okay, you're on for a jump!'

There was no turning back. Gingerly I stepped forward until my toes were right at the end of the platform. The countdown began. 'Five … four … three …' the voice of fear in my head was replaced by another:

'How can you face your friends if you wimp out?' I jumped.

For the next three and a half seconds I was weightless – like an astronaut in orbit, but at a fraction of the cost. And what a rush it was! It's hard to describe because it was like being in another world – with no sound and nothing supporting me. My body felt utterly vulnerable, powerless to prevent itself smashing into the water surface at over 100 kilometres an hour.

Panic was beginning to swamp my consciousness, but just then came the reassuring tug on my ankles from the cord above. I knew bungee ropes could stretch four times their natural length, but it seemed to extend forever. I could feel the tension building up and had to close my eyes. And then I felt the splash of water on my face as my head dipped briefly under, before I was pulled up again. Suddenly I was an enormous yo-yo, reaching fully half way back to the top. Gradually the bounces grew smaller and smaller until I came to rest above the waiting jet boat. The sense of relief was overwhelming.

Bungee jumping is not a new sport. For hundreds of years, men of the island of Pentecost, off Papua New Guinea, have leaped from wooden towers with jungle vines attached to their ankles (Figure 1.48). For them it's a test of their courage – the closer they swoop to the ground, the greater their bravery. Inevitably there have been deaths, one of which happened while Queen Elizabeth II and the Royal Family were watching the ceremony. So how has bungee jumping evolved into a relatively safe activity?

It's got a lot to do with two guys from New Zealand. One is a daredevil businessman called A. J. Hackett, and the other, Henry van Asch, a physicist. Back in the mid-1980s they planned some outrageous stunts to bring bungee jumping to the world's attention, including a leap from the Eiffel Tower. A policeman arrested Hackett after the jump (presumably once he'd stopped bouncing). A. J. Hackett Bungee became the world's first professional bungee operation in Queenstown, New Zealand. Since then, bungee jumping has become established worldwide.

Figure 1.48 The origins of bungee jumping

Analysing a bungee jump

In Activity 29 you are going to put yourself into the position of a bungee designer: given a piece of bungee rope, how do you choose the length to stop someone just short of the ground? Make it too long and they're history; too short and you'll remove a lot of the excitement. Before you do, you will see how you can apply the physics of this chapter to working out the correct length of a bungee rope.

One approach to solving many problems involving motion is to use forces. The forces on a bungee jumper are shown in Figure 1.49. Because the tension in the rope varies with extension, the force on the jumper is not constant, which makes it difficult to apply the equations of motion.

F — Tension in rope varies with extension

Weight = mg

Figure 1.49 The forces on a bungee jumper

Another approach is to use energy conservation. Figure 1.50 shows the energy transfers at different stages during a jump. As the jumper falls, he loses gravitational potential energy. At first, during free fall, he gains kinetic energy. Then the rope begins to stretch, reducing the jumper's kinetic energy. Then the rope begins to stretch, reducing the jumper's kinetic energy. Energy is transferred to the rope as it stretches, so the rope now has potential energy that we will call **elastic energy** (and symbolise E_{el}). When the jumper comes momentarily to rest at the bottom of the first 'bounce', his kinetic energy is again zero; he has lost gravitational potential energy and the rope has gained elastic potential energy. Energy conservation tells us that the gravitational energy lost ($mg\Delta h$) must be equal to the elastic energy gained by the rope.

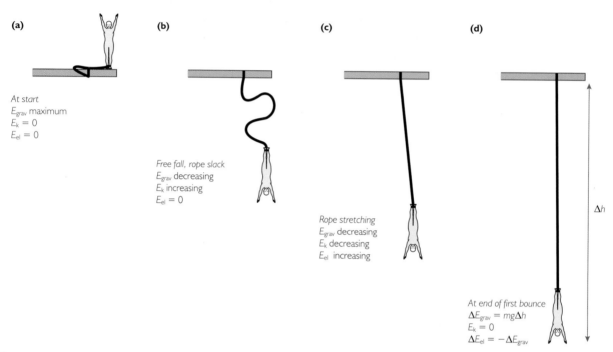

(a)

At start
E_{grav} maximum
$E_k = 0$
$E_{el} = 0$

(b)

Free fall, rope slack
E_{grav} decreasing
E_k increasing
$E_{el} = 0$

(c)

Rope stretching
E_{grav} decreasing
E_k decreasing
E_{el} increasing

(d)

Δh

At end of first bounce
$\Delta E_{grav} = mg\Delta h$
$E_k = 0$
$\Delta E_{el} = -\Delta E_{grav}$

Figure 1.50 Energy in a bungee jump

Measuring elastic energy

You will remember we used the definition of work to derive formulae for gravitational and kinetic energy. Force–extension graphs give us a way to find the work done in stretching a bungee rope. Figure 1.51 shows a typical force–extension graph. The force varies, so we cannot simply multiply force by distance to find the work done in stretching the sample. But for a small increase in extension, Δx, the stretching force F is very nearly constant, so the work done is equal to the area of the strip:

$$\Delta W = F \times \Delta x$$

The total work done in stretching the sample can be found by adding up all the areas of the narrow strips (each with a different height). In other words, the elastic energy stored can be found from the **area under a force–extension graph**.

Study note

Notice that this is very similar to finding the distance travelled from a velocity-time graph. See section 1.3.

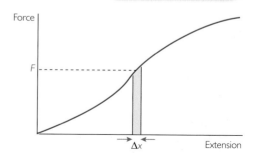

Force

F

Δx

Extension

Figure 1.51 A force–extension graph

For a material that obeys Hooke's law, the force–extension graph is a straight line
(Figure 1.52) and there is a simple formula for E_{el}:

$$E_{el} = \tfrac{1}{2}Fx \qquad\qquad (24)$$

where F is the force needed to produce an extension x. Using Equation 14 we can
write an expression involving the stiffness k, which is constant for a sample that obeys
Hooke's law:

$$E_{el} = \tfrac{1}{2}kx^2 \qquad\qquad (25)$$

However, with a rubber bungee cord, which does not obey Hooke's law, the graph is
curved. We therefore estimate the area by counting squares of graph paper under the
curve. In Figure 1.53, the values for force and extension are marked on the graph.
On the scale shown, the area of each square represents 10 N × 0.1 m = 1 J. You can
estimate the area by counting only the squares where at least half the area is under the
curve, giving 25 squares, i.e. the elastic energy stored is 25 J.

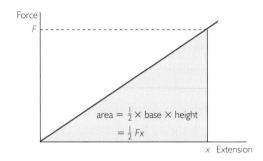

Figure 1.52 A force–extension graph for a material that obeys Hooke's law

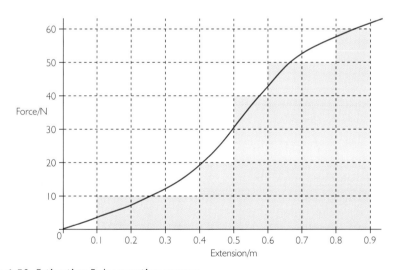

Figure 1.53 Estimating E_{el} by counting squares

Activity 29 Bungee challenge

Set up a model bungee jump, using a piece of elastic. By calculating elastic energy
from a force–extension graph, you can work out the height from which a given object
can 'jump' with a given piece of elastic, so that it will just miss the floor.

Questions

36 In a bow and arrow, the wire stretches by 60 cm when the bow is pulled back. If the stiffness of the wire is 0.4 N m^{-1}, how much elastic energy is stored?

37 In most bungee jumps, several cords are used, just in case one of them fails. *The number of cords depends on the jumper's body weight.* Explain the sentence in italics.

38 In February 1992 Greg Rifti set a world bungee record jumping from a helicopter with a rope 250 m long. His cord stretched to 610 m (2000 feet). Calculate (a) the gravitational potential energy lost as he fell (take his mass to be 75 kg), (b) the stiffness of the cord assuming it obeyed Hooke's law and (c) the extension of the cord when he finally came to rest.

4.2 Pole vaulting

Pole vaulters are truly the astronauts of the stadium (Figure 1.54). The Russian pole vaulter, Sergei Bubka, who has steadily pushed the record up past 6 metres, is effectively jumping over three people, one on top of the other. In the 1960s, there was a sudden rise in the pole vault record heights (Figure 1.55). It was the result of fibreglass poles being introduced, replacing bamboo and aluminium poles. Fibreglass totally changed the event, allowing the pole to bend nearly into a half circle during the swing.

Can we expect the pole vault record to keep on increasing, or is there a limit? Answer Questions 39 and 40 to help you decide.

Figure 1.54 Pole vaulting

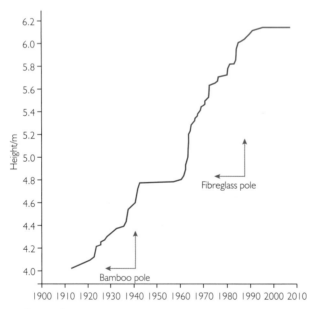

Figure 1.55 Pole vault records

Questions

39 Write down the labels that you would need to add to Figure 1.56 so that it shows the energy in a pole vault (like Figure 1.50 showed for a bungee jump).

40 (a) If a pole vaulter can sprint at 10 m s^{-1} (the world sprint record is 10.27 m s^{-1}), through what height can he raise his entire centre of gravity if all his kinetic energy is used to supply additional gravitational potential energy?

(b) Assuming that the vaulter's centre of gravity is 1 m above the ground when running, estimate the height of the vault.

(c) Say what, in practice, might (i) prevent the vaulter from clearing this height or (ii) enable him to exceed this height.

(d) Say whether you think the graph in Figure 1.55 will keep on rising. (Use $g = 9.8$ N kg^{-1} = 9.8 m s^{-2}.)

(a) The run-up   **(b)** Bending the pole   **(c)** Vaulting over the bar

Figure 1.56 Stages in a pole vault

Further investigations

Investigate the energy transfers in a model pole vault. Devise a way to measure the work done in deforming a springy rod, and the kinetic energy that it imparts to a catapulted mass. Compare the efficiency of the energy transfer obtained with rods of different materials. Scale up the results from you model and compare them with real pole vault records.

4.3 Summing up Part 4

In this part of the chapter you have seen how to apply the idea of energy conservation and how to measure elastic energy using a force–extension graph. In doing so, you have used and extended ideas from earlier in the chapter. Questions 41 to 44 give you further practice in using ideas about energy.

Questions

Use $g = 9.8$ N kg^{-1} = 9.8 m s^{-2} in these questions.

41 An overhead electricity cable between two pylons stretches 50 cm under its own weight. How much elastic energy does it store? (Assume the cable obeys Hooke's law and has $k = 10^7$ N m^{-1}.)

42 Itaipu, one of the world's biggest hydroelectric power stations, lies near one of the world's natural wonders – the Iguacu Falls in South America. It supplies a large fraction of Brazil's and Paraguay's energy needs. When the flow of water is 10 000 m^3 s^{-1} the power station generates 12 GW of power (one gigawatt, 1 GW = 10^9 W).

    (a)   Calculate the mass of water flowing per second. (Density of water = 1000 kg m^{-3}).

    (b)   Estimate the height through which the water drops. What assumption did you make? Will this lead to an underestimate or an overestimate of the height?

43 (a)   A swimmer moving at constant speed through the water uses a force to do work, but does not increase her kinetic energy. How can this be?

    (b)   Figure 1.57 is a graph of the energy cost of swimming a kilometre for aquatic animals. For humans, a typical energy cost while swimming is 80 kJ min^{-1} and it takes many minutes to swim a kilometre. Approximately where on the graph would the 'human' data point lie? Comment on this.

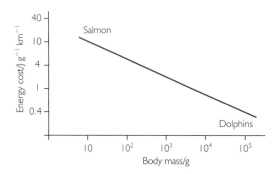

Figure 1.57 The energy cost of swimming

44 Here is an energy puzzle. A rock climber slips vertically, gradually stretching her rope until it supports her, i.e. the tension in the rope is equal to her weight ($F = mg$). She falls through a height Δh and her rope is stretched by the same amount. She has lost gravitational potential energy $\Delta E_{grav} = mg\Delta h$. But the elastic energy stored in the rope (which obeys Hooke's law) is $\Delta E_{el} = \frac{1}{2}Fx = \frac{1}{2}mg\Delta h$. What has happened to the missing energy?

5 Jumping and throwing

Many events in the Olympics involve launching objects into the air. The jumping events involve throwing your own body as a **projectile** and the throwing events involve launching another object.

The four Olympic throwing events are discus, shot, javelin and hammer. Despite the varying masses and shapes of all these objects thrown, there are two main aspects that all throwers have to master: a speed building phase, and a throwing position angle.

The main components of long jump are horizontal speed and vertical lift. International long jumpers should be fast enough to earn a place in their national relay squad. Jesse Owens and Carl Lewis were the world's fastest sprinters in their time and were also the best long jumpers. Heike Drechsler, one of the world's best ever women long jumpers, was joint world record holder for the 200 m and has a personal best of 10.91 s for the 100 m sprint. Coaching for the long jump involves coaching for sprinting. The long jump also involves achieving sufficient height to stay in the air for a long time. The longer the time in the air and the greater the horizontal speed, the greater the horizontal distance travelled.

5.1 Ski jumping

The Winter Olympic sport of ski jumping is similar to the long jump – the aim is to leave the ramp at high speed in order to travel as far as possible before landing. In this section, you will explore the relationship between the launch speed and the length of the 'jump' and so reach some general conclusions about projectiles that move freely under gravity.

Activity 30 Ski jump

Use the arrangement shown in Figure 1.58 to see how the launch height h_1, and the height of the vertical drop h_2 affect the horizontal distance travelled.

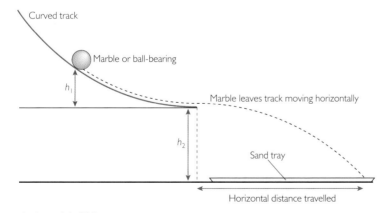

Figure 1.58 A model ski jump

Activity 30 shows that increasing the launch height and the vertical drop both increase the length of the jump. How can these results by explained more precisely? Activity 31 provides a clue.

Activity 31 Projectile motion

Release two squash balls side by side at the same time. Do they hit the floor at the same time? Repeat, but this time launch one squash ball horizontally. Do they hit the floor at the same time?

Figure 1.59 shows strobe photographs of two balls moving as in Activity 31. Notice that they both hit the floor at the same time. It does not matter that one of them has got a horizontal velocity as well as a vertical acceleration. Notice too, that the sideways-moving ball covers equal horizontal distances in equal time intervals – its horizontal velocity remains constant and is unaffected by its vertical motion. The vertical and horizontal components of the motion are *independent of each other*, so we can treat them quite separately.

Projectile motion

The displacement, velocity and acceleration of the projectile (the ball or the skier) are all vectors. We can treat the horizontal components of these vectors completely separately from the vertical components of velocity and acceleration. We will use x and y to indicate horizontal and vertical components of velocity and acceleration.

Figure 1.59 Strobe photographs of projectiles

If the ball, or skier, leaves the ramp horizontally, the initial velocity vector has no vertical component: $u_y = 0$. In the vertical direction, the force of gravity provides a uniform acceleration, so the vertical component of velocity increases. The vertical motion is described by Equations 3b and 4a with $a_y = g = 9.8$ m s^{-2}, taking downwards as positive:

$$v_y = gt$$
$$\text{and } y = \frac{1}{2}gt^2 \tag{26}$$

In the horizontal direction there is no component of the gravitational force, so there is no acceleration: $a_x = 0$. The horizontal velocity remains equal to its initial value u_x, and the same as Equation 4a with $a = 0$:

$$x = u_x t \tag{27}$$

The horizontal and vertical components combine to give a projectile a **trajectory** (path through the air) that has the shape of a **parabola**. Activity 32 illustrates this.

Activity 32 Parabolic trajectory

Use Equations 26 and 27 to complete Table 1.5 and hence to plot the trajectory of a ball thrown sideways at 2 m s^{-1}.

Time t/s	Vertical displacement y/m	Horizontal displacement x/m
1		
2		
3		
4		
5		

Table 1.5 Table for Activity 32

The following example shows how analysis of projectile motion can be useful in police forensic work.

Worked example

Q A car ran off a mountain road and landed 80 m from the foot of a cliff face having fallen 100 m. Was the driver going too fast? The speed limit was 30 mph (about 13 m s^{-1}).

A If air resistance is ignored, the horizontal component of the velocity will not change as the car leaves the road and travels through the air. The horizontal velocity can be estimated if the horizontal distance travelled (80 m) is divided by the time that the car is in the air.

Dealing with vertical motion first to find t:

    displacement vertically, y = 100 m

    initial velocity, u_y = 0 m s^{-1}

    acceleration downwards due to gravity, g $\approx$ 10 m s^{-2}

Using Equation 26: $y = \frac{1}{2}gt^2$, so

$$t = \sqrt{\left(\frac{2y}{g}\right)} = \sqrt{\left(\frac{2 \times 100 \text{ m}}{10 \text{ m s}^{-2}}\right)} = 4.47 \text{ s}$$

Now the horizontal velocity can be found using Equation 27

$$u_x = \frac{x}{t} = \frac{80 \text{ m}}{4.47 \text{ s}} = 17.897 \text{ m s}^{-1} \approx 18 \text{ m s}^{-1}$$

So the driver was well over the speed limit and this probably contributed to the crash.

5.2 Throwing

In throwing events such as the discus and shot, and in games such as football or golf, the projectile is not launched horizontally. But what launch angle gives the maximum range?

A top British shot putter (ranked third in the UK with a distance of 17.90 m) took part in biomechanical tests to find the limiting factors in his performance. The digitised information from a video of his throwing was analysed by computer. Data on height of release h, projection speed V, and projection angle θ (see Figure 1.60) were all recorded. By calculating the effect of changing the angle and launch speed, he found that he needed to increase the speed and alter the launch angle. By changing his technique, he increased his personal best by 0.4 m and set a new Scottish National record.

The range of a projectile

The motion of a shot can be analysed by resolving the initial velocity into horizontal and vertical components:

$$u_x = V\cos\theta$$

$$\text{and } u_y = V\sin\theta$$

(28)

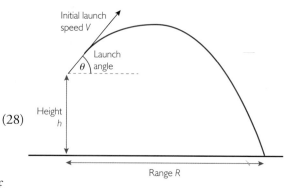

Figure 1.60 Measurements on a shot putt trajectory

and then treating the horizontal and vertical motions independently.

For a projectile launched from ground level ($h = 0$), the time of flight can be found by considering the vertical component of its motion and using Equation 4a:

$$s = ut + \frac{1}{2}at^2$$

with $s = 0$ (the projectile falls back to the ground), $u = u_y = V\sin\theta$ and a = $-g$, we have:

$$0 = V\sin\theta t - \frac{gt^2}{2}$$

so:

$$\frac{gt^2}{2} = Vt\sin\theta$$

$$t = \frac{2V\sin\theta}{g}$$

(29)

The range R (i.e. the horizontal displacement x) can then be found using Equation 27:

$$x = u_x t$$

$$= \frac{V\cos\theta \times 2V\sin\theta}{g}$$

$$= \frac{2V^2\sin\theta\cos\theta}{g}$$

(30)

In this simple situation, the range is greatest for $\theta = 45°$.

However, real life is less straight forward – for a start, h is rarely zero. Also, air resistance significantly affects the motion of many projectiles, so the launch angle varies according to the object in question.

The angle of release in a hammer throw is close to 45° because it is launched from near the ground and because air resistance has little effect on its motion. The shot putt requires an angle less than 45° because it is launched from a position higher than where it lands. Because of their shape, a discus and javelin 'float' on the air: the angle of release depends on release speed and headwind, but is always less that 45°. The effects of 'spin' and air resistance make the trajectories of cricket and golf balls even more complicated – again, their best launch angle is less than 45°.

Using a computer model

In Activity 33, you use a computer to calculate the range of a projectile launched from above the ground. The main power of the computer is that it can perform routine calculations very quickly. You, as the scientist, just need to tell it the correct calculations to perform. Once you get an equation of the physics correct and set up a working computer program based on correct physical equation, you can design problems that may be very difficult or expensive to run for real. If the results do not match up with what happens in a real-world experiment, this tells you that the equation inserted in the program is probably wrong. Refinements can then be made until the computer simulation is closer to reality. Effectively, you can perform a computer experiment and obtain results quickly, cheaply and safely. Engineers design aircraft, cars, buildings and bridges in this way.

Activity 33 Range of a projectile

By carrying out some algebra, derive an equation for the **range** of a projectile launched from above the ground. Ignore air resistance. Use a spreadsheet to calculate the range for various launch speeds and angles.

5.3 Summing up Part 5

In this part of the chapter you have revisited ideas about vectors and about uniformly accelerated motion and used them to study projectiles. Questions 45 to 47 and Activities 34 and 35 use these ideas in a variety of sporting and non-sporting situations – real and imaginary!

Questions

45 A skier slides down the slope shown in Figure 1.61, leaving the ramp at X horizontally at 20 m s^{-1}.

(a) She drops through 20 m before hitting the mountainside at Y. Calculate the time this takes.

(b) Using the time from (a), calculate the horizontal displacement R.

(c) Explain how each of the following actions affects the distance R:

   (i) the skier pushes herself off horizontally when leaving X

   (ii) the skier tries to jump upwards at X before leaving the ramp.

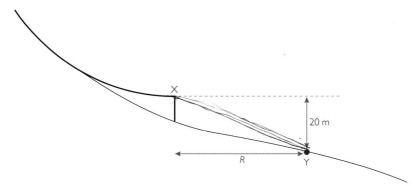

Figure 1.61 Ski jump diagram for Question 45

46 A cartoon character (like 'Roadrunner' for example) is sometimes shown running off the edge of a cliff (Figure 1.62). It is only when he looks down and realises that there is no ground at his feet that he stops moving forwards and starts to drop vertically.

(a) Explain what *should* happen to his horizontal component of his velocity and the vertical component of his velocity as soon as he runs off the edge of the cliff.

(b) Sketch the cartoon trajectory of his motion and also the correct physical parabolic trajectory.

Figure 1.62 A cartoon character runs over the edge of a cliff

47 Figure 1.63 is from a 16th century book, and shows the supposed trajectory of a cannon ball.

(a) What evidence have you seen that shows that such a trajectory is incorrect?

(b) Sketch a more realistic path for the cannon ball (ignoring air resistance).

(c) If the cannon ball is launched from ground level at a speed of 50 m s^{-1}, calculate its range for launch angles of 30°, 45° and 60°.

Figure 1.63 The supposed trajectory of a cannon ball, drawn in 1561

Activity 34 What happens next?

A James Bond-type super-hero has just spotted a wicked villain climbing over the balcony of a building opposite, intent on some dastardly deed (Figure 1.64). He takes aim and fires. Just as he does so, the villain notices what's happening and lets go, hoping to drop to the ground and escape. What happens next? Does the villain escape the bullet? Use your knowledge of projectile motion to complete the story.

Figure 1.64 What happens next?

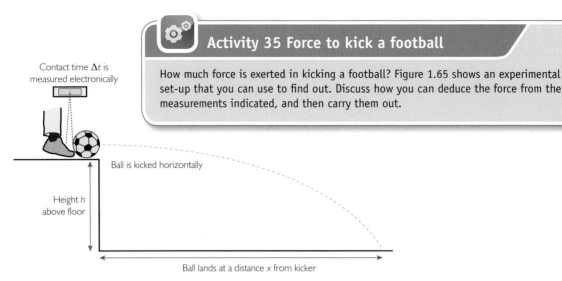

Activity 35 Force to kick a football

How much force is exerted in kicking a football? Figure 1.65 shows an experimental set-up that you can use to find out. Discuss how you can deduce the force from the measurements indicated, and then carry them out.

Ball is kicked horizontally

Height h above floor

Ball lands at a distance x from kicker

Figure 1.65 Apparatus for Activity 35

Further investigations

Use computer models and experimental measurements to investigate the effect of air resistance on the range of projectiles with various shapes.

6 Last lap

6.1 Summing up the chapter

In this chapter you have studied some aspects of motion, balanced and unbalanced forces, and energy. This concluding session is intended to help you to look back over the whole chapter and consolidate your knowledge and understanding.

Activity 36 Sports consultant

Look back through the chapter and make sure you know the meanings of the key terms printed in bold.

Use what you have learned in this chapter to write a brief guide for a coach, or sports teacher to help them integrate physics into their training. Choose any sport you like which involves one or more of the concepts you have studied in the chapter. Here are a few possibilities: high jump, skiing, paragliding, sailing.

Include explanations of the relevant physics principles, and how to apply them to the sport. Write in an appropriate style for your readers.

Throughout this chapter, you have seen how graphs can be just as useful as equations in calculating physical quantities. You have used the gradient of a graph and the area under a graph.

Activity 37 Using graphs

Summarise in a table all the different quantities you know how to find using gradients and areas of graphs along with the graph they are measured from.

Activity 38 Advertising

A tyre manufacturer introduced a new tyre designed to improve fuel consumption – see Figure 1.66.

Imagine you are in the marketing department. You have been asked to write the text of a short newspaper advertisement for the 'Energy Tyre', designed to show how much further a tank of petrol will take you, and why – simply. The technical department has supplied the graph shown in Figure 1.67, and the data given below.

Tyre data

- Fuel consumption of average family car with normal tyres = 9.5 litres per 100 km
- Reduction in fuel consumption with 'Energy Tyres' = 3%
- The 'Energy Tyre' is the first example of 'low rolling resistance technology'. Rolling resistance: the absorption of energy and its dissipation in the form of heat, by the tyre, as a result of low-frequency deformation
- High-frequency, microscopic deformation of the tyre also occurs, and results in grip

Figure 1.66 A new design of tyre

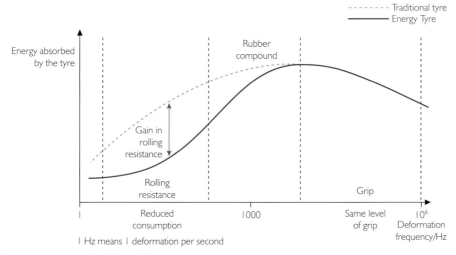

Figure 1.67 Tyre data for Activity 38

6.2 Questions on the whole chapter

Questions

48 A fielder throws a cricket ball of mass 160 g towards the stumps. The wicket keeper catches the ball, which is travelling at 15 m s⁻¹. In bringing the ball to rest, his hands move 0.4 m backwards. What is the average stopping force applied to the ball?

49 In 1976, basketball player Darrell Griffith's standing vertical jump measured 1.20 m. Estimate the speed at which he left the ground.

50 A karate expert sets up a block of wood, mass 500 g, to break with his fist. The energy needed to smash the block is 40 J. Estimate the minimum speed of the karate expert's fist to break it, assuming that all its kinetic energy is transferred to the block.

51 Before a match, a trainer makes his players take in 640 g of carbohydrate food. If 100 g produces 100 kJ of useful energy, estimate the average power expended over a 90 minute game.

52 A ski-lift can carry 100 people at a time up a slope at 20° to the horizontal, at 3 m s⁻¹. Estimate the minimum power of the ski-lift motor.

53 When an aeroplane lands on an aircraft carrier, it can be stopped by a huge steel 'arresting wire', stretched across the deck. A plane of mass 20 t lands at 50 m s⁻¹ and the wire has a stiffness $k = 3.3 \times 10^8$ N m⁻¹. (1 t = 1 × 10³ kg). Find

    (a)   the kinetic energy of the plane landing

    (b)   the amount the arresting wire stretches (assuming the wire obeys Hooke's law).

54 A car skidded off the road and hit a low stone wall. Glass from the windscreen was found 4.0 m in front of the car windscreen in the field.

    (a)   If the height of the middle of the windscreen was 1.2 m, estimate the speed of the car as it hit the wall.

    (b)   Why is this value likely to be lower than the actual speed of the car when it had been on the road?

55   (a)   In a sprint race a sprinter of mass 55 kg had an initial horizontal acceleration of 10 m s⁻², produced by the force with which her feet pushed against the starting blocks. Calculate the size of this accelerating force.

    (b)   An average net force of 200 N is exerted on a tennis ball, which has a mass of 0.055 kg, while it is in contact with the racket. Calculate the ball's average acceleration during this contact time.

    (c)   A golf club strikes a golf ball with an average force of 6.00 kN, producing an average acceleration of 1.40×10^5 m s⁻². What must be the mass of the ball?

6.3 Achievements

Now you have studied this chapter you should be able to achieve the outcomes listed in Table 1.6.

Table 1.6 Achievements for the chapter *Higher, Faster, Stonger*

	Statement from examination specification	Section(s) in this chapter
1	use the equations for uniformly accelerated motion in one dimension: $v = u + at$ $s = ut + \frac{1}{2}at^2$ $v^2 = u^2 + 2as$	1.2, 1.3, 5.1, 5.2
2	demonstrate an understanding of how ICT can be used to collect data for, and display, displacement/time and velocity/time graphs for uniformly accelerated motion and compare this with traditional methods in terms of reliability and validity of data.	1.3, 1.4
3	identify and use the physical quantities derived from the slopes and areas of displacement/time and velocity/time graphs, including cases of non-uniform acceleration	1.3
4	investigate, using primary data, recognise and make use of the independence of vertical and horizontal motion of a projectile moving freely under gravity	5.1, 5.2
5	distinguish between scalar and vector quantities and give examples of each	1.2
6	resolve a vector into two components at right angles to each other by drawing and by calculation	2.1, 3.2
7	combine two coplanar vectors at any angle to each other by drawing, and at right angles to each other by calculation	2.1
8	draw and interpret free-body force diagrams to represent forces on a particle or on an extended but rigid body, using the concept of *centre of gravity* of an extended body	2.1
9	investigate, by collecting primary data, and use $\Sigma F = ma$ in situations where m is constant [Newton's first law of motion ($a = 0$) and second law of motion]	1.4, 2.1
10	use the expressions for gravitational field strength $g = F/m$ and weight $W = mg$	1.4, 4.1, 4.2
11	identify pairs of forces constituting an interaction between two bodies [Newton's third law of motion]	1.4
12	use the relationship $E_k = \frac{1}{2}mv^2$ for the kinetic energy of a body	3.1
13	use the relationship $\Delta E_{grav} = mg\Delta h$ for the gravitational potential energy transferred near the Earth's surface	3.1, 4.4, 4.2
14	investigate and apply the principle of conservation of energy including use of work done, gravitational potential energy and kinetic energy	3.1, 4.1, 4.2
15	use the expression for work $\Delta W = F\Delta s$ including calculations when the force is not along the line of motion	3.1, 3.2
16	understand some applications of mechanics e.g. to safety or to sports	all
17	investigate and calculate power from the rate at which work is done or energy transferred	3.3

Answers

1 $v = u + a\Delta t$

$= 2.0 \text{ m s}^{-1} + 1.5 \text{ m s}^{-2} \times 3.0 \text{ s}$

$= 2.0 \text{ m s}^{-1} + 4.5 \text{ m s}^{-1} = 6.5 \text{ m s}^{-1}.$

2 $v = u + a\Delta t.$

Assuming he is falling from rest,

$u = 0 \text{ m s}^{-1}.$

Taking downwards as positive, $a = g = +9.8 \text{ m s}^{-2}$,

$\Delta t = 2.5 \text{ s}$, and so

$v = +9.8 \text{ m s}^{-2} \times 2.5 \text{ s}$

$= +24.5 \text{ m s}^{-1}.$

3 $a = \dfrac{(v - u)}{\Delta t}$

$= \dfrac{(0 \text{ m s}^{-1} - 9.0 \text{ m s}^{-1})}{0.003 \text{ s}}$

$= \dfrac{-9.0 \text{ m s}^{-1}}{0.003 \text{ s}}$

$= -3000 \text{ m s}^{-2}.$

4 $a = \dfrac{(v - u)}{\Delta t}$

$v = -25 \text{ m s}^{-1}$, $u = 5.0 \text{ m s}^{-1}$, $\Delta t = 0.012 \text{ s}$

$a = \dfrac{(-25 \text{ m s}^{-1} - 5.0 \text{ m s}^{-1})}{0.012 \text{ s}}$

$= -2500 \text{ m s}^{-2}.$

Note that the velocity directions are carefully given positive and negative signs, positive for rightwards velocity and negative for leftwards.

5 (a) $\Delta s = 4.0 \text{ m}$ (s has increased from 2.0 m to 6.0 m) and

$\Delta t = 4.0 \text{ s}$ so $v = 4.0 \text{ m}/4.0 \text{ s} = 1.0 \text{ m s}^{-1}.$

(b) If you draw any other triangle on this graph you should get the same velocity.

6 (a) The velocity would be more than 1.0 m s^{-1} – displacement increases by more than 1.0 m in each second.

(b) The velocity would be negative i.e. the motion is in the negative direction so the displacement in the positive direction decreases with time – or (if the graph goes below the horizontal axis) the displacement becomes larger in the negative direction.

7 Using the triangle shown in Figure 1.5(b),

$\Delta v = 7.5 \text{ m s}^{-1} - 3.5 \text{ m s}^{-1} = 4.0 \text{ m s}^{-1}$ and

$\Delta t = 3.0 \text{ s} - 1.0 \text{ s} = 2.0 \text{ s}:$

$a = \dfrac{4.0 \text{ m s}^{-1}}{2.0 \text{ s}} = 2.0 \text{ m s}^{-2}.$

Using any other triangle would give the same answer.

8 The graph would slope downwards from left to right, showing that the velocity in the positive direction is decreasing, or (if the graph goes below the horizontal axis) that the velocity is increasing in the negative direction.

9 Using Equation 3,

final velocity

$v = u + a\Delta t = 8.00 \text{ m s}^{-1} + 0.70 \text{ m s}^{-2} \times 3.00 \text{ s}$

$= 8.0 \text{ m s}^{-1} + 2.1 \text{ m s}^{-1} = 10.1 \text{ m s}^{-1}.$

displacement

$s = ut + \tfrac{1}{2}at^2$

$= 8.00 \text{ m s}^{-1} \times 3.00 \text{ s} + \tfrac{1}{2} \times 0.70 \text{ m s}^{-2} \times (3.00 \text{ s})^2$

$= 24.0 \text{ m} + 3.15 \text{ m} = 27.15 \text{ m}.$

10 A is ahead (there is a larger area under A's graph).

A's total displacement can be found by adding together the areas (1–4) of the triangles and rectangles as shown in Figure 1.68:

displacement $= 37.5 \text{ m} + 75 \text{ m} + 5 \text{ m} + 65 \text{ m}$

$= 182.5 \text{ m}.$

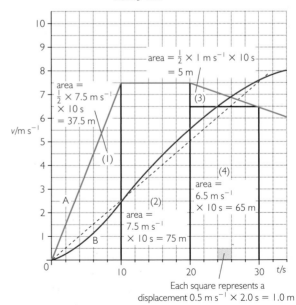

Figure 1.68 See the answer to Question 10

Alternatively, use Equation 4 – but remember to treat each 10 s interval separately, since the acceleration is different in each case.

B's total displacement can be found by counting the squares under the curved line. Each large square represents a displacement of 1.0 m (see Figure 1.68). Alternatively (and quicker) notice that B's graph can be approximated to a straight line giving a triangle of 'height' 7.5 m s^{-1} and 'base' 30 s, that represents a displacement of approximately

½ × 7.5 m s^{-1} × 30 s = 112.5 m.

This puts A about 70 m ahead of B after 30 s.

(You could refine this by estimating the number of graph-paper squares above and below the curved line, and adding and subtracting them from 112.5 m. Using this approach, B's actual displacement is about 5 m greater than 112.5 m – say 118 m altogether.)

11 Using Equation 6,

net force $F = ma$

$$= 65 \text{ kg} \times 2.0 \text{ m s}^{-2}$$

$$= 130 \text{ kg m s}^{-2} = 130 \text{ N}.$$

12 Using Equation 6a,

$$F = \frac{m\Delta v}{\Delta t}$$

Taking the initial direction of motion as positive:

$u = +5.0$ m s^{-1}, $v = -25.0$ m s^{-1},

$\Delta v = v - u = -25.0$ m s^{-1} – 5.0 m s^{-1} = –30.0 m s^{-1}

(notice the negative sign) so

$$F = \frac{0.120 \text{ kg} \times (-30.0 \text{ m s}^{-1})}{0.015 \text{ s}}$$

$$= -240 \text{ N}.$$

The negative sign shows that the force was acting in the opposite direction to the initial motion – which it would need to be in order to stop and reverse the motion of the ball.

13 The answer will depend on your mass – but whatever the numbers, you should give your answer in newtons and say that it acts downwards. As you are asked for an estimate, you can use an approximate value for $g \approx 10$ N kg^{-1}. For example, suppose $m = 64$ kg. Taking downwards as positive,

$g \approx +10$ N kg^{-1}, $W = mg \approx 64$ kg × 10 N kg^{-1} = 640 N acting downwards.

14 There are three 'Newton's third law' pairs of forces:

- contact force of person on the ground and ground on the person

- contact force of person on the javelin and javelin on the person

- gravitational force of person on the Earth and the Earth on the person.

15 Taking the initial direction of the ball's motion as positive, $a = -12$ 200 m s^{-2}.

$F = ma$

$$= 0.024 \text{ kg} \times (-12 \text{ 200 m s}^{-2})$$

$$= -292.8 \text{ N} (\approx 300 \text{ N}).$$

The ball exerts a force of equal size on the racket, in the same direction as its initial motion.

16 (a) (i) There is a pair of gravitational forces attracting the diver and earth towards each other (as in Figure 1.14).

(ii) In addition to the gravitational forces, as the diver enters the water there is a pair of forces involving the diver and the water – the diver exerts a downward force on the water and the water exerts an upward force that decelerates the diver.

(b) In an awkward splash landing, the diver comes to rest rapidly, experiencing a large change of velocity in a short time i.e. a large acceleration, so a large, painful, force must be exerted on the diver by the water. If the diver enters the water smoothly, the change of velocity is much more gradual i.e. the acceleration (and hence the force exerted by the water) is smaller.

(c) The two forces in the diagram are a third-law-force pair, so they are equal in size:

$$F_{\text{Earth}} = -F_{\text{diver}}$$

Using $F = ma$:

$$m_{\text{Earth}} a_{\text{Earth}} = -m_{\text{diver}} a_{\text{diver}}$$

$$a_{\text{Earth}} = \frac{-m_{\text{diver}} a_{\text{diver}}}{m_{\text{Earth}}}$$

acceleration of Earth:

$$a_{Earth} = \frac{(70 \text{ kg} \times 9.8 \text{ m s}^{-2})}{6 \times 10^{24} \text{ kg}}$$

$$= 1.1 \times 10^{-22} \text{ m s}^{-2}.$$

17 (a) Acceleration was uniform from 0 s to 40 s.

(b) Magnitude of uniform acceleration

$$a = \frac{\Delta v}{\Delta t}$$

$$= \frac{20 \text{ m s}^{-1}}{40 \text{ s}}$$

$$= 0.5 \text{ m s}^{-2}.$$

(c) Magnitude of the instantaneous acceleration at $t = 80$ s is found from slope of graph at $t = 80$ s.

$$a = \frac{\Delta v}{\Delta t}$$

$$= \frac{20 \text{ m s}^{-1}}{130 \text{ s}}$$

$$= 0.15 \text{ m s}^{-2}.$$

(Your answer may differ by ± 0.03 m s^{-2}, depending on exactly how you drew your tangent line.)

(d) Net force

$$F = ma$$

$$= 200\,000 \text{ kg} \times 0.5 \text{ m s}^{-2}$$

$$= 100\,000 \text{ N}.$$

(e) Using area under section of graph between 0 s and 40 s:

$$\text{displacement} = \frac{1}{2} \times 20 \text{ m s}^{-1} \times 40 \text{ s}$$

$$= 400 \text{ m}.$$

(Or you could use Equation 4 with $u = 0$ m s^{-1}, $a = 0.5$ m s^{-2}, $t = 40$ s, which gives the same answer.)

18 The three forces W (the weight acting vertically), H (the wind acting horizontally) and T (the force exerted by the support arm at an angle θ to the vertical) are in equilibrium so must form a triangle (Figure 1.69). Measurement on Figure 1.69 shows that $T \approx 25\,500$ N and $\theta \approx 11°$.

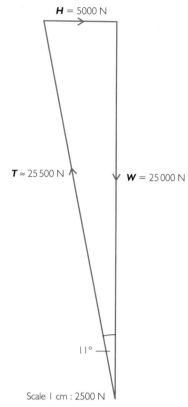

Figure 1.69 Vector diagram for the answer to Question 18

19 (a) The plane has constant velocity (which means no acceleration) therefore the forces must be in equilibrium: $\Sigma F = 0$. See Figure 1.70(a).

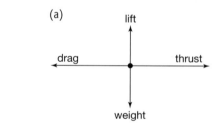

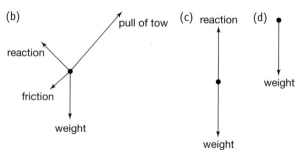

Figure 1.70 The answers to Question 19(a) a plane with constant velocity (b) a skier being towed up a slope (c) a friction-free ice puck with constant velocity (d) a ball after being thrown into the air

(b) As the skier is moving at constant velocity, the net force is zero $\Sigma \boldsymbol{F} = 0$.

(c) The only forces are vertical: the weight of the puck and the upward reaction (contact force) exerted by the ice.

(d) The only force acting on the ball is gravity. (A common misconception is to also draw the force from the hand that has just released it.)

20 See Figure 1.71. From Figure 1.25, $\tan\theta = \frac{1}{3}$.

Horizontal force

$F = W \tan\theta$

$= 600 \text{ N} \times \left(\frac{1}{3}\right)$

$= 200 \text{ N}.$

By Pythagoras,

$T^2 = F^2 + W^2$

$= (200 \text{ N})^2 + (600 \text{ N})^2$

$= 4.00 \times 10^5 \text{ N}^2$

so $T = 632 \text{ N}.$

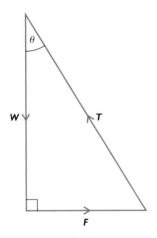

Figure 1.71 Vector diagram for the answer to Question 20

21 (a) $T = \dfrac{W}{2\sin\theta}$, where W is the climber's weight (600 N). As θ becomes small, then $\sin\theta$ approaches zero and T becomes very large.

(b) Rearranging the expression from (a):

$\sin\theta = \dfrac{W}{2T} = \dfrac{600 \text{ N}}{(2 \times 15 \times 10^3 \text{ N})}$

$= 2 \times 10^{-2}.$

$\theta = \sin^{-1}(2 \times 10^{-2}) = 1.1°$

22 (a) Figure 1.72 shows a free-body diagram and a force vector diagram for the problem.

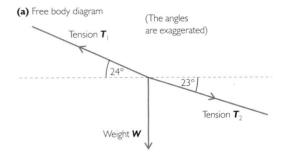

(a) Free body diagram

Tension $\boldsymbol{T}_1$

(The angles are exaggerated)

24°

23°

Tension $\boldsymbol{T}_2$

Weight $\boldsymbol{W}$

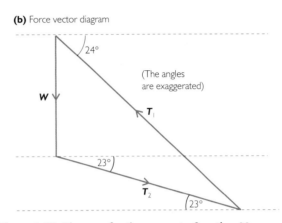

(b) Force vector diagram

24°

(The angles are exaggerated)

$\boldsymbol{W}$

$\boldsymbol{T}_1$

23°

$\boldsymbol{T}_2$

23°

Figure 1.72 Diagrams for the answer to Question 22

(b) The horizontal components of T_1 and T_2 must be equal and opposite:

$T_1 \cos24° = T_2 \cos23°$ (i)

and the vertical component of T_1 acting upwards must equal W and the vertical component of T_2 acting downwards:

$T_1 \sin24° = W + T_2 \sin23°$ (ii)

Rearranging (i) gives:

$T_2 = T_1\left(\dfrac{\cos24°}{\cos23°}\right)$ (iii)

Substituting (iii) in (ii) gives:

$T_1 \sin24° = W + T_1\left(\dfrac{\cos24°}{\cos23°}\right)\sin23°$

Rearranging:

$T_1\left(\sin24° - \left(\dfrac{\cos24°}{\cos23°}\right)\sin23°\right) = W$

$T_1 = W \div \left(\sin24° - \left(\dfrac{\cos24°}{\cos23°}\right)\sin23°\right)$

$= 5.27 \times 10^6 \text{ N}.$

Substituting this value back into (i) gives:

$$T_2 = T_1 \frac{\cos 24°}{\cos 23°} = 5.23 \times 10^6 \text{ N}.$$

(c) If T_1 and T_2 both acted along the same direction, then it would be impossible to draw the closed vector triangle in Figure 1.72. It is the difference in the vertical components of T_1 and T_2 that supports the weight of the cable car.

23 (a) See Figure 1.73. By Pythagoras,

$$R^2 = (500 \text{ N})^2 + (100 \text{ N})^2 = 2.60 \times 10^5 \text{ N}^2$$

so $R = 510$ N.

$$\tan\theta = \frac{100}{500} = 0.2, \text{ so } \theta = 11.3°$$

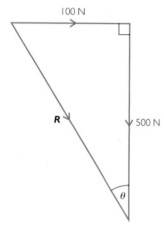

Figure 1.73 Diagram for the answer to Question 23

(b) $R = \Sigma F = ma$

$$a = \frac{R}{m} = \frac{510 \text{ N}}{100 \text{ kg}} = 5.1 \text{ m s}^{-2}$$

The direction of acceleration is that of the resultant force i.e. at 11.3° to the vertical.

24 The rope does *not* obey Hooke's law. The load in the second test is three quarters that in the first (600 N/800 N = 0.75), but it does not produce three quarters the extension (0.012 m/0.020 m = 0.60).

In the first case,

$$\text{stiffness} = \frac{800 \text{ N}}{0.020 \text{ m}}$$

$$= 4.00 \times 10^4 \text{ N m}^{-1}$$

$$= 40.0 \text{ kN m}^{-1}.$$

In the second case,

$$\text{stiffness} = \frac{600 \text{ N}}{0.012 \text{ m}}$$

$$= 50.0 \text{ kN m}^{-1}.$$

25 (a) From Equation 14, extension $x = \dfrac{F}{k}$

$$x = \frac{650 \text{ N}}{(60 \times 10^3 \text{ N})} = 1.1 \approx 10^{-2} \text{ m (11 cm)}$$

(b) Imagine that the 4 m rope is two 2 m ropes joined end to end. Each rope would be subject to the same force (650 N) so each extends by 11 cm, giving a total extension of 22 cm.

26 200 J is 20% of the total energy, so

total energy = 5 × 200 J = 1000 J.

27 $E_k = \frac{1}{2}mv^2$

$$= \frac{1}{2} \times 200 \text{ kg} \times (44 \text{ m s}^{-1})^2$$

$$= 1.94 \times 10^5 \text{ J}$$

28 The skier has slightly more kinetic energy than the car (if you guessed 'the same' you were not far out).

Car: $E_k = \frac{1}{2} \times 600 \text{ kg} \times (22 \text{ m s}^{-1})^2 = 1.45 \times 10^5 \text{ J}$

Skier: $E_k = \frac{1}{2} \times 80 \text{ kg} \times (66 \text{ m s}^{-1})^2 = 1.74 \times 10^5 \text{ J}$

29 (a) $\Delta E_{\text{grav}} = mg\Delta h$

$$= 65 \text{ kg} \times 9.8 \text{ N kg}^{-1} \times 35 \text{ m}$$

$$= 2.23 \times 10^4 \text{ J}$$

$E_k = 2.23 \times 10^4$ J

Rearranging Equation 20, $v^2 = \dfrac{2E_k}{m}$

so $v = \sqrt{\left(\dfrac{2E_k}{m}\right)}$

$$= \sqrt{\left(\frac{2 \times 2.23 \times 10^4 \text{ J}}{65 \text{ kg}}\right)}$$

$$= 26 \text{ m s}^{-1}.$$

(b) The mass makes no difference to the speed. This can be shown algebraically:

$$v = \sqrt{\left(\frac{2E_k}{m}\right)} = \sqrt{\left(\frac{2\Delta Erav}{m}\right)}$$

$$= \sqrt{\left(\frac{2mg\Delta h}{m}\right)} = \sqrt{(2g\Delta h)}$$

An alternative argument is to say that all free-falling objects have the same acceleration so will fall at the same rate and so reach the same speed.

30   Angle between slope and *vertical* is $\theta = 40°$.

Refer back to Question 28: kinetic energy of 80 kg skier moving at 66 m s^{-1} is 1.74×10^5 J.

$$\Delta W = F\Delta s \cos\theta$$

so $\Delta s = \dfrac{\Delta W}{(F \cos\theta)}$

$\Delta W = 1.74 \times 10^5$ J

F = downward gravitational force on skier

$\quad = 80$ kg $\times 9.8$ N kg^{-1}

so $\Delta s = \dfrac{1.74 \times 10^5 \text{ J}}{(80 \text{ kg} \times 9.8 \text{ N kg}^{-1} \times \cos 40°)}$

$\quad\quad = 290$ m.

Alternatively, start with some algebra to eliminate mass as in Question 29:

$$\tfrac{1}{2}mv^2 = mg\Delta h = mg\Delta s \cos\theta$$

so

$$\Delta s = \frac{v^2}{(2g \cos\theta)}$$

$$= \frac{(66 \text{ m s}^{-1})^2}{(2 \times 9.8 \text{ m s}^{-1} \cos 40°)}$$

$\quad\quad = 290$ m.

31   Combining Equations 19 and 23:

$$P = \frac{mg\Delta h}{\Delta t}$$

so $t = \dfrac{mg\Delta h}{P}$

$$= \frac{60 \text{ kg} \times 9.8 \text{ N kg}^{-1} \times 0.60 \text{ m}}{150 \text{ W}}$$

$\quad\quad = 2.4$ s.

32   Power $= \dfrac{\text{gain in kinetic energy}}{\text{time taken}}$

$$= \frac{\tfrac{1}{2}mv^2}{\Delta t}$$

$$= \frac{\tfrac{1}{2}\, 60 \text{ kg} \times (10 \text{ m s}^{-1})^2}{3 \text{ s}}$$

$\quad\quad = 1000$ W

33   Car must gain kinetic energy $\tfrac{1}{2}mv^2$.

Power $P = \dfrac{\Delta E}{\Delta t}$

so $\Delta t = \dfrac{\Delta E}{P} = \dfrac{\tfrac{1}{2}mv^2}{P}$

$$= \frac{\tfrac{1}{2} \times 800 \text{ kg} \times (30 \text{ m s}^{-1})^2}{60 \times 10^3 \text{ W}}$$

$\quad\quad = 6.0$ s.

34   1 day $= 24 \times 60 \times 60$ s $= 8.64 \times 10^4$ s

so energy $= 90$ W $\times 8.64 \times 10^4$ s

$\quad\quad = 7.8 \times 10^6$ J.

35   2 weeks $= 14 \times 8.64 \times 10^4$ s

so energy $= 14 \times 8.64 \times 10^4$ s $\times 10^{-4}$ W $= 121$ J

36   $E_{el} = \tfrac{1}{2}kx^2 = \tfrac{1}{2} \times 0.4$ N m$^{-1} \times (0.6$ m$)^2 = 7.2 \times 10^{-2}$ J

37   Each cord contributes just part of the force that acts on the jumper. For example, if a 600 N jumper is supported at rest by six cords, then a 700 N jumper will stretch a seven-cord rope by the same amount – each cord provides a force of 100 N.

38   (a) $\Delta h = 610$ m

$\quad\quad \Delta E_{grav} = mg\Delta h = 75$ kg $\times 9.8$ N kg$^{-1} \times 610$ m

$\quad\quad\quad\quad = 4.5 \times 10^5$ J.

(b)   Extension of cord, $x = 610$ m $- 250$ m $= 360$ m.

$\quad\quad E_{el} = \tfrac{1}{2}kx^2$

$\quad\quad$ so $k = \dfrac{2E_{el}}{x^2} = 6.9$ N m^{-1}.

(c)   Now the cord must just support his weight i.e. upward force exerted by cord is $F = kx = mg$

$\quad\quad$ so $x = \dfrac{mg}{k} = \dfrac{75 \text{ kg} \times 9.8 \text{ N kg}^{-1}}{6.9 \text{ N m}^{-1}}$

$\quad\quad = 107$ m.

39 See Figure 1.74.

E_k maximum
E_{grav} minimum
E_{el} = 0

(a) The run-up

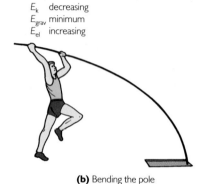

E_k decreasing
E_{grav} minimum
E_{el} increasing

(b) Bending the pole

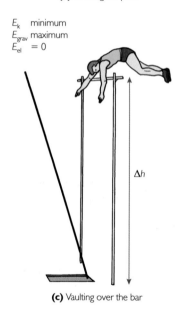

E_k minimum
E_{grav} maximum
E_{el} = 0

Δh

(c) Vaulting over the bar

Figure 1.74 The answer to Question 39

40 (a) Initial $E_k = \frac{1}{2}mv^2$. Increase in gravitational energy, $\Delta E_{grav} = mg\Delta h$. If all E_k becomes E_{grav}, then:

$$mg\Delta h = \frac{1}{2}mv^2$$

$$\Delta h = \frac{v^2}{2g} = \frac{(10 \text{ m s}^{-1})^2}{2 \times 9.8 \text{ m s}^{-2}} = 5.1 \text{ m}.$$

(b) If his centre of gravity just clears the bar, then the height will be 6.1 m.

(c) (i) The energy conversion via bending the pole will not be 100% efficient so not all the initial E_k will become E_{grav}. Also, the vaulter still has some E_k at the top of the vault – he is moving over the bar.

(ii) The vaulter is not simply catapulted over, but can push himself further upwards using the pole.

(d) Probably not. Improvements in materials are unlikely to make much difference as the process must already be close to 100% efficient. Athletes may be able to increase the speed of the run-up (though probably not much) and refine their vaulting technique, but these will probably make only a slight difference to the height of the vault. (Between 1994 and the time of writing (2007) there was no increase in the world record for pole vaulting.)

41 $E_{el} = \frac{1}{2}kx^2 = \frac{1}{2} \times 10^7 \text{ N m}^{-1} \times (0.5 \text{ m})^2$

$= 1.25 \times 10^6 \text{ J}$

42 (a) The mass flow rate is $10\,000 \text{ m}^3 \text{ s}^{-1} \times 1000 \text{ kg m}^{-3} = 10^7 \text{ kg s}^{-1}$.

(b) Assume that all gravitational energy lost by the falling water eventually provides electrical energy i.e. that the overall process is 100% efficient. In 1 second, falling water must transfer $\Delta E = 12 \times 10^9 \text{ J}$. This is achieved by a mass $m = 10^7 \text{ kg}$ falling through a height Δh where $\Delta E = mg\Delta h$, so

$$\Delta h = \frac{\Delta E}{mg} = \frac{12 \times 10^9 \text{ J}}{(10^7 \text{ kg} \times 9.8 \text{ N kg}^{-1})}$$

$= 122 \text{ m}.$

In practice the efficiency will be (much) less than 100%, so the water will have to fall through a (much) greater height in order to generate the same amount of electrical power.

43 (a) She is opposed by an equal force due to the water so she does not accelerate. She transfers energy by heating the water and herself, rather than increasing her kinetic energy.

(b) The data point for humans would lie slightly to the left of the 'dolphin' point at 10^5 g (100 kg). A human of mass 80 kg would expend about 1 kJ kg^{-1} min^{-1} which is equivalent to 1 J g^{-1} min^{-1}. Swimming 1 km would take many minutes, and so a human swimmer would expend many J g^{-1} km^{-1}, putting the data point well above the plotted line. Humans transfer considerably more energy while swimming than would an aquatic animal of similar body mass. This is because our bodies are a less suitable shape (less streamlined) so we experience a much greater opposing force than if we had evolved for swimming.

44 Some of the energy must have been transferred elsewhere. If she slipped 'gradually' by sliding against a rock face, then friction between her body and the rock would have given rise to heating, which would account for the apparently 'missing' energy. If she fell and 'bounced' before coming to rest (as in a bungee jump) she would initially have some kinetic energy, then heating of air, air esistance and internal heating in the rope would account for the 'missing' energy.

45 (a) Her initial vertical velocity is zero. Using Equation 26: $y = \frac{1}{2}gt^2$ so:

$$t = \sqrt{\left(\frac{2y}{g}\right)} = \sqrt{\left(\frac{2 \times 20 \text{ m}}{9.8 \text{ m s}^{-2}}\right)} = 2.0 \text{ s.}$$

(b) Using Equation 27:

$x = u_x t = 20 \text{ m s}^{-1} \times 2.0 \text{ s} = 40 \text{ m.}$

(c) (i) This would increase u_x, so she would travel further in the time she is in the air.

(ii) If she is initially moving upwards, she would take longer to reach the ground so would travel further.

46 (a) The horizontal velocity remains constant and the vertical velocity increases steadily with a uniform acceleration.

(b) See Figure 1.75.

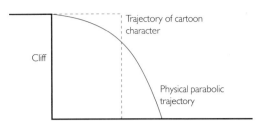

Figure 1.75 The answer to Question 46(b)

47 (a) You might have seen: long-exposure photographs of projectiles, or stop-frame video or film. Also, if you sketch a parabola on a whiteboard, you can throw a small projectile so that it follows the same path – you cannot do this if you sketch the path shown in Figure 1.63.

(b) See Figure 1.76.

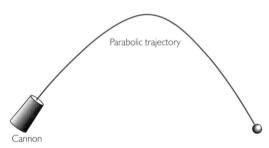

Figure 1.76 The answer to Question 47(b)

(c) $\theta = 30°$:

$$\text{range} = \frac{2 \times (50 \text{ m s}^{-1})^2 \sin30° \cos30°}{9.8 \text{ m s}^{-2}}$$

$$= 221 \text{ m}$$

$\theta = 45°$:

$$\text{range} = \frac{2 \times (50 \text{ m s}^{-1})^2 \sin45° \cos45°}{9.8 \text{ m s}^{-2}}$$

$$= 255 \text{ m}$$

$\theta = 60°$:

range = 221 m (same as $\theta = 30°$ because $\sin30° = \cos60°$ and $\cos60° = \sin30°$)

CHAPTER 2

Good Enough to Eat

Why a chapter called *Good Enough to Eat*?

Everyone has to eat, and for the most part, we enjoy the taste and textures of food as well as the social aspects of eating. Foods have to be manufactured and packaged so that they can be transported and stored safely and affordably, and labelled so that we know what we are eating.

The food production industry is enormous. In 2006, Nestlé, the world's largest food, drink and confectionery producer, was making over eight and a half thousand different products (Figure 2.1) to be sold in more than a hundred countries, and employing over two hundred thousand people directly. In 2006, Nestlé's sales amounted to £41 thousand million.

Physicists have a part to play in most stages of food production. Physics principles are used to assess the raw materials' quality and condition. Ingredients must be weighed and mixed and brought together to form a homogeneous mass at a specified temperature. Food manufacture often involves the product flowing along pipes, being pushed through orifices or shaped in moulds, each of which is affected by the physical properties of the materials. The product must be tested to check that it has the desired properties. Finally the product needs to be packaged for safe storage and distribution, and so the physical nature of the packaging is important. Labelling, date stamping and detecting contaminants during processing all involve important elements of physics.

Figure 2.1a Confectionery products

Overview of physics principles and techniques

In this chapter you will see how the flow properties of liquids are affected by concentration and temperature, and how they can be measured or compared. Products need to be tested (for example, to ensure that a biscuit provides a suitable 'crunch'), and in this part of the chapter you will learn about materials testing and such factors as hardness, brittleness and toughness. Material properties feature again in the final part of the chapter, on packaging, where you will also see how physics relates to aspects of health and safety.

During the course of this chapter you will learn about instrumentation and calibration, and be introduced to some important techniques and measuring instruments. You will learn how to read a vernier scale and how to use a micrometer screw gauge, and how to treat experimental errors and uncertainties.

In this chapter you will extend your knowledge of:

- forces and motion and using graphs from *Higher, Faster, Stronger*.

In other chapters you will do more work on:

- forces and motion in *Transport on Track*
- bulk properties of materials in *Spare Part Surgery* and *Build or Bust?*

In *Good Enough to Eat* you will see how physics principles are used in the confectionery industry – in particular, the manufacture of biscuits and chocolates.

Figure 2.1b Confectionery products

1 Physics in the food industry

Stephen Beckett is a professional physicist working at the Nestlé Research and Development Centre in York. Here he provides an introduction to the place of physics in the food and confectionery industry.

1.1 Physics in the food and confectionery industry

Chocolate making, and indeed the food industry as a whole, is not at first sight an obvious place to need physics. Yet closer inspection shows that the industry needs and uses physics to an ever-increasing degree. Not only is food processing the UK's biggest industry, but it is also one that is currently in the middle of big changes from an essentially craft-based industry to a highly automated one requiring critical control. This makes the challenge and the opportunity for physicists even greater.

Food has the advantage over many industrial products in that if it is processed well, and to the customer's liking, then a repeat purchase is likely in the very short term, unlike other industries whose products may last for many years. It does however have an extra challenge in that the products must be absolutely safe to eat and obey the food laws. You cannot for instance help chocolate to melt in the mouth by something that tastes nasty or, even worse, makes the consumer ill. It is always worth remembering that food is bought because of how it looks and tastes, not because of the clever science that has been used to make it. Science can, however, help to manipulate the taste and texture of a product, ensure that it is relatively consistent from day to day (not an easy task when your raw ingredients are always varying) and indeed help to ensure a product's safety.

In addition, science can help to make the industry more efficient by optimising processing and other factors like extending shelf-life so that a product can be made all the year round, rather than having operators and expensive machinery employed for only a few weeks.

Further research

When a box of chocolate assortments becomes old, the centres containing nuts are usually the first ones to turn a white colour. This is known as 'bloom' and is when the fat from the sweet comes to the surface and sets there.

The reason why it is worse in the nut sweets is because the nuts contain a fat, which is mainly liquid at room temperature, whereas most of the fat within chocolate (cocoa butter) is solid. The soft nut fat reacts with the cocoa butter and softens it and also migrates through the sweet to the surface. In order to obtain more information about this process, magnetic resonance imaging (as used in body scanners) has been used at the University of Cambridge to monitor the changes in the position of the nut fat taking place within the sweets.

So, every time you buy a KitKat, Mars Bar or Crunchie it has been produced with the aid of physics. The chocolate industry, however, still has a lot of physics that remains to be done. Not only is a simple method required to detect plastic in chocolate, but other problems remain, such as the measurement of the three-dimensional contraction of chocolate as it sets in the mould, or of its stickiness as it melts in the hand. Perhaps you have the solution!

Stephen Beckett has written further comments to accompany some other sections as you progress through the chapter.

Activity 1 Food web

Use the Internet to find some more background information about the food industry. If you want to find out more about the variety of food products, or if you are wondering whether your future might be in this field of employment, use the Search facility on your web browser to look up some famous names. Some relevant sites are listed on www.shaplinks.co.uk.

The Answers website provides a good introduction to the food industry and if you would like to find out more about Nestlé, then go to their website, which gives you and insight into the company, its research, products, sales, staffing and history.

The Cadbury's website (Figure 2.2) provides details of the company history and the history of chocolate.

If you have a particular interest in chocolate, then visit the Chocolate Encyclopedia at the Exploratorium website.

Click first on 'Chocolate Encyclopedia' and then on 'The production process'.

The website 'Wayne's this and that' also contains information about chocolate.

The URLs for all these websites can be found at www.shaplinks.co.uk

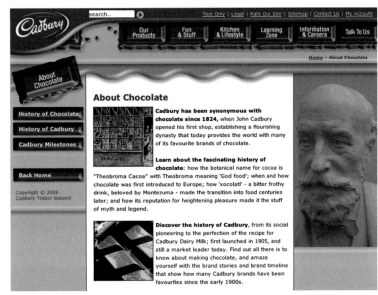

Figure 2.2 Cadbury's website

Activity 2 How big?

Manufacturers need to monitor their production, and they need to know how precise their measurement is. Make measurements to determine the volume of different sweets, and estimate the uncertainty in your measurements.

2 Going with the flow

In many industrial processes it is essential to get liquids flowing at speeds to match the needs of the process. In the manufacture of chocolates or chocolate biscuits, the chocolate needs to flow at a rate that allows the correct thickness to be placed on a biscuit or soft centre, or to fill a mould, and to keep up the required rate of production. The description, understanding, measurement and control of the flow all involve physics.

2.1 Flowing chocolate

Activity 3 Flowing chocolate

Read the following short article in which Stephen Beckett describes aspects of chocolate manufacture. Then make notes of how physics is, or might be, used in the circumstances that he mentions. Expand on any points you can, suggesting techniques and equipment that might be used for sensing, testing or control.

Chocolate differs from most other foods in that it is solid at room temperature and yet easily melts in the mouth. This is because the fat it contains has a melting point below that of blood temperature. In confectionery manufacture, chocolate is produced as a liquid to pour into moulds, or to pour over (enrobe) a sweet centre, before being cooled to enable the fat to set (Figure 2.3). Incorrect flow properties will result in a poor-quality product. This may take the form of mis-shapes (Figure 2.4) where the chocolate runs down the sweet but, instead of flowing through the open grid on which the sweet is enrobed, it sticks to it, forming a sort of foot.

Figure 2.3 The enrobing process

In aerated products, such as an Aero bar, the flow properties of the chocolate will affect the size of the air bubbles (Figure 2.5). Weight control also becomes difficult, with thick chocolate sticking on top of the centre and sides and making it overweight. Too thin a chocolate may run off the centre altogether, allowing it to pick up or lose moisture and hence be more likely to deteriorate.

Figure 2.4 Mis-shapen chocolates

The flow property of chocolate is in fact very complex. Only about one-third is made up of fat which is capable of melting at moderate temperatures and helping chocolate to flow. The remainder consists of solid particles (sugar, cocoa and milk solids) that must be coated with fat for them to flow smoothly past one another when the chocolate is being processed or melted in the mouth. This high solids content makes the chocolate flow in what is known as a non-Newtonian manner. In other words its viscosity depends on how quickly it is moving. It is in fact a bit like tomato ketchup or non-drip paint, in that it becomes runnier when stirred or mixed quickly.

Figure 2.5 Incorrect aeration of an Aero bar

In addition, the more fine particles there are, the bigger is the surface to be coated by fat to enable it to flow and hence the thicker the chocolate becomes. Once again physics becomes important in that it is used to measure the size distribution of these solid particles. This is done by dispersing the chocolate in a liquid and then shining a laser through the dispersion. The distribution can be calculated from the relative intensity of the light scattered at different angles.

2.2 Viscosity

How fast liquids flow is partly dependent on their **viscosity**, a factor controlling their resistance to flow. Loosely speaking, the lower the viscosity of a fluid (a liquid or gas), the 'runnier' it is. Tar and treacle are very **viscous**, while water has low viscosity. Devices for comparing and measuring viscosity are called **viscometers**.

Comparing viscosities

One of the simplest tests for comparing viscosities is known as the line spread test and uses a device known as a consistometer (Figure 2.6). A fixed quantity of liquid is allowed to flow out of a container and spread out on a flat surface. How far or fast it spreads provides a measure of its viscosity. Another instrument, the Redwood viscometer (Figure 2.7), involves allowing the liquid to flow through a narrow tube driven by its own head of pressure. (This instrument was first developed by French physiologist and physicist Jean Léonard Poiseuille (1799–1869), who used it to study blood flow.) Both these tests are used in the food industry.

Figure 2.6 A consistometer used to perform a line spread test

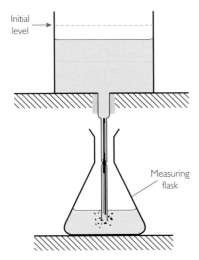

Figure 2.7 Redwood viscometer

Activity 4 Comparing viscosities

Use a model line spread test and Redwood viscometer to compare the viscosities of sugar solutions, syrups and honeys. Rank your samples in order of increasing viscosity. Comment on the reliability and ease of use of the two techniques.

Defining viscosity

The line spread and Redwood tests are useful for comparing the behaviour of fluids, but they do not give an absolute measurement of viscosity. For this we can use a falling ball viscometer. This instrument, developed by Irish physicist George Gabriel Stokes (1819–1903), involves timing a ball falling at constant speed through a fluid. The instrument has been adapted for use in the chocolate industry; as chocolate is opaque, a rod is attached to the ball so that its movement can be monitored (Figure 2.8).

Figure 2.8 Chocolate falling ball viscometer

Stokes was able to quantify viscosity by studying the force exerted on a spherical object as it moves through a fluid, or when a fluid flows past it. (There is an equivalence between a ball-bearing moving through a still fluid and a fluid moving past a stationary ball-bearing.) This **viscous drag** force is described by the relationship known as **Stokes's law**:

$$F = 6\pi\eta v \qquad\qquad\qquad (1)$$

where r is the radius of the sphere, v the velocity of the fluid relative to the sphere and η the coefficient of viscosity of the fluid (or commonly, just 'the viscosity'). The direction of the force is opposite to that of the velocity. Equation 1 can be rearranged to get an expression for η:

$$\eta = \frac{F}{6\pi r v} \qquad\qquad\qquad (1a)$$

From this we can see that the SI units of η are N s m^{-2}. To give you an idea of typical values, the viscosity of water at 20 °C is 1.000×10^{-3} N s m^{-2} and that of air at 27 °C and a pressure of 1 atmosphere is 18.325×10^{-6} N s m^{-2}. (Note that the temperatures are quoted – the viscosities of most fluids are highly dependent on temperature.)

> **Maths reference**
>
> Manipulating units
> See Maths note 2.2

Drag force

For a small sphere moving slowly through a fluid, the viscosity described by Stokes's law is the main contribution to drag. But there is also a drag force that depends on the density of the fluid (the moving object has to push fluid out of its path). For large objects moving at high speed this drag force is much larger than the viscous drag, but in the falling-ball viscometer it is much smaller and can be ignored in comparison to the viscous drag.

Archimedes in the balance

Figure 2.9 shows the forces acting on a sphere falling through a fluid. The sphere's weight acts downwards, and in addition to the drag force there is another upwards force: the **upthrust** or **buoyancy force**. This force always acts on an object immersed in a fluid, and arises because the object displaces some of the fluid around it. When getting in to or out of a bath or swimming pool we feel lighter or heavier, and so become aware of the upthrust provided by the water. A similar force acts in air, too, and the weight we measure on bathroom scales is very slightly less than it would be in a vacuum. Indeed anything that is completely or partially immersed in a fluid will experience an upthrust or buoyancy force, be it a skydiver, a ship or a falling ball-bearing.

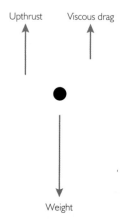

Figure 2.9 Forces on a falling ball

Activity 5 Archimedes' principle

Demonstrate Archimedes' principle by weighing an object in air and immersed in water.

The size of the upthrust is described by **Archimedes' principle**: 'When a body is partially or totally immersed in a fluid, the upthrust is equal to the weight of the fluid it displaces.'

Legend has it that Archimedes (*c.* 287–212 BCE) discovered this while in his bathtub and then ran through the streets of Syracuse in Sicily where he lived shouting 'Eureka' (meaning 'I found it'). He was at the time trying to develop a method of checking that the king's crown really was made of pure gold as the maker had claimed.

A simple way of looking at upthrust is to consider a floating object. Its weight must still be acting downwards but, with an upthrust provided upwards, the forces are in equilibrium and the object remains at rest. Now think of an object fully immersed in fluid. If the upthrust exceeds the weight, there will be a net upward force, which will push the object upwards until it is only partially immersed and displaces exactly its own weight of fluid. If, on the other hand, the upthrust on the fully immersed object is less than the object's weight, there is a net downward force and so the object sinks.

When an object falls through a fluid, it first accelerates due to the net downward force (weight minus upthrust). But it also experiences a drag force that increases with speed, so the net downward force is reduced as speed increases. When the net force reaches zero, the object can no longer accelerate and it falls with a constant downward velocity called its **terminal velocity**.

Questions 1 to 4 take you through some calculations of forces involved in a falling ball viscometer.

Questions

These questions refer to a ball-bearing of radius $r = 1.0 \times 10^{-3}$ m, made of steel with a density ρ_{steel} falling through oil with density ρ_{oil} and viscosity eta.

$\rho_{steel} = 7.8 \times 10^3$ kg m^{-3}

$\rho_{oil} = 920$ kg m^{-3}

$\eta = 8.4 \times 10^{-2}$ Ns m^{-2}

The gravitational field strength, g, is 9.8 N kg^{-1}.

A sphere with radius r has volume $V = \dfrac{4\pi r^3}{3}$.

1. Assuming that Stokes's law applies, calculate the viscous drag on the ball-bearing when it is travelling through the oil at a speed of 2.0×10^{-2} m s^{-1}.

2. Calculate (a) the volume of the ball-bearing, (b) its mass and (c) its weight.

3. When the ball-bearing is immersed in the oil, what are (a) the volume, (b) the mass and (c) the weight of the oil that it displaces?

4. Using your answers to Questions 1 to 3, (a) state the size of the upthrust acting on the ball-bearing and hence (b) calculate the size and direction of the net force acting on the ball-bearing.

Measuring viscosity

In a falling ball viscometer, the ball-bearing is selected so that it reaches it terminal velocity after travelling only a short distance through the fluid.

Referring back to Questions 1 to 4, we can then write down an expression for the forces acting on the ball-bearing, which must combine to give a resultant force of zero – in other words, the magnitudes of the forces must be related as follows:

upthrust + drag forces = weight

Ignoring the drag force that depends on density, we can write:

$$\frac{4\pi r^3 \rho_{fluid} g}{3} + 6\pi r \eta v = \frac{4\pi r^3 \rho_{steel} g}{3} \qquad (2)$$

To learn more about viscous drag visit the Brookfield Engineering website (see www.shaplinks.co.uk for details of the URL).

Activity 6 Measuring viscosity

Use a falling ball viscometer to determine the viscosity of honey or syrup at a particular temperature. Plan your experiment carefully and decide how best to analyse your measurements, taking account of any experimental uncertainties.

By comparing results with other students who have used different temperatures, explore the relationship between viscosity and temperature.

Calibration

A falling ball viscometer allows us to measure viscosities of fluids fairly directly. Once this has been done, then fluids of known viscosity can be used to **calibrate** other viscometers – that is, to establish a relationship between the performance of a particular viscometer and the actual viscosity of any fluids used. For example, your results from Activity 6 could be used to calibrate the consistometer or the Redwood viscometer that you used earlier; you could relate the flow of a fluid in a given time to its viscosity.

Figure 2.10 shows another type of viscometer popular in the food industry – the viscous drag viscometer. Here the liquid under test is contained within the outer cylinder, which is rotated at constant speed. As a result of the viscous properties of the liquid, this drags the inner cylinder round against the force of the spring S, moving a pointer over a scale. The position of the pointer indicates the liquid's viscosity, but there is not a simple relationship between the movement of the pointer and the viscosity of the liquid with the viscous drag viscometer.

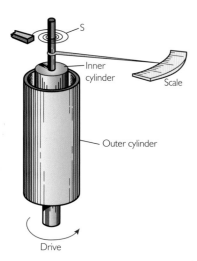

Figure 2.10 A viscous drag viscometer

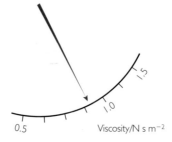

Figure 2.11 Pointer on scale

Activity 7 Calibration

Discuss how you would go about calibrating a viscous drag viscometer. Suggest a value for the viscosity of the liquid which moved the pointer to the position shown in Figure 2.11.

Questions

5 If you were provided with a forcemeter, a measuring cylinder, some water, a piece of cotton and an object that could be immersed in the measuring cylinder, outline how you could check Archimedes' principle. (You will need to know that $g \approx 10 \text{ N kg}^{-1}$ and that the density of water is 1000 kg m^{-3}.)

6 The pressure P at a depth h within a fluid is given by $P = \rho g h$, where ρ is the fluid's density and g is the gravitational field strength. By considering the pressures at the top and bottom surfaces of the rectangular object of area A shown in Figure 2.12, show that the upthrust must be equal to the weight of the fluid displaced.

7 Suppose that when Archimedes compared the weight of the king's crown immersed in water with the same crown in air, the ratio came to 0.948. If the density of gold is 19.3 times that of water, was the crown solid gold?

8 Some motor oils are labelled 'viscostatic'. What behaviour might you expect these oils to have over a wide range of temperatures?

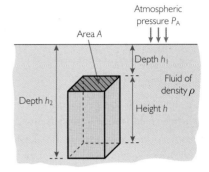

Figure 2.12 Rectangular block immersed in fluid

2.3 More about flow

Laminar/streamlined flow or turbulence?

Stokes's law relies on the flow of the liquid past the ball-bearing being **streamlined** or **laminar** and not **turbulent**. In streamlined or laminar flow the liquid does not make an abrupt change in direction or speed and adjacent layers of within the fluid only mix on a molecular scale – the word 'laminar' means 'layered'. With turbulent flow there is a lot of mixing and a series of eddies (little whirlpools) are produced along the object's path. See Figure 2.13.

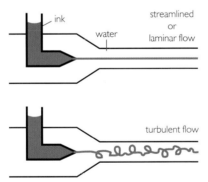

Figure 2.13 Streamlined and turbulent flow patterns

Turbulence is the unsteadiness that we observe with smoke billowing away from bonfires, the flapping of sails on yachts and of a flag on a flagpole, or the buffeting that one occasionally feels on an aeroplane. Turbulence gives rise to heating as energy is transferred to the fluids; this is usually unwanted and reduced the efficiency of a process.

Laminar or streamlined flow is what one hopes for around a vehicle as it lessens the fuel consumption. Similarly a professional speed skier will purchase clothing and adopt a posture that aims to achieve streamlined flow of air past the body and so allow him to go faster. Likewise less energy will be needed to move fluids in a factory if turbulence is avoided. The flow of chocolate onto centres needs to be as laminar as possible in order to produce a fairly even coating without air bubbles.

Activity 8   Laminar and turbulent flow

In this short activity you will be able to see the difference between laminar and turbulent flow. Connect a piece of transparent plastic tubing to a laboratory water tap and arrange for a length of it to be horizontal before going into a sink. Fill a syringe with ink and pass the syringe needle into the tube just where it becomes horizontal. Turn the tap on slowly and squeeze some ink into the tube.

Compare, and comment on, the pattern of movement of the ink as the flow of the water is increased from slow to fast.

Thixotropy

A number of foods display the interesting property of **thixotropy**. Margarine in its container at normal temperatures will not flow. However, on exerting a force with a knife in order to spread it, the margarine's viscosity lessens and it flows. When that force is removed, the margarine's viscosity again rises and it acts like a solid on the bread. This behaviour identifies it as a thixotropic material. Similar effects are to be seen with chocolate spread, tomato ketchup and mayonnaise (Figure 2.14).

Figure 2.14 Some thixotropic foods

Activity 9   Stirring custard

In this activity you will be able to observe behaviour known as negative thixotropy (rheopexy).

Put two heaped teaspoons of custard powder into a cup. Mix in up to two spoonfuls of water until, when the custard is stirred slowly, it is just runny. Now stir quickly and note what happens.

Further investigations

If you let honey or syrup pour from a spoon or jar it forms a small hill with a few spirals at the summit as shown in Figure 2.15. Might this perhaps provide a means of determining viscosity? What variables might be worth investigating?

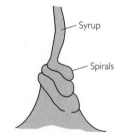

Figure 2.15 Syrup hill

2.4 Measuring flow rates

The measuring of flow rates is vital in the confectionery industry if one is to maintain consistency of the product – the same thickness of chocolate on the sweet's centre or in the mould for a chocolate rabbit or Easter egg (Figure 2.16). One way to do this would be to measure the mass or volume collected in a given time interval and hence determine the **mass flow rate** (the mass per unit time) or **volume flow rate** (the volume per unit time). However, this would interrupt the process, and it is preferable to use a flowmeter that can be left in position all the time. There is a tutorial that explains this very clearly at the Omega website, details at which can be found at www.shaplinks.co.uk.

Figure 2.16 Chocolate moulds

The light-gate flowmeter

One simple type of flowmeter is the light-gate flowmeter. When conducting experiments with dynamics trolleys you will probably have used light-gates to time the movement of a known length of card past a fixed point. The light-gate consists of a source of light (a bulb or a light-emitting diode (LED)), shining on to a photodiode. When the beam is interrupted, the change of illumination triggers an electronic timer, which is stopped when the illumination is restored.

The same principle can be used in a flowmeter (Figure 2.17). Here, a rotating propeller blade, driven by the flowing liquid, repeatedly interrupts the illumination. The meter registers the frequency of interruptions, i.e. the number of 'darkenings' per second. To be useful, such a meter first has to be calibrated using known flow rates. When this has been done, a calibration certificate is attached to the instrument (see Figure 2.18).

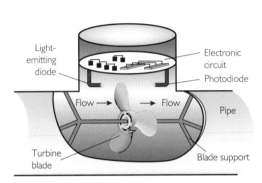

Figure 2.17 Cutaway of light-gate flow meter

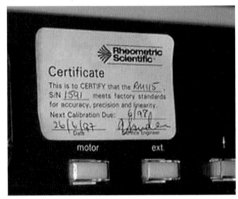

Figure 2.18 Calibration certificate

You will calibrate a flowmeter in Activity 10. The following Worked example and Questions 9 and 10 illustrate what is involved.

While frequency can be measured directly on a frequency meter, you will often need to calculate it from an oscilloscope screen trace as shown in Figure 2.19.

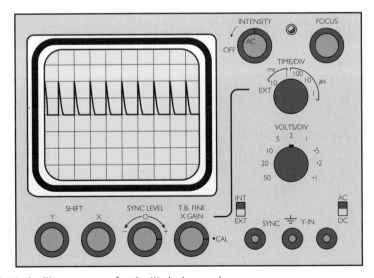

Figure 2.19 Oscilloscope trace for the Worked example

Worked example

Q Calculate the frequency of the output signal shown in Figure 2.19 if (a) the time-base setting was 10 ms/div and (b) 5.0 µs/div.

A The spacing between each repeated part of the trace is 1 division

(a) With a time-base setting of 10 ms/div the time period of the signal is

$$T = 10 \text{ ms} = 10 \times 10^{-3} \text{ s} = 1.0 \times 10^{-2} \text{ s}$$

frequency $f = \dfrac{1}{T}$

so

$$f = \frac{1}{(1.0 \times 10^{-2} \text{ s})} = 100 \text{ Hz}$$

(b) With a time-base setting of 5.0 µs/div the time period is:

$$T = 5.0 \text{ µs} = 5.0 \times 10^{-6} \text{ s}$$

$$f = \frac{1}{T} = \frac{1}{(5.0 \times 10^{-6} \text{ s})} = 2.0 \times 10^{5} \text{ Hz} = 200 \text{ kHz}$$

Questions

9 Calculate the frequency of the signal shown on the oscilloscope trace in Figure 2.20. The time-base setting was 10 ms/div.

Maths reference

SI prefixes
See Maths note 2.4

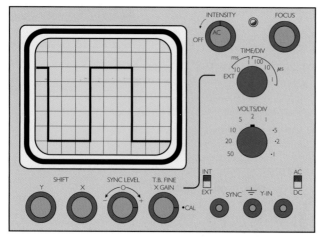

Figure 2.20 Oscilloscope trace for Question 9

10 Figure 2.21 shows a calibration graph for an RS flowmeter.

(a) (i) When the output frequency was 400 Hz, what was the volume flow rate of the fluid (in litres per minute)?

   (ii) If the fluid was water, with density 1 kg m⁻³, what was the mass flow rate (in kilograms per second)?

(b) What would be the output frequency for a volume flow rate of 3 litres per minute?

(c) How would you describe the relationship between output frequency and flow rate for this particular flowmeter?

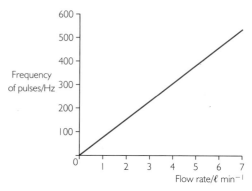

Figure 2.21 Calibration graph from RS flow transducer

Activity 10 Calibrating a flowmeter

Calibrate a light-gate flowmeter by investigating how its output frequency changes with volume flow rate. Compare your results with those shown in Figure 2.21.

Many types of flowmeter

Flowmeters are used widely in industry. For example, in the oil industry they may monitor flow from oil fields to refineries or from storage depots into tankers; they measure milk flow from farms into collection tankers; and they are vital for process control in the brewing industry. If you visit any industry, look out for flowmeters in use. Figure 2.22 shows meters used to record domestic gas consumption.

There are many types of flowmeter, most of which make ingenious use of physics in their design. The names of some common types of meter give a clue to their diversity; there are the diaphragm meter, vortex meter, rotameter, hinged plate or gate meter, turbine meter, ultrasonic Doppler effect meter, Pitot tube, orifice meter, V-notch, and the electromagnetic meter, to mention but a few.

Figure 2.22 Domestic gas meters

Activity 11 How does it work?

Figure 2.23 shows the sequence of operations in a type of meter known as a diaphragm meter, widely used to measure gas flow. Write a paragraph explaining how it works. Imagine you are writing for a book about how things work, suitable for readers about 15 years old who are studying science.

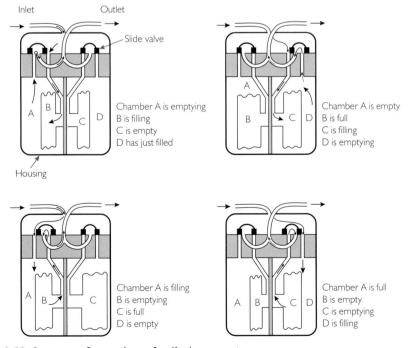

Figure 2.23 Sequence of operations of a diaphragm meter

2.5 Controlling the flow

Being able to measure rates of flow is not enough; their control is essential too. Mostly this is achieved using valves and pumps, but a rather more novel method is being considered at Michigan State University in the USA. In 1996 researchers in their Agricultural Engineering Department found molten chocolate to be an electro-rheological fluid, that is, one in which electric fields affects its viscous properties. The stronger the electric field, the more viscous the fluid. An extremely strong field can make the fluid solid.

An electric field is a region in which a charged object experiences a force. Electric fields are produced when objects become charged. You may have generated an electric field yourself by rubbing a comb on your clothes and seeing how it can then pick up small pieces of paper or make your hair stand up a little. More controllably, an electric field can be produced by connecting a potential difference between a pair of conducting plates.

The electro-rheological effect was first discovered in 1948. Since then a number of applications have been considered and developed. These include a clutch, for a motor vehicle, in which the coupling between the engine, clutch and finally the wheels is controlled by the viscosity of the material in the clutch. The more viscous the fluid, the greater the coupling and the faster the vehicle will go. The viscosity is controlled by an electric field placed across the material.

Question

11 Suggest how you think electro-rheology might be used in the making of a chocolate-coated product.

2.6 Summing up Part 2

This part of the chapter should have given you some insight into the flow properties of materials and how they can be measured. Use Activities 12 and 13, and Questions 12 to 15, to check your progress and understanding.

Activity 12 Summing up Part 2

Look back through your work and ensure that your notes include a clear definition, explanation or description of each of the terms printed in bold type.

Activity 13 Going with the flow

Draw a series of annotated sketches to illustrate the following:
(i)   very viscous flow
(ii) flow with little viscosity
(iii) a thixotropic material
(iv) streamlined flow
(v)   turbulent flow.
In the cases of the first three, list some materials that would behave as described.

Questions

12 In a test of some motor oil, a ball-bearing of radius 0.5×10^{-3} m was dropped down the centre of a wide container of the oil and quickly reached a terminal velocity of 0.03 m s^{-1}.

(a) What is meant by *terminal velocity* and what can be said of the forces acting on the ball-bearing when it has reached its terminal velocity?

(b) Calculate the upthrust or buoyancy force on the ball-bearing. The density of this motor oil is 900 kg m^{-3} at this temperature.

(c) Calculate the weight of the ball-bearing. The density of the steel from which it was made is 7860 kg m^{-3} and g, the gravitational field strength, can be taken as 9.8 N kg^{-1}.

(d) The fall through the oil was such that Stokes's law could be applied. Calculate the viscosity η of the oil.

13 The data in Table 2.1 were collected from a flowmeter. Explain whether they indicate that output frequency is proportional to flow rate for this type of meter.

Mean output frequency / Hz	Mean flow rate / l s^{-1}
12.5	0.15
13.9	0.17
16.0	0.20
18.9	0.23

Table 2.1 Flowmeter data for Question 13

14 It is often more useful to measure mass flow rate than volume flow rate (for example, when charging customers for water or gas).

(a) Suggest a reason for this.

(b) Explain how mass flow rate can be calculated from volume flow rate.

15 Figure 2.24 shows a vortex-shedding flowmeter. When a non-streamlined obstruction is placed in a section of pipe, a series of vortices or eddies are produced non-symmetrically by the fluid flowing past. Mounted within the obstruction are two heated NTC thermistors. The eddies cool the thermistors, which increases their resistance.

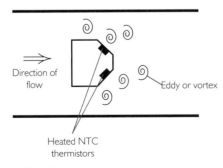

Figure 2.24 Vortex-shedding flowmeter

(a) Tables 2.2 and 2.3 show data collected for this type of meter for obstructions of diameters 10 mm and 5 mm.

Speed of flow / m s⁻¹	Time for 20 vortices to pass / s	Frequency of vortices / Hz
0.5	2.00	
1.0	0.95	
1.5	0.65	
2.0	0.48	
2.5	0.38	

Table 2.2 Data for a vortex flowmeter: 10 mm obstruction

Speed of flow / m s⁻¹	Time for 20 vortices to pass / s	Frequency of vortices / Hz
1.0	0.49	
1.5	0.32	
2.0	0.24	
2.5	0.19	

Table 2.3 Data for a vortex flowmeter: 5 mm obstruction

(i)   Copy and complete the 'Frequency of vortices' column in each table.

(ii) What appears to be the relationship between the speed of flow and frequency of vortices for an obstruction of given size?

(iii) For any one speed of flow, what appears to be the relationship between the diameter of the obstruction and the frequency of the vortices?

(b) Suggest a problem that might occur with this type of meter if the vortex frequency became very high.

3 Testing, testing ...

3.1 Good enough to eat?

It is no use developing a product that no-one will buy. You are likely to try a new food or sweet just to see what it tastes like, and may purchase it again if it is nice. But what makes a popular product, and how do manufacturers know this?

What qualities do we look for in a chocolate, biscuit or sweet? The flavour is very important, and this will be determined by the chemical composition of the food. Another important factor is the texture, which, for some people, and especially young children, can make all the difference between liking and disliking the food.

Consider the texture of the food; when we put something in our mouth, a number of things can happen. To discuss this with others and explain what it is that we like or dislike, we need to be able to describe what happens so that others understand. We use a large number of words for this. A sweet manufacturer may employ tasting panels (Figure 2.25) to find out what the people like and dislike. Below, a member of a tasting panel talks about the need to agree on exactly what each word means.

Figure 2.25 A tasting panel

The company has tasting panels to comment on texture, flavour and smell. When testing for flavour we have a number of references to compare with the test flavours, but we don't do that for texture. We were recruited through an advertisement in the local press and we spend one day a week testing. We did a lot of training and have regular training sessions. We have to discuss the product and come up with a description that we all agree with, although of course, no opinion is wrong and people do taste things very differently sometimes.

When the panel formed we spent a long time defining different words. For example chewy means partly how much it sticks to your teeth – how flexible it is. Some sweets are compact, they go into a ball and move around your mouth more than others. Brittle means it breaks into shards, sharp edged pieces, easily and quickly when you bite it. Crunchy means it makes a noise when you chew it. A sticky sweet sticks to your teeth and leaves a residue. It might be creamy or grainy or something else – we have defined all these descriptions.

When we do a profile for a product we start by saying what it looks like and what it feels like. Then we bite into it. Some products have a very different surface to the inside, so we comment on that, and on the initial bite. We time things like how long we suck before we bite, and how long the sweet lasts in the mouth. Finally we comment on the aftertaste.

Once tasters are experienced enough to give consistent descriptions, they can be employed in rating the attributes of various products. Frequently some 30 to 60 attributes will be judged on scales of 1–5, 1–7 or 1–9. These data are fed into a computer, which then constructs a star or spider diagram (Figure 2.26) to display the so-called sensory profile of the product.

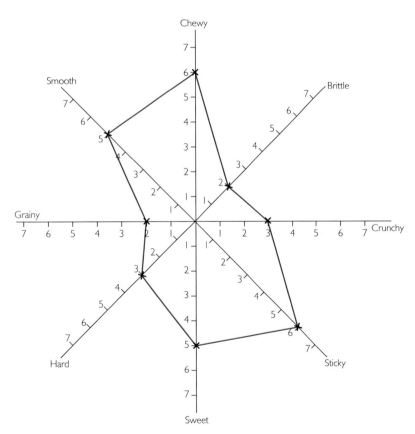

Figure 2.26 A star or spider diagram

Activity 14 Describing food

Make a list of the words you might use to describe the texture of some foods or confectionery. Combination products such as KitKat, Lion bar, Mars bar, Picnic and Snickers will give you opportunities to deal with a number of textures and tastes. Construct star/spider diagrams for your chosen products.

Human tasters give a subjective definition of what happens inside their mouth, described in words, with no measurement of how hard one has to chew to break food. The taster speaking earlier does measure the time to suck a sweet, to melt it and the time before it is soft enough to bite, but on the whole tasting panels are not concerned with measurement. However, research has been done to match sensory factors, such as chewiness, stickiness and hardness, to the tester (Figure 2.27), and the minute electrical signals are recorded as the product is eaten. These are then commented on.

The need for objective measurement arises when foods are to be produced on a large scale. The manufacturer must ensure that the product is always the same. You would be very surprised if you bought a food product one week and then the next week you bought the same product and found that the taste or texture was different. Is it possible to measure qualities like brittleness or chewiness? To describe behaviour quantitatively we need to be more definite about the meaning of the words, replacing verbal description with measurements. We must investigate and measure how the food responds to deformation, to force being applied to food.

Figure 2.27 Recording electrical signals while eating

3.2 Hardness

It is essential that sweets that are designed to be bitten can indeed be bitten without breaking your teeth – they must not be too **hard**. A hard material is one that is not readily scratched or indented. There is no absolute measure of hardness, but it is easy to rank samples in order of hardness according to whether they scratch, or can be scratched by, one another.

For minerals, an Austrian mineralogist Friedrich Mohs (1773–1839) devised the Mohs scale of hardness. He selected ten minerals arranged them from the softest to the hardest. Talc (Figure 2.28) is 1 on the Mohs scale and diamond (Figure 2.29) the hardest known mineral, is 10. Any mineral will scratch all the materials below it in the scale, and none of those above. The intervals of the Mohs scale are not regular; a fingernail has a hardness of about 2.5 and a copper coin has a hardness of about 3.5.

> **Study note**
>
> The terms used to describe food are applied to other materials as well. You will meet them in the chapter *Spare Part Surgery*.

Figure 2.28 Talc

Figure 2.29 Diamond

Activity 15 Sorting sweets

Choose a variety of sweets and cut them so that each has a rough surface. Use a scratch test to put them in order of hardness, and record that order.

Another way to define hardness is to apply a known force to a surface and measure the indentation, but each test of this type defines its own scale. There is no such thing as the hardness of a material – the value is only defined for a particular test. One common test of this type is the Brinell hardness test in which a hardened steel ball is pressed into the surface of the material for 10 or 15 s, and the surface area of the indentation measured (Figure 2.31). A commercially made hardness-measuring machine is shown in Figure 2.31. You might see similar machines if you visit an industrial or research laboratory.

The Brinell hardness number (BHN) is defined as:

$$\text{BHN} = \frac{\text{mass of applied load in kg}}{\text{surface area of indentation in mm}^2} \div \qquad (3)$$

The area of the curved surface can be calculated from the diameter, D, of the indenting ball and the diameter, d, of the indentation, which leads to:

$$\text{BHN} = m \div \left\{ \left(\pi \frac{D}{2} \right) \times [D - \sqrt{(D^2 - d^2)}] \right\} \qquad (4)$$

where D is the diameter of the indenting ball in millimetres, d the diameter of the indentation in millimetres, and m the mass of the applied load in kilograms. An alternative expression is:

$$\text{BHN} = m \frac{m}{\pi D H} \qquad (5)$$

where D is the diameter of the ball indenter in millimetres, h the depth of penetration in millimetres (not exceeding the diameter of the ball) and m the mass of the applied load in kilograms.

Expressions 3 to 5 are unlike most of the expressions you meet in physics; they are defined only for a particular situation and they only 'work' in specified units.

In the food industry, once a tasting panel has agreed on a product they like, the hardness can be measured to establish a standard. Quality control testing on each batch of the product can then ensure that the hardness falls within an accepted range. Compared with some other materials, such as metals and ceramics, foodstuffs are fairly soft. However, they tend to be referred to as 'hard' if they are made from compacted material that has little in the way of air- or liquid-filled gaps.

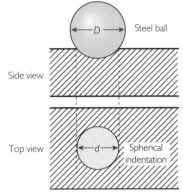

Figure 2.30 Indentation produced by a steel ball

Figure 2.31 A commercial hardness tester

Activity 16 Can you bite it?

Use a Brinell hardness test to measure the BHN for a variety of mints. There are various questions that you could investigate.

Are the mints in one packet as hard as those in another of the same type?

Do you get the same hardness all over the mint? Does the hardness change if the mints are hot, cold or even frozen?

Does the hardness change if the mints are heated or cooled and then returned to room temperature?

Do you get the same Brinell hardness number regardless of the diameter of the indenter or the size of the load?

To carry out hardness tests on very soft materials, for example plastics, or very hard materials such as hard steel used to make tools, different indenters can be used. In commercial testing there are other hardness tests, for example the Vickers and the Rockwell. In the Rockwell hardness test there are different sizes of hardened steel balls which may be used to make the indentation, or various diamond indenters. The choice depends on the hardness of the material, and there are a number of scales, one for each indenter and the force used.

Question

16 An extra strong mint was tested with a ball indenter of diameter 2.00 mm. An applied load of 5.3 kg produced an indentation of mean diameter 1.25 mm. What was the Brinell hardness number for this mint?

3.3 Crunching and chewing

Hardness is not the only property of interest to food and confectionery makers. The way foods deform, stretch or break under an applied force is related to the sensations we experience when eating them. Much of the testing in the food industry concerns the behaviour of foods under compression (Figure 2.32), because the manufacturers are interested in what happens to food when we chew or bite it, but tensile ('pulling') tests are also used (Figure 2.33). Some sweets, such as strawberry or cola laces (see Activity 19), are often gripped with the teeth and pulled to break them, so their tensile behaviour is of interest.

Figure 2.32 Compression testing of a cake

Figure 2.33 Tensile testing of pasta

The results of tensile and compressive tests are often displayed as graphs. From a load–extension graph such as that in Figure 2.34, we can measure the stiffness of a material.

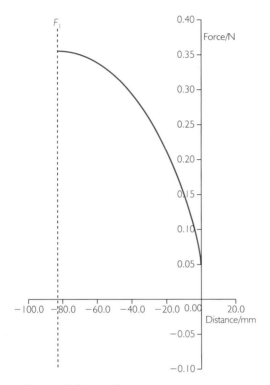

Figure 2.34 Load–extension graph for noodles

As you saw in the earlier chapter *Higher, Faster, Stronger*, the **stiffness** k is defined as:

$$k = \frac{F}{x} \qquad (6)$$

where F is the applied force and x the resulting extension. The stiffness depends on the size and shape of the sample, as well as on the material from which it is made. A thick sample will be stiffer than a thin one of the same material, and a long sample will be less stiff than the short one. However, provided the samples are all of a standard size and shape, graphs like that in Figure 2.34 provide a quick visual way to compare stiffness. A **stiff** material is one such as seaside rock (Figure 2.35) that does not easily change shape when a force is applied, so its force–extension (or force–compression) graph will be steep – a large load produces only a small deformation.

Figure 2.35 Seaside rock, a stiff foodstuff

When the load is removed, some materials (such as jelly cubes or sweets, Figure 2.36) will spring back to their original shape; such materials are said to be **elastic**. Others, described as **plastic** materials, will remain deformed like well-chewed chewing-gum (Figure 2.37). Many materials are elastic under small loads, up to their **elastic limit**, but deform plastically when subjected to larger loads.

Figure 2.36 Jelly, an elastic foodstuff

Figure 2.37 Chewing gum, which is plastic when well chewed

There are various ways in which a material can be deformed plastically. A **ductile** ('drawable') material is one that can readily be pulled out by a tensile force into a longer, thinner shape (like well-chewed gum), and a **malleable** ('hammerable') material can be deformed under compression. All ductile materials are malleable, but not all malleable materials are ductile; some, like fudge (Figure 2.38) may tear apart under tension.

Figure 2.38 Fudge, a malleable foodstuff that breaks apart under tension

Tensile or compression testing can also measure the force needed to break a sample. Again, the actual force needed depends on the size and shape, but compression of standard samples gives information about strength; a **strong** material is one that requires a large force to make it break.

The way in which foodstuffs break is of great interest to manufacturers and consumers. A **brittle** material is one that easily cracks, like a boiled sweet or hard toffee, or a biscuit; such foods might loosely be described as 'crunchy'. In brittle materials, an applied force is unable to deform the material; brittle materials are usually stiff, and show essentially no plastic deformation. Instead, the load causes small cracks to spread rapidly. Sometimes, as with wafers, this behaviour arises because there are many small air gaps within the structure which are unable to stop cracks spreading.

In contrast to brittle materials, a **tough** material deforms plastically and can withstand dynamic loads such as shock or impact. A tough material requires a large force to produce a small deformation – in other words, a large amount of **work** must be done on the material in order to produce a small plastic deformation. Few foodstuffs can be described as tough according to this definition (apart, perhaps, from some meats); it is a term more usually applied to materials such as Kevlar, which is used to make bullet-proof vests (Figure 2.39).

Figure 2.39 Bullet-proof vest made of Kevlar

Activity 17 Brittle and ductile toffee

Investigate how modifying the structure of toffee changes its physical properties.

Activity 18 Tough cookies

Compare the toughness of different sweets (such as CurlyWurly bars) chocolate bars and biscuits.

Activity 19 Stretchy sweets

Investigate the behaviour of strawberry or cola laces under tension. Display your results using graphs, and use appropriate technical terms to describe the behaviour of your samples.

Question

17 Table 2.4 lists some measurements obtained from hanging masses on a sweet called a Glow-worm. When the load is removed, the Glow-worm slowly goes back to its original length. It was 8.0 cm long at the start of the test, and its cross-section was an equilateral triangle with sides of 1 cm.

    (a) Plot a graph to display the data from Table 2.4.

    (b) Use appropriate words to describe and explain what was happening to the Glow-worm during this test.

    (c) Using your answers to parts (a) and (b), describe what the texture of a Glow-worm would be like as you ate it.

Mass of load / g	Length / cm
20	8.6
30	8.8
40	9.1
50	9.5
60	9.9
70	10.0
90	10.6
150	11.6
200	12.6
250	13.4
300	14.2
400	15.0

Table 2.4 Tensile tests on a Glow-worm

More testing

Standard tests in industry are normally carried out by machines that automatically apply loads and record the resulting deformation. Figure 2.40 shows one widely used machine, the Instron Universal Testing Machine. The moving part of the machine is driven vertically to compress or stretch the sample, and the drive is connected to a chart recorder so the force and distance can be recorded. It can perform tests in tension, compression or bending, and can be used for many materials including wood, plastic and adhesives. There are different models, some developed specifically for the food industry, and accessories for performing different tests can be attached to the machine as required. The model shown here can apply a force in the range from 2 N to 5 kN and the speed of deformation can vary from 0.02 to 50 cm min^{-1}.

Food manufacturers use a variety of tests in addition to straightforward compressive and tensile testing. One common test is the three-point bend test, shown in Figure 2.41, testing the 'bendiness' of a wafer. This test is used in many industries – steel rails and PVC window frames, for example, are tested in this way.

Figure 2.40 Instron Universal Testing Machine

Figure 2.41 Three-point bend test

Activity 20 Bendy wafer

Devise and carry out a three-point bend test on a wafer.

Another test used to monitor the crispness of wafers is shown in Figure 2.42. A wafer is snapped in front of a microphone, and its waveform and frequency spectrum displayed. The sound produced, and hence the shape of the waveform and frequency spectrum, depends on the crispness or bendiness of the wafer.

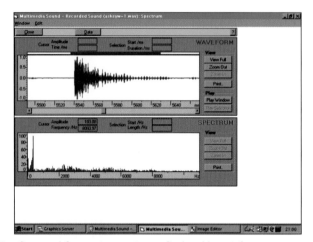

Figure 2.42 Waveform and frequency spectrum of a breaking wafer

Study note

You will find out more about sound and frequency spectra in the chapter *The Sound of Music*.

In 1988 Michael McIntyre and James Woodhouse of Cambridge University suggested that Chladni figures, the patterns on resonant surfaces, might by used to analyse material properties. This was followed up for wafers by Simon Livings, now working at the Centre Recherche Nestlé in Switzerland, as part of his doctoral project. He placed glitter on the top surface of wafers, placed them near a loudspeaker and observed the patterns as the wafer resonated.

Further investigations

Materials testing provides scope for a variety of investigations, using foodstuffs or other materials. You could explore the sounds produced by snapping brittle foods, and try to relate the results to other properties of the materials such as stiffness or strength, or perhaps you could investigate Chladni figures on resonating wafers. Or you might like to devise a completely different 'crispness' test for wafers. You could investigate how the 'crispness' varied with water content of wafers left exposed to damp air, or how the stretchiness and strength of spaghetti or noodles varied with the quantity of water absorbed.

You could try to devise tests to measure other physical aspects of food materials, such as 'stickiness'. A good test must by easy to use and should give reproducible results, i.e. the same test applied on different occasions to the same sample should give results within a relatively small range of uncertainty.

3.4 Summing up Part 3

In this part of the chapter, you have learned about so-called **mechanical properties** of materials. You have focused on the food industry, but the tests, and the terms used to describe the materials, are applicable wherever materials are tested and used.

Activity 21 and Questions 18 to 21 are intended to help you look back through this part of the chapter and check what you have learned.

Activity 21 Pick and mix

Select a variety of sweets or other foodstuffs, and use the words malleable, ductile, elastic, plastic and brittle, as appropriate, to describe their properties.

Questions

18 Why will the Brinell test not work for very hard or very soft materials?

19 In a Brinell test of a sample of steel, a 3000 kg load produced an indentation of diameter 3.85 mm with a ball indenter of diameter 10.00 mm. Calculate the Brinell hardness number for this sample.

20 Figure 2.43 shows the results of compression tests on Gouda cheeses.

(a) Describe what happens to the stiffness of Gouda cheese as it ages.

(b) What happens to the strength of the cheese as it ages? (Strength as in physical property, not as in taste!)

(c) What effect do the cumin seeds have on the physical properties of the cheese? Suggest a reason for this effect.

21 Imagine that you are working in the quality control area of a textile factory. You are asked to select four threads from a large number to demonstrate to some visitors threads that are:

(a) strong but not stiff

(b) stiff but not strong

(c) both strong and stiff

(d) neither strong not stiff.

If you tested the threads by hand, what would you look for to identify each of (a) to (d)?

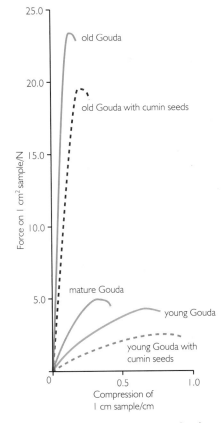

Figure 2.43 Compression tests on Gouda cheeses

4 Wrapping up

In this final part of the chapter, you will look first at some other aspects of physics in the food industry and then review the work that you have done in earlier parts of the chapter.

4.1 Food quality and safety

It is important to ensure that foods, including confectionery, are fit to eat. Government regulations have a role to play in this. The Food Safety Act (1990) and The Food Safety (Temperature Control) Regulations (1995) are two important pieces of legislation covering the food chain from the farm to the shop.

At each stage, the manufacture and processing of food must conform to high standards of quality. The relevant techniques, procedures and management systems are specified in 'British Standards'. The procedures for certification are independently inspected, and a company would lose its certificate if it failed to reach the required standards. Stephen Beckett of Nestlé Research and Development comments here on how physics in involved.

> *The development and use of instrumentation is perhaps one of the most important areas for physicists in the food industry. Some of this is for safety and must be carried out on-line, e.g. foreign material detectors; other instruments are for quality control, e.g. temperature and humidity. Even these are not as simple as they may at first appear. An item containing a metal object must be detected and rejected when the product is passing at the rate of several hundred per minute, and a simple, robust device for detecting pieces of plastic within chocolate has still to be invented. In addition the chocolate's temperature must be monitored and controlled to within a fraction of a degree to ensure that it is glossy and has a good 'snap' when broken.*

As you would expect, hygiene and safety are paramount. To ensure that any potential hazards are identified in a food production system, a Hazard and Critical Control Points (HACCP) system was developed. The first such system was design in the USA in the 1960s to ensure the safety of the food for astronauts (Figure 2.44) – getting food poisoning in space would be quite a problem! The key principles of HACCP involve:

- identifying steps in food production where significant hazards occur
- identifying critical control points (CCPs) where it is essential that the hazards are removed
- establishing critical limits which describe the difference between safe and unsafe
- establishing the means of monitoring and controlling the hazards
- keeping records
- verifying that the system is working correctly.

The hazards could be biological ones, such as salmonella in chicken, but they could also be chemical, related to the cleaning materials or lubricants used on the production line, or physical hazards, such as pieces of glass, metal, stones or wood getting into the food.

Figure 2.44 NASA Shuttle astronaut Rhea Seddon having a meal on the Space Shuttle

Activity 22 Physical hazards in food

In a small group, discuss your suggestions and ideas concerning the following:

- the types of physical hazards that could potentially contaminate food in the course of production
- the possible source(s) of these contaminants
- the techniques you might be able to use to detect such contaminants
- the techniques you might use to keep out such contaminants.

Make notes on your discussions as you will need them for Activities 25 and 26.

If you have the opportunity to visit a food production company, you may be able to identify some of their critical control points, together with what is done to ensure monitoring and control at these points.

4.2 Packaging

Most confectionery products are wrapped to preserve them in good condition. There is a vast number of wrapping materials in use, mainly plastic based, including cellulose acetate, polyester, polyethylene, polypropylene, polyvinyl chloride, nylon-6 and polyvinylidene chloride amongst others; the aluminium foil wrap of many chocolate bars is a good example of a non-plastic material. Stephen Beckett puts the final production stage (Figure 2.45) into perspective:

> *Packaging plays an important part in the chocolate confectionery industry, as it is often the only thing that a would-be purchaser sees. It is important to ensure therefore that the wrapper has been put on correctly. Major chocolate brands are produced at a rate of several million per day, and it is impossible to see moving parts of the wrapping machines to determine what happens when something goes wrong. In such cases high-speed video cameras can be of great benefit in being able to slow down the motion and help determine the cause of the fault.*

Figure 2.45 A confectionery product going through a wrapping machine

Activity 23 Which packaging?

Your task here is to make a recommendation on the ideal wrapping material(s) for a white chocolate covered Crunchie bar which is wrapped at a rate of 1000–1500 bars per minute. Write a report giving reasons for your recommendations.

Activity 24 Too much packaging?

Identify some of the issues involved in food packaging. Write a summary report with some recommendations for future practice.

Activity 25 Food miles

Research what is meant by 'food miles'. Then discuss whether shops should stock foods that have been transported through long distances (Figure 2.46), and whether customers should buy them.

Figure 2.46 Some food is transported through long distances

4.3 The final product

The following activities are designed to help you look back over your work in this unit; in doing so, make sure you have a full record of your work and that you are familiar with the key terms printed in bold earlier in the unit.

Activity 26 Coping with hazards

Design a double page spread (perhaps using computer graphics) for a magazine showing how some physical hazards could be detected and coped with in the food industry. Accompany your diagrams by a short paragraph outlining the physics involved in the techniques employed. Use your discussion notes from Activity 22 to help you.

Activity 27 Sweet physics

Write a short educational pamphlet 'Sweet physics' about physics in the confectionery industry, which might be useful for visitors to a confectionery company. Use your notes on the whole chapter to help you with this.

4.4 Questions on the whole chapter

Questions

22 A student wants to measure the density of the sugar mix used to make aniseed balls. He measures the diameters of 5 different balls with a micrometer screw gauge and obtains the following results 10.2, 10.7, 10.1, 10.3, 10.5 mm. He weighs
10 balls, and obtains a mass of 11.16 g.

(a) (i) What is the mean diameter and its uncertainty?

(ii) What is the percentage uncertainty in the diameter, and in the radius?

(iii) What is the volume of one ball? Give your answer in mm^3 and in cm^3.

(b) If the balance reads to ± 0.01 g, what is the percentage uncertainty in the total mass, and in the mass of 1 ball?

(c) What is the density of the sweet? Give your answer in $g\ cm^{-3}$ and in $kg\ m^{-3}$.

(d) (i) What is the uncertainty in the density?

(ii) How many significant figures should you give in the value for the density?

(iii) Write the density with the appropriate number of significant figures and its uncertainty.

23 A student wanted to build a model surveyor's ultrasound 'tape measure'. She connected a signal generator to a speaker, next to a microphone. The output from the microphone was connected to a storage oscilloscope. The timebase of the oscilloscope was set at $2\ ms\ cm^{-1}$. The speed of sound was taken as $335\ m\ s^{-1} \pm 10\ m\ s^{-1}$. The student switched the generator on and off very quickly, and obtained a trace similar to that in Figure 2.46.

(a) (i) Estimate the time between the signal received from the speaker and the echo.

(ii) Calculate the distance travelled by the sound in that time, and hence the distance away of the wall that reflected the sound.

(b) (i) If the position of each pulse can be judged ± 0.1 cm, calculate the uncertainty in the time for the signal to return.

(ii) Calculate the percentage uncertainty in the time for the signal to return.

(c) Calculate the percentage uncertainty in the speed of sound.

(d) (i) Calculate the actual uncertainty in the distance to the wall.

(ii) State the distance to the wall with the correct number of significant figures, and its uncertainty.

24 When water vapour in the atmosphere cools, it condenses to form droplets of liquid water. This forms clouds. If the droplets cool further, they freeze to form ice pellets, which fall from the cloud. By the time they reach the ground they have usually melted again to form rain, but sometimes they reach the ground as hail. When the ice pellets melt, they break up into small water droplets, which reach their terminal velocity before they hit the ground. Hailstones can cause a lot of damage because they are larger and more massive and travel faster than raindrops.

(a) Explain, with the aid of a diagram, how the forces on an object falling through a viscous fluid bring it to a terminal velocity.

(b) Write down an expression for the forces on a raindrop of radius r and density ρ_{water}, when it has reached a terminal velocity v while falling through air of density ρ_{air} and viscosity η.

(c) Given that $\rho_{air} \ll \rho_{water}$, rearrange your answer to (b) to obtain an expression for v.

(d) Given that $\rho_{ice} \approx \rho_{water}$, explain why hailstones reach the ground travelling faster than raindrops.

4.5 Achievements

Now you have studied this chapter you should be able to achieve the outcomes listed in Table 2.5.

Table 2.5 Achievements for the chapter *Good Enough to Eat*

Statement from examination specification	Section(s) in this chapter
18 understand and use the terms *density, laminar flow, streamline flow, terminal velocity, turbulent flow, upthrust* and *viscous drag*, e.g. in manufacturing	2.2, 2.3
19 recall, and use primary or secondary data to show that the rate of flow of a fluid is related to its viscosity	2.2, 2.3, 2.5
20 recognise and use the expression for Stokes's law, $F = 6\pi\eta r v$ and upthrust = weight of fluid displaced	2.2
21 investigate, using primary or secondary data, and recall that the viscosities of most fluids change with temperature. Explain the importance of this for industrial applications	2.2
25 investigate elastic and plastic deformation of a material and distinguish between them	3.3
26 explore and explain what is meant by the terms *brittle, ductile, hard, malleable, stiff* and *tough*. Use these terms, give examples of materials exhibiting such properties and explain how these properties are used in a variety of applications, e.g. foodstuffs	3.2, 3.3, 4.2

Answers

1 Using Equation 1,

$$F = 6\pi\eta rv$$

so

$$F = 6\pi \times 8.4 \times 10^{-2} \text{ N s m}^{-2} \times 1.0 \times 10^{-3} \text{ m} \times 2.0 \times 10^{-2} \text{ m s}^{-1}$$

$$= 3.2 \times 10^{-5} \text{ N}.$$

2 (a) $V = \dfrac{4\pi \times (1.0 \times 10^{-3} \text{ m})^3}{3}$

$$= 4.2 \times 10^{-9} \text{ m}^3.$$

(b) Mass m_{steel}

$= \rho_{\text{steel}} V$

$= 7.8 \times 10^3 \text{ kg m}^{-3} \times 4.2 \times 10^{-9} \text{ m}^3$

$= 3.3 \times 10^{-5} \text{ kg}$

(c) Weight of ball-bearing

$= m_{\text{steel}} g$

$= 3.3 \times 10^{-5} \text{ kg} \times 9.8 \text{ N kg}^{-1}$

$= 3.2 \times 10^{-4} \text{ N}.$

3 (a) Volume $= 4.2 \times 10^{-9} \text{ m}^3$ as the ball-bearing will displace its own volume.

(b) Mass of oil $m_{\text{oil}} = \rho_{\text{oil}} V$

$= 920 \text{ kg m}^{-3} \times 4.2 \times 10^{-9} \text{ m}^3$

$= 3.9 \times 10^{-6} \text{ kg}.$

(c) Weight of displaced oil $= m_{\text{oil}} g$

$= 3.9 \times 10^{-6} \text{ kg} \times 9.8 \text{ N kg}^{-1}$

$= 3.8 \times 10^{-5} \text{ N}.$

4 (a) Upthrust = weight of displaced oil

$= 3.8 \times 10^{-5} \text{ N}.$

(b) Upward force = upthrust + viscous drag force

$= 3.8 \times 10^{-5} \text{ N} + 3.2 \times 10^{-5} \text{ N}$

$= 7.0 \times 10^{-5} \text{ N}.$

Weight of ball-bearing $= 3.2 \times 10^{-4}$ N acting downwards.

There is therefore a net downwards force of magnitude F, where

$$F = 3.2 \times 10^{-4} \text{ N} - 0.7 \times 10^{-4} \text{ N}$$

$$= 2.5 \times 10^{-4} \text{ N}.$$

5 Weigh the object in air by suspending it from the forcemeter by the cotton.

Pour some water into the measuring cylinder to about half-way and note the volume.

Lower the object, still attached to the forcemeter, completely into the water but don't let it touch the bottom.

Note the new forcemeter reading and the new level of water in the measuring cylinder.

The difference between the two forcemeter readings gives the upthrust.

The difference between the two levels in the measuring cylinder gives the volume water displaced (which is also the volume of the object).

Calculate the mass of water displaced and hence its weight, which should be equal to the upthrust.

6 See Figure 2.12.

Pressure exerted by water on top of block

$= P_A + h_1 \rho g$

Downward force on top of block

$= (P_A + h_1 \rho g)A$

Pressure exerted by water at bottom of block

$= P_A + h_2 \rho g$

Upward force on bottom of block $= (P_A + h_2 \rho g)A$

So net upward force on block $= (h_2 - h_1)\rho g A$

$= h\rho g A.$

Volume of fluid displaced = volume of block $= Ah$

Mass of fluid displaced $= \rho Ah$

Weight of fluid displaced $= \rho Ahg$, which is the same as the net upward force.

7 Using symbols:

C_a = weight of crown in air

C_w = weight of crown in water

W_d = weight of water displaced

ρ_w = density of water

ρ_c = density of crown

V = volume of crown = volume of water displaced

By Archimedes' principle, we know that

$$C_w = C_a - W_d$$

and we are told that

$$\frac{C_w}{C_a} = 0.948$$

so we can write

$$\frac{(C_a - W_d)}{C_a} = 0.948$$

and so

$$1 - \frac{W_d}{C_a} = 0.948$$

$$\frac{W_d}{C_a} = 1 - 0.948 = 0.052$$

But we can also say that

$$C_a = V\rho_c g$$

and

$$W_d = V\rho_w g$$

so

$$\frac{\rho_w}{\rho_c} = \frac{W_d}{C_a}$$

or

$$\frac{\rho_c}{\rho_w} = \frac{C_a}{W_d} = \frac{1}{0.052} = 19.3$$

This is the density ratio of pure gold to water, so the crown must be made of pure gold.

8 You might expect them to have the same coefficient of viscosity over a range of temperatures – visco (viscosity) and static (the same). (In reality this is not actually so, but the changes of viscosity are such as to enable the vehicle to function satisfactorily regardless of the temperature changes. For extremes of temperature, oils are specially formulated to provide suitable viscosity.)

9 With 6 divisions between repeated parts of the trace the time period is $T = 60$ ms $= 60 \times 10^{-3}$ s.

Frequency $f = \dfrac{1}{T}$

$$= \frac{1}{(6.0 \times 10^{-2}\ \text{s})}$$

$$= 16.7\ \text{Hz}.$$

10 (a) (i) About 6 ℓ min^{-1}

(ii) 6 ℓ = 6×10^{-3} m^3. This volume of water has a mass of 6 kg. This gives a mass flow rate of 6 kg min^{-1}. 1 min = 60 s, so mass flow rate = 0.1 kg s^{-1}.

(b) About 200 Hz

(c) The graph is a straight line: output frequency is directly proportional to the volume flow rate.

11 By changing the electric field at certain points in the flow, the rate of flow of the chocolate could be changed in order to alter the thickness of a coating or take account of the change of speed of a production line.

12 (a) The terminal velocity is the greatest steady velocity reached by the falling ball-bearing. It is reached when the forces acting on the ball-bearing are balanced.

(b) Upthrust

$$= \frac{4\pi r^3 \rho_{oil}\, g}{3}$$

$$= \tfrac{4}{3}\pi \times (0.5 \times 10^{-3}\,\text{m})^3 \times 900\ \text{kg m}^{-3} \times 9.8\ \text{N kg}^{-1}$$

$$= 4.62 \times 10^{-6}\ \text{N}.$$

(c) Weight $= \dfrac{4\pi r^3 \rho_{steel}\, g}{3}$

$$= \frac{4\pi}{3} \times (0.5 \times 10^{-3}\ \text{m})^3 \times 7860\ \text{kg m}^{-3} \times 9.8\ \text{N kg}^{-1}$$

$$= 4.03 \times 10^{-5}\ \text{N}.$$

(d) From Equation 2:

viscous drag force = weight − upthrust

$$6\pi \eta v = 4.03 \times 10^{-5}\ \text{N} - 4.62 \times 10^{-6}\ \text{N}$$

$$\eta = \frac{(4.03 \times 10^{-5}\ \text{N}) - (4.62 \times 10^{-6}\ \text{N})}{6\pi \times 0.5 \times 10^{-3}\ \text{m} \times 0.03\ \text{m s}^{-1}}$$

$$= 1.26 \times 10^{-1}\ \text{N s m}^{-2}.$$

13 If you plot the data on a graph, the points lie almost on a straight line, suggesting that the output frequency is proportional to the flow rate.

14 (a) The fluid will expand and contract as temperature changes, so customers would be charged incorrectly.

(b) Mass = density × volume, so mass flow rate = density × volume flow rate. (Some flowmeters incorporate densitometers which automatically measure the density and compensate for any changes.)

15 (a) (i) See Tables 2.5 and 2.6.

(ii) The speed of flow is directly proportional to the frequency of vortices. As one doubles so the other also doubles.

(iii) The diameter of the obstruction is inversely proportional to the frequency of vortices. As one doubles, the other halves; as one is reduced to a third, so the other triples.

Speed of flow / m s⁻¹	Time for 20 vortices to pass / s	Frequency of vortices / Hz
0.5	2.00	10
1.0	0.98	20
1.5	0.66	30
2.0	0.50	40
2.5	0.40	50

Table 2.5 Table 2.2 completed (10 mm obstacle)

Speed of flow / m s⁻¹	Time for 20 vortices to pass / s	Frequency of vortices /Hz
1.0	0.50	40
1.5	0.33	61
2.0	0.25	80
2.5	0.20	100

Table 2.6 Table 2.3 completed (5 mm obstacle)

(b) If the frequency of vortices was very high there might not be enough time for the thermistors to heat up again before the next vortex cooled it. Then the vortices would not be detected.

16 Using Equation 4:

$$BHN = 5.3 \div \left\{ \left(\pi \times \frac{200}{2} \right) \times [2.00 - \sqrt{(2.00^2 - 1.25^2)]} \right\}$$

$$= 5.3 \div \{\pi \times (2.00 - 1.56)\} = 3.84.$$

17 (a) See Figure 2.47. Notice that this graph is plotted using g and cm; if you worked out the load in newtons and the extension in m, your graph would have the same shape but the numbers on the axes would be different; a mass of 10 g has a weight of approximately 0.1 N, so when the mass is 20 g, the load is 0.2 N. You could also have plotted extension rather than total length, in which case your graph would start at the origin.

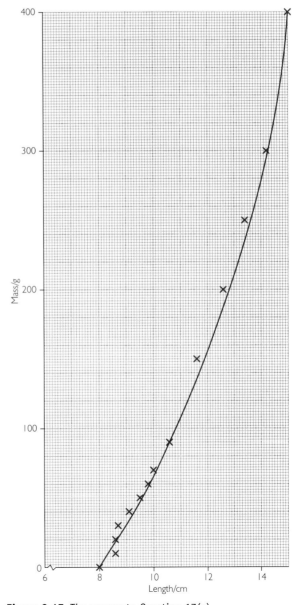

Figure 2.47 The answer to Question 17(a)

(b)   The behaviour was elastic throughout (the Glow-worm returned to its original length when the load was removed). It was not very stiff (despite being quite thick, it doubled its length under a moderate load). As it extended it became slightly stiffer (the graph curves upwards, so load ÷ extension is greater when the extension is large).

(c)   The Glow-worm would feel quite 'rubbery' and stretchy when chewed; it would deform and spring back.

18   For very hard materials the ball may make an indentation too small to measure. For very soft materials the ball will penetrate to a depth greater than the radius of the ball. The diameter of the indentation will then not increase any further and the indentation will just get deeper.

19   Using Equation 4:

$$BHN = 3000 \div \left\{\left(\pi \times \frac{10.00}{2}\right) \times [10.00 - \sqrt{(10.00^2 - 3.85^2)}]\right\}$$

$$= 3000 \div \{15.71 \times (10.00 - 9.23)\}$$

$$= 3000 \div 12.09$$

$$= 248$$

20 (a)   As it ages the cheese gets stiffer.

(b)   It also becomes stronger with age.

(c)   With cumin seeds, the cheese is less stiff and breaks more easily. Maybe the seeds act as small cracks, which spread when the cheese is deformed.

21 (a)   Stretches easily, but requires a large force to break it.

(b)   Does not stretch much, and breaks under a small force.

(c)   Does not stretch much, and also does not break easily.

(d)   Easily stretches and breaks.

CHAPTER 3

Spare Part Surgery

Why a chapter called *Spare Part Surgery*?

You probably know someone who has had spare part surgery. Perhaps an older relative or friend has a hip replacement such as in Figure 3.1(a), or new heart valve or a pacemaker. You might know someone who has had an operation for cataracts, in which the eye lens has been replaced by an artificial implant. Figure 3.1(b) shows surgery in progress, while Figure 3.1 (c) shows a modern prosthetic limb.

All of these spare parts help people to lead more pleasant and active lives, though there can be problems when a spare part wears out or is simply not as good as the original. Sometimes failures are dramatic and make the headline news. There have been cases involving heart valves that stick, and replacement hips that fail within months – unpleasant and possibly dangerous for the users, and expensive for the makers. Physics plays a key role in the development of good spare parts. To design a replacement joint you need to know about forces and the behaviour of materials in order to match the joint with natural bone.

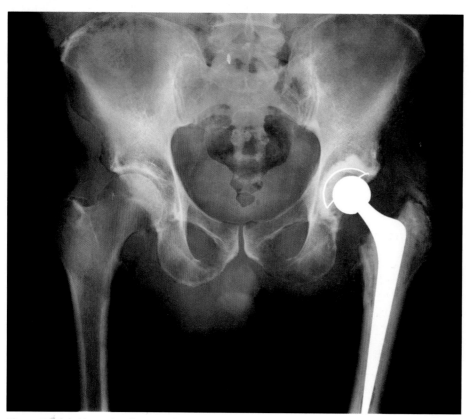

Figure 3.1(a) X ray showing a replacement hip

Figure 3.1(b) Spare part surgery in progress

Figure 3.1(c) Modern prosthetic limbs allow athletic movement

Overview of physics principles and techniques

In this chapter you will study the physics of materials. Some of this work will build on and extend ideas that you have already met earlier, but you will also meet some new ideas – for example you will learn how large-scale properties can be related to small-scale structure. There are many opportunities in this chapter for practical work, and to develop your skills in communication – both very important in medical uses of physics.

In this chapter you will extend your knowledge of:

- Bulk properties of materials from *Higher, Faster, Stronger* and *Good Enough to Eat*.

In other chapters you will do more work on:

- Bulk properties of materials in *Build or Bust?*
- Microscopic properties of materials in *Technology in Space*, *Digging Up the Past* and *The Medium is the Message*.

1 Spare parts

1.1 Informed consent

Most people are somewhat nervous about any sort of hospital treatment, whether it is an operation or some kind of examination – they want to know what is involved, and how it will affect them. At the end of this chapter, you will be asked to use what you have learned to produce some information for patients about one aspect of spare part surgery.

When deciding whether someone should have spare part surgery, several things need to be taken into account. One of these is the benefit to the patient, but there are other factors too, such as the cost of the operation, which need to be considered.

The following activities provide a general introduction to this chapter, and should also help you to plan what you would tell a patient.

Activity 1 The patient's view

Use a digital camera, tape recorder or mobile phone to record an interview with someone who has had some type of spare part surgery. Ask them how it has affected their life. Ask them about the advantages and also ask if there have been any problems. Would they recommend it to someone else? Is there any information they would like to pass on to other people having a similar operation?

Activity 2 Spare parts

Look up information on the internet on the price of various spare parts. Suggest reasons why some of the parts (e.g. an electric heart pump) are so much more expensive than others (e.g. a knee ligament). The actual cost of spare part surgery is much more than the cost of the part; suggest reasons for this. Choose one of the parts and identify the aspects of physics that were probably involved in its development.

1.2 Decisions

Activity 3 The doctor's view

In the last decade there has been a large increase in the demand for prosthetic body parts, especially arm and leg parts. List any reasons you can think of for this increase.

As a doctor, you would have to assess the needs of each patient for prosthetics and prioritise them, based upon your available resources.

Discuss with one or two other students how you and why you might prioritise the treatment of the following patients:

- a soldier requiring two artificial legs
- a pensioner requiring a hip replacement
- an eight-year-old child needing a new electronic arm
- a professional athlete in need of a powered artificial knee joint and leg
- a 50-year-old businessman needing an artificial heart pump.

You might find it helpful to refer to the information below on ethical frameworks.

The discussion in Activity 3 is an example of ethical decision-making – deciding what is the right course of action. Ethics is an academic discipline in its own right, and ethicists have identified various **ethical frameworks** that characterise the thought processes that we go through, and the reasons we give, in deciding how to act in a given situation. Four of the main frameworks are summarised below.

Utilitarianism

This way of thinking is based on the principle that the best course of action is the one that gives rise to the greatest amount of happiness (and the greatest reduction in the amount of unhappiness). A utilitarian might argue that a young child should have priority for spare part surgery, as they are likely to benefit for a much longer period than an older person. The likely effects on others (such as family members and carers) would also be taken into account in weighing up the priorities. A utilitarian might also argue that it is better to carry out two relatively low-cost operations rather than a single expensive one, as more people would benefit.

Divine command

For many people with a religious faith, writings such as the Bible or the Qur'an are the first source of guidance on ethical issues. While such scriptures are unlikely to say anything explicitly about spare part surgery, they may offer some guidance on the relative importance to be attached to the well-being of various types of people. Some interpretations of scripture present firm rules about medical procedures (for example, forbidding blood transfusions and organ transplants), and these will be considered of paramount importance by some people.

Virtue ethics

This way of thinking is based on the notion of a virtuous person – which is difficult to define but possibly less difficult to recognise. Here, the right course of action is that which is considered 'virtuous'. In the context of spare part surgery, it is probably quite difficult to say which decision is any more virtuous than any other, though it could be argued that an administrative decision to fund spare part surgery might be more virtuous than using the money for higher salaries for administrators.

Rights and duties

This framework uses the principle that people have certain rights (some of which might be formalised – for example, in the European Declaration of Human Rights, or in the constitution of the USA). And if someone has a right, other people have a duty to ensure that right is granted to them. Within this framework, it might be argued that someone has a right to a particular quality of life (e.g. a right to mobility, a right to be out of pain) in which case the doctor may have a duty to grant that right.

2 Boning up

As we age, our joints begin to wear out, making movement difficult and painful. Joints can also be affected by disease (such as arthritis), or may become injured. Nowadays, damaged or worn-out joints can be replaced relatively easily, and hip replacement is one of the commonest types of spare part surgery. This has become possible partly through the development of new materials and partly through improvements in surgical techniques.

In this chapter, you will study the properties of bones and the materials used to replace them. In doing so, you will advance your knowledge of the mechanical properties of materials.

2.1 Bone and joint replacement

In order to understand what is involved in making a good replacement joint, it is important first to know something about bones and joints and their function in our bodies. The article below outlines some key features of bones and joints, discusses some of the issues involved in bone replacement and then describes a recent development in artificial bone technology.

Activity 4 Bone and joint replacement

Read the following article about bones and joints and their replacement, and then answer Questions 1 to 6.

Bones and joints

Imagine what we would look like without a skeleton (Figure 3.2) – we would certainly not be able to move! Our bony skeleton enables us to move against two important forces: weight (due to gravity) and the drag of the medium through which we are moving – this is usually air but may also be water. The cell is the basic building block of living material. Plants' cells have rigid walls, so plants do not need skeletons, but animal cells are surrounded by a weak membrane that cannot be used for support. Our cells excrete materials to build up an internal skeleton (endoskeleton). Some animals have an external skeleton (exoskeleton).

Figure 3.2 Artist's impression of a person without a skeleton

Human skeletons are made of bone formed when proteins such as collagen are hardened by calcium and phosphorus salts excreted by cells. Bone is mainly crystalline calcium phosphate and calcium carbonate. About 20% of the bone is made up of living cells, which are fed with blood vessels through cavities in the bone.

The bones form levers in the body. They are held together by ligaments. Muscles are attached to the bones by tendons. A place where two bones meet is called a joint. Joints which allow free movement are called synovial joints (Figure 3.3).

The capsule holds the joint together and the synovial fluid acts as a lubricant. The cartilage on the bone ends provides a smooth surface to allow the bones to move over each other with the minimum of friction. There are two main types of joint: ball-and-socket joints (such as the hip) and hinge joints (such as the knee). (See Figure 3.4.)

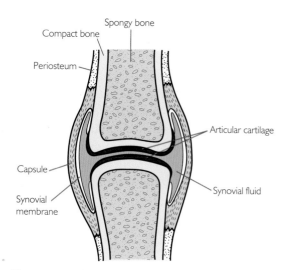

Figure 3.3 Schematic diagram of a synovial joint

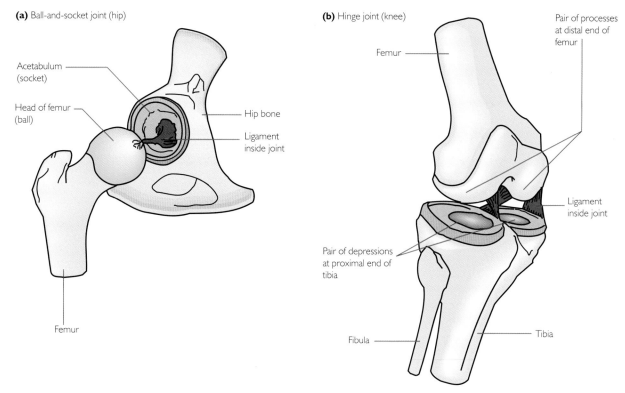

(a) Ball-and-socket joint (hip)

Acetabulum
(socket)

Head of femur
(ball)

Hip bone

Ligament
inside joint

Femur

(b) Hinge joint (knee)

Femur

Pair of processes
at distal end of
femur

Ligament
inside joint

Pair of depressions
at proximal end of
tibia

Fibula

Tibia

Figure 3.4 Exploded sketches of knee and hip joints

Replacement surgery

Like any mechanical device, joints undergo wear. Usually our body is able to repair any damage, but if the joint surfaces wear too quickly then osteoarthritis occurs. Disease can lead to a destruction of the surfaces; this is called rheumatoid arthritis. In severe cases replacement of the joint is necessary. The commonest joint that requires replacing is the hip joint.

Bone cancer used to require amputation but now the bone and joint can usually be replaced.

Replacement is rarely as a result of fracture since when this occurs blood vessels grow into the clot and a repair tissue develops which holds the bones together. This hardens and calcium salts are deposited to form bone. The role of an orthopaedic surgeon is normally to make sure the bones are aligned correctly. In the rare occasions that healing fails, then bone grafts from another part of the body can be used. This exposes the patient to the stress of two operations and an increased risk of complications and pain. It also costs a lot to perform a double operation and the long recovery time keeps the patient in an expensive hospital bed. Taking donor bone increases the risk of rejection or transmission of diseases such as hepatitis B or HIV.

New materials

Introducing new materials to the body can cause problems. Materials that are too weak may themselves break too easily. If the replacement material is much stiffer than the natural bone to which it is attached, then forces become concentrated in the replacement material. Bone regrows naturally when subjected to forces, but otherwise it wastes away – astronauts who live for long periods in weightless conditions can suffer problems with their bones if they do not undertake enough exercise. Not least, the new materials must be sterile and also must not cause the body's immune system to reject them.

'Of his bones are coral made'

Coral is a natural material, similar in many respects to bone. It is the subject of research to see whether it might make a good bone substitute. The title of this subsection comes from Shakespeare's The Tempest, *so even then the similarity between bone and coral was recognised.*

Coral reefs are the largest structures on Earth. Tiny marine animals called corals feed off microscopic plankton and turn the consumed carbon into calcium carbonate. When they die their mineral skeletons become part of a reef, which we call coral (Figure 3.5). It is similar in structure to bone, and a few species of coral have the necessary porous inner structure to allow new blood vessels to develop into the graft which will allow the body to start the growth of new bone and tissue around the coral.

Figure 3.5 A coral reef

Two researchers at the French Research Institute (CNRS) in Paris, Drs Patat and Guillemin, have pioneered the use of medical coral. They have mended shattered limbs, backbones and jaws with, they claim, 'as good as new' results (Figure 3.6).

Jonathan Knowles, a researcher at the Interdisciplinary Research Centre in Biomedical Materials, based at Queen Mary and Westfield College, London, believes that coral will be especially useful for treating more difficult fractures. 'When a bone breaks it normally heals within eight weeks. But there are some cases where the two ends don't grow back together. If you put in a metal plate, it still doesn't heal. A stimulus is usually needed to get a new bone to grow. Coral actively promotes bone formulation and healing in a way that materials like steel don't,' explains Knowles.

Figure 3.6 A patient's shattered spine is repaired using coral

As well as being tough and less likely to be rejected by the body's immune system, coral carries no risk of infection, unlike grafts of human bone. The only drawback is that coral has the potential to absorb toxic metals such as nickel, cadmium and mercury.

Medical coral is harvested from warm reefs on the South Pacific islands of New Caledonia, east of Australia, under strict rules that minimise environmental damage.

Questions

1 Name a common external skeleton (exoskeleton).

2 Write down at least two functions of a skeleton.

3 How might bone behave if it was without living fibres?

4 Describe the main difference in the movement of a hinge joint compared with that of a ball and socket.

5   List at least two advantages of using a bone substitute rather than using human bone.

6   What material is missing from coral that is present in bone?

2.2 The right stuff

To make a good bone substitute, a material must be strong enough to exert and withstand the forces involved in normal movement; if it is to be used in a hip replacement, it must support a person's weight. Its **mechanical properties** (the way it behaves when subjected to forces) must be similar to those of real bone. In this section you will see how materials are tested, described and compared.

If you have studied the chapter *Good Enough to Eat*, you will have met several terms used to describe materials and their behaviour. Some of these are listed in Table 3.1, along with two extra terms (**smooth** and **durable**) not used in that chapter. In the course of this chapter, the meanings of some of these terms will be refined.

elastic	returns to its original shape when the load is removed
plastic	remains deformed when the load is removed
brittle	cracks and breaks without plastic deformation
tough	does not readily crack; can withstand dynamic loads such as shock or impact
ductile	can be deformed plastically under tension; can be pulled into a long thin shape
malleable	can be deformed plastically under compression; can be hammered into a sheet
hard	not readily scratched or indented
stiff	requires a large force to produce a small deformation
smooth	low friction surface
durable	properties do not worsen with repeated loading and unloading

Table 3.1 Terms used to describe the behaviour of materials

Question

7   Use terms from Table 3.1 to describe the ideal properties of the following spare parts: (a) blood vessel, (b) teeth and (c) thigh bone.

Stress and strain

In *Higher, Faster, Stronger*, you saw that the extension of a rope or cord under a given load depends on its length and thickness as well as on the material from which it is made. One way to compare the behaviour of different materials is to use samples of a standard size. Another way is to define and measure properties in such a way that they depend only on the material and not on the size and shape of the sample; as bones do not come in standard sizes, it makes sense to compare their properties in this way.

Bones are normally subjected to compressive forces – that is, to forces that tend to squash them. Under a relatively small force, bone will deform slightly (though not much, because it is stiff), but if the force becomes large the bone may break. If two samples of different thickness are both subjected to compressive forces, the thinner one will break under a smaller force. If one sample has twice the cross-sectional area of the other, then it will withstand twice the force, but for each sample the force divided by the area is the same. Applied force, F, divided by area of cross-section, A, is called the stress and is represented by σ (the Greek letter sigma):

$$\sigma = \frac{F}{A} \qquad\qquad (1)$$

The same definition and symbols are used for **tensile stress** (when the sample is pulled) and for **compressive stress** (when the sample is squashed). The SI unit of stress is the pascal (Pa); $1\ \text{Pa} = 1\ \text{N m}^{-2}$. You have probably met the unit before – it is also used for pressure. 1 Pa is a very small stress; the SI prefixes kilo and mega are usually needed when measuring stresses in practical situations.

The stress needed to break a material is called its **ultimate compressive** (or tensile) **stress**, or simply the **breaking stress**, and is a measure of the **strength** of the material that does not depend on the size of the sample. The ultimate compressive stress of bone, or bone substitute, is clearly important to the user, as you will see in Questions 8 and 9 and Activity 5.

> **Maths reference**
>
> SI prefixes
> See Maths note 2.4

Questions

8 Estimate the compressive stress in your leg bones when you are standing still. Explain why their ultimate compressive stress needs to be much greater than this.

9 A 70 kg man jumps from a wall 1.5 m high, lands on both feet together and takes 0.1 s to come to rest. The cross-sectional area of the bones in each of his lower legs is 30 cm².

(a) How fast is he moving just as he reaches the ground?

(b) What is his average deceleration?

(c) What is the average force exerted as he comes to rest?

(d) What is the stress in his lower legs as he comes to rest?

Use $g = 9.8\ \text{m s}^{-2} = 9.8\ \text{N kg}^{-1}$. You might need to look back at the work you did in *Higher, Faster, Stronger*. Note that you need to express the areas in m².

Activity 5 Crunchie bones?

The inside of a Crunchie bar is very similar to bone in structure and appearance. Measure the breaking stress for a piece of Crunchie bar, and hence decide whether it would make an acceptable bone substitute.

The amount by which a sample deforms also depends on the stress. In the chapter *Higher, Faster, Stronger*, you learned that, for a material that obeys **Hooke's law**, the extension is proportional to the applied force. Figure 3.7 shows two samples that obey Hooke's law. They are both made of the same material and are the same length but one has twice the cross-sectional area of the other. If the same force is applied to each, then the thicker sample is deformed by only half as much as the thinner one, but if they are both subject to the same stress, then they both suffer the same deformation. So, for samples of the same length, compression (or extension) Δx is proportional to stress, and Hooke's law can be expressed as

$$\Delta x \propto \sigma \qquad\qquad (2)$$

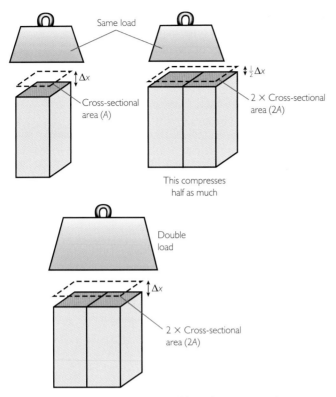

Figure 3.7 Samples of different thickness being subjected to compression

The extension, or compression, of a sample, depends on its length as well as on its area of cross-section. As shown in Figure 3.8, doubling the length of the sample doubles the compression. For each sample, the ratio of compression, or extension Δx to original length l is the same. This ratio is defined as the strain and is represented by the symbol ε (the Greek letter epsilon):

$$\varepsilon = \frac{\Delta x}{l} \qquad\qquad (3)$$

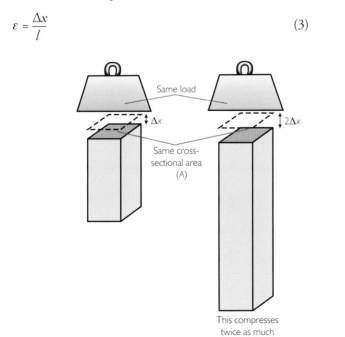

Figure 3.8 Samples of different length being subjected to compression

The same symbols are used for **compressive strain** and for **tensile strain**. Notice that strain is a ratio of two lengths and so it has *no units*. Strain is often expressed as a percentage – that is, the compression or extension as a percentage of the original length.

Questions

10 In a test of a sample of possible bone substitute, a sample of 40 cm long is compressed by 2.5 mm. What is the compressive strain expressed as a decimal and as a percentage?

11 A metal wire is given a tensile strain of 0.15%. If the original length of the wire was 50 cm, by how much did it extend? Express your answer in mm. If a wire ten times as long is given the same strain, by how much will it extend?

Maths reference

Fractions, decimals and percentages
See Maths note 3.1

Hooke's law and the Young modulus

In previous work, you have used force–extension graphs to show how a material sample deforms under a load. If, instead of force against extension, we plot stress against strain, then we get a graph that depends only on the nature of the material and not on the size and shape of the sample. Figure 3.9 shows such a **stress–strain graph** (b) plotted for a material sample, alongside two force–extension graphs (a) for different samples of the same material. Notice all the graphs are the same shape, although they have different numbers on the axes.

The labels on the graphs in Figure 3.9 are related to the mechanical behaviour of the sample. In Figure 3.9(a), from O to H each graph is straight; H is the **limit of proportionality**. Force is proportional to extension, i.e. the sample obeys Hooke's law. In Figure 3.9(b), between O and H, stress is proportional to strain – which is another way of expressing Hooke's law.

Figure 3.9 the results of testing a material sample displayed (a) (i) and (ii) as force against extension and (b) as stress against strain

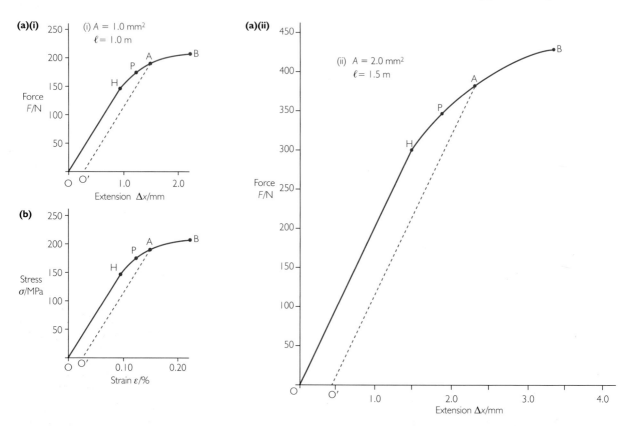

Elastic deformation occurs up to P, the **elastic limit**. This means that the material returns to its original length when released. Beyond P the material gains permanent extension and is therefore said to behave plastically. P is also known as the **yield point**. It is worth noting that if released at A (i.e. beyond P) the material recovers along AO′, which is parallel to HO, giving it a permanent extension of OO′. If stress is reapplied then the curve O′AB is followed. At B the sample breaks. In Figure 3.9(b), the stress at B is the breaking stress or ultimate tensile stress.

In the chapter *Higher, Faster, Stronger* we defined the stiffness, *k*, of a sample such that:

$$k = \frac{\Delta F}{\Delta x} \qquad (4)$$

The constant *k* is equal to the gradient of the force–extension graph. Rather similarly, we can measure the gradient of the stress–strain graph. Now, though, we get a value that depends only on the material and not on the size or shape of the sample – it is a measure of the **stiffness of the material**. This value is known as the **Young modulus**, *E*, of the material:

$$E = \frac{\sigma}{\varepsilon} \qquad (5)$$

Since strain has no unit, the SI unit of the Young modulus is that of stress, i.e. pascals, Pa.

The Young modulus is important when designing materials for replacement joints. Figure 3.10 shows a typical hip replacement. Bone typically has a Young modulus of about 1×10^{10} Pa. If the replacement has a smaller Young modulus than the natural bone into which it is inserted, then most if the load will be supported by the natural bone, which might then be under too much stress and become damaged. (You can model this using two springs side by side, as in Figure 3.11.

> **Study note**
>
> The Young modulus is given the symbol *E* because it is a so-called 'elastic modulus'. There are other elastic moduli which you will meet in the chapter *Build or Bust?*

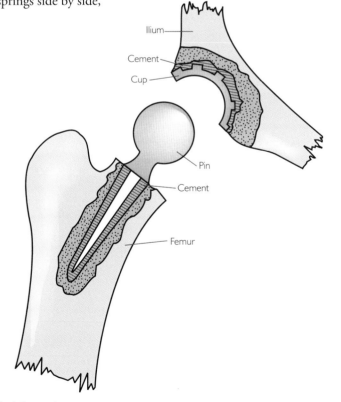

Figure 3.10 A typical hip joint replacement

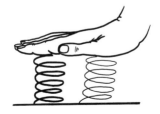

Figure 3.11 Squashing two springs of different stiffness

If one spring is stiffer than the other, then as you squash both springs you feel the stiffer spring exerting a greater force.) However, if the Young modulus of the replacement is too high, then it supports most of the load. Far from protecting the natural bone, this causes it to waste away because it is deprived of the stress that stimulates regrowth.

2.3 By design

Modern prosthetics

Modern prosthetics can now do much more than just replace a limb. Advances in miniaturising microprocessors have led to the development of the C-leg by Otto Bock Healthcare (Figure 3.12). The advanced knee joint samples data 50 times a second and processes it to give the user a more stable limb, for better balance and a more controlled gait. Two strain gauges measure pressures on the leg and note how often the heel strikes the ground. Magnetic sensors can detect changes in knee angle. It can recognise different actions (e.g. walking up stairs, running, hopping) and adjust itself accordingly so the wearer doesn't need to think about it. These new prosthetics and give the users a better quality of life and allow them greater mobility, to take part in a variety of sporting and leisure activities once more. For more information, see the Otto Bock website.

Figure 3.12 The C-leg

Figure 3.13 shows sportsman Cameron Clapp. As a teenager, a train accident resulted in him becoming a triple amputee. He uses a variety of advanced prosthetics, including the C-legs, to take part in numerous sporting events. You can read more about his remarkable achievements and his various prostheses on his website.

For details of both these websites, visit www.shaplinks.co.uk

Figure 3.13 Cameron Clapp

New materials

Replacement hip joints used to be made of a metal ball in a polyethylene and titanium socket. They last for about ten to fifteen years, after which time small, abrasive particles from the metal ball begin to wear away the plastic parts, and they then need to be replaced. For this reason, hip replacements are not generally given to younger people since they would need several operations in the course of a lifetime.

In 2003 approval was given for the development of ceramic on ceramic replacement hip joints. Ceramic materials are far more durable – they produce much less debris and therefore last a lot longer. Golfer Jack Nicklaus received an experimental ceramic hip joint in 1999. One of the industry standard hip replacement joints used worldwide is the BHR – Birmingham Hip Replacement (Figure 3.14(a)), and as the name suggests, it was developed in Birmingham. The metal alloy hip joint has a special surface cast called Porocast™ that helps resist particulate wear, by making the porous surface an integral part of the metal casting process (Figure 3.14(b)). It also encourages bone in-growth, helping to fix the new joint better.

The metal Porocast™ is covered with a hydroxyapatite ($Ca_{10}(PO_4)_6(OH)_2$) ceramic coating. This prevents the body from rejecting the metal implant, and allows bone minerals to grow into the hydroxyapatite.

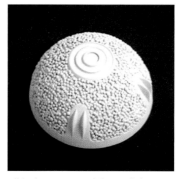

Figure 3.14 (a) The Birmingham Hip Replacement (b) the porous surface of the BHR

Research and development

Prosthetics is a major area of research and development where materials scientists and engineers work on the design of the prosthetics and the materials used to make them. Some of this work takes place in universities and hospitals, and some is sponsored by manufacturers. It is important that new ideas are tested, and that research findings are communicated accurately to other people working in the same field.

Tests usually involve laboratory work or (for designs at an advanced stage) clinical trials with patients. Potential users of new materials and new designs (surgeons, manufacturers, patients) want to know details of the tests, and these may be reported in specialist journals. A paper describing the work is sent to the editors of the journal and goes through a process of **peer review**. That is, the editors ask other experts in the field to read the paper and say whether the work appears to be of good quality and accurately reported – for example, the reviewers might look for evidence that tests were properly conducted, and that other relevant work is referred to and acknowledged. If the reviewers agree, the paper is published, and other workers in the field may then draw on it to help them make further progress.

Sometimes a new development may be commercially sensitive. A manufacturer working on a revolutionary design will not want to report details too soon, as rival manufacturers may use the same idea. Once the details of the new design are fully worked out and tested, the manufacturer usually takes out a **patent**. This means that a technical description of the design is legally registered and made public, then anyone else who wants to use the design can do so if they pay a fee to the patent holder. After the patent is registered, the developer may then publish details of tests and trials.

Editors of technical and research journals sometimes commission an expert to write a **review article** on a particular chapter. Such an article does not report new research, but summarises all the recent and important work on the topic and provides information about where the work was originally reported.

As well as sharing information through peer-reviewed journals, researchers hold conferences at which they report their findings and discuss them with others working in the same field. Proceedings of conferences are often written up and published to provide a permanent record.

You can find more about academic publishing and the relevant procedures on Wikipedia the online encyclopaedia. Details of the URL can be found at www.shaplinks.co.uk.

Activity 6 Replacement hip joints

Use the Internet to research information for a review article about replacement hip joints and various materials that could be used. You could start by looking at the BHR site (type 'hip resurfacing' into a search engine).

Try to find information that offers a genuine and reliable comparison of different hip joint materials and designs. Look for evidence of tests and trials, and for references to work carried out by others. Also look for any indication that the work has been peer-reviewed.

Write a 500-word report of your key findings and opinions.

Developing a new material

You have read of one possible substitute bone material: coral. The development of completely artificial bone materials is the subject of much current research. These have the advantage that they can be made to precise specifications in controlled sterile conditions. The most promising new materials are **polymers**, since these can be designed to have properties very similar to real bone and they are also very strong. One such material, HAPEX, has been developed at the Biomedical Materials Interdisciplinary Research Centre (IRC) at Queen Mary and Westfield College, London, under the leadership of Professor Bonfield. HAPEX is a composite material, consisting of a polymer (Polyethylene) and a ceramic (hydroxyapatite), which is quite similar to natural bone in structure and chemical composition. (Coral, incidentally, is a natural hydroxyapatite.)

HAPEX has been used to replace the tiny bones in the middle ear which are responsible for the conduction of sound (Figure 3.15). Stronger versions of the material are being developed in collaboration with the Polymer IRC at Leeds, Bradford and Durham Universities and it is hoped to use it for hip replacements. As the material closely resembles real bone the real bone will grow into it and attach the implant very firmly. It is anticipated that HAPEX implants will last about thirty years and will therefore be feasible for use with young sports-active people.

Figure 3.15 A tiny HAPEX artificial bone for ear implants compared in size with a 5p coin

As well as the replacement bone itself, the material to make the outer parts of the replacement joint must be carefully chosen to have suitable mechanical properties. Again, polymer materials are proving ideal. In Activity 7, you are asked to test a sample of UHMWPE (ultra-high-molecular-weight polythene) to check that its properties are suitable. This material is used to make the cup in replacement ball-and-socket joints. The main reason for its use is its low friction, but the material must also be able to withstand the large stresses exerted on the joint.

Polymers

Polymers consist of molecules that are long chains made up of fairly simple arrangements of carbon and other atoms repeated many times over. Polyethylene is a polymer whose molecules are long chains of C_2H_2 units joined together (Figure 3.16). Its official chemical name is poly(ethene) (C_2H_2 is ethene) and it is known colloquially as polythene. The length of the chains depends on how the polymer is made. UHMWPE has very long chains – 'ultra-high-molecular-weight' means that the (average) mass of each molecule is large.

Ethene

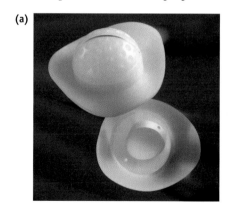

Polyethylene

Figure 3.16 Ethene and poly(ethene)

UHMWPE has a very smooth surface, and is also very tough, so it is used for many other purposes as well as for replacement joints. These include:

- coating of hoppers in the steel industry to allow coke to flow easily
- rear of boring heads for tunnel building to allow earth and clay to fall off easily
- sliders in railway points so that ice will not stick to them
- harbour fenders
- rail guides for conveyor belts.

Perplas Medical Ltd in Lancashire specialise in the production of UHMWPE. They can produce rods, sheets and machined components using computer-aided design machining equipment and presses. Their machines work in an environment controlled by electrostatic filtration to submicrometre size to ensure that airborne contaminants are kept to an absolute minimum. Any material introduced into the body must be free of microbes. This medical-grade material is used in hip joints (Figure 3.17(a)), and is supplied with a certificate that allows total traceability and figures to show test results of its important mechanical properties to its user (Figure 3.17(b)).

(a)

(b)

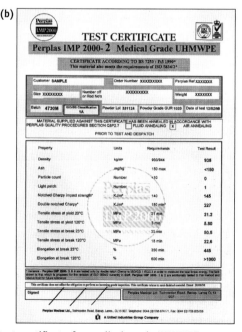

Figure 3.17 (a) Hip joints made with UHMWPE (b) a test certificate for medical-grade UHMWPE

Activity 7 Is it good enough?

Test a sample of UHMWPE and see how well it compares with the medical-grade sample shown in Figure 3.17(b). Figure 3.18 shows two possible versions of the apparatus.

Figure 3.18 Apparatus for Activity 7: (a) simple version and (b) materials testing kit

Stress–strain graphs, and the Young modulus, are widely used in all areas of materials science, not just those concerned with medical applications, as Questions 12 to 14 illustrate.

Questions

12 A steel wire of cross-sectional are 0.5 mm² and length 8.0 m is found to stretch 6.0 mm when the tension on it is increased by 75 N. It returned to its original length when the extra tension was released.

(a) What was its cross-sectional area in square metres?

(b) What stress had been applied?

(c) What was the resultant strain?

(d) What is the Young modulus for steel?

(e) Bone typically has a Young modulus of about 1×10^{10} Pa. What problems would there be in using steel as a replacement bone material?

(f) A lift cable consists of 100 strands of wire. The lift is limited to 10 people of maximum mass 85 kg each. When the lift is full, by how much will the 90 m cable stretch? ($g = 9.8$ N kg^{-1}.)

13 A steel rod was tested to destruction. It had a cross-sectional area of 10^{-4} m² and a length of 65×10^{-3} m. The results are listed in Table 3.2 in the order in which they were obtained.

(a) Use Table 3.2 to draw up a table of stress and strain and plot a stress–strain graph. Join the plotted points with a smooth curve, being careful to join them in the order in which they are listed. (You might need two graphs, one for small strains and one for large strains.)

(b) From your graph:

(i)   What was the yield stress of the steel?

(ii) What was the ultimate tensile stress of the steel?

(iii) What was the percentage strain when it broke?

Tension / 10^3 N	Extension / 10^{-3} m
0.0	0.0
35.0	0.1
38.0	0.2
36.5	0.4
36.0	0.7
37.0	1.0
40.0	1.4
45.0	2.1
50.0	3.1
55.0	5.1
57.5	7.2
60.0	9.1
60.0	13.1
57.5	14.7
55.0	15.2
52.5	15.8 (broke)

Table 3.2 Data for Question 13

14 A nylon guitar string has a diameter of 0.4 mm. The length of the string from its fixed point to the tension key is 840 mm. Turning the tension key once extends the string by 4.0 mm. Calculate the tension in the string when the key has been turned eight times. (Young modulus for nylon = 3.0×10^9 Pa.)

2.4 Take the strain

Elastic energy

Look back at Figure 3.1, and think again about the different types of materials used to make spare parts such as bones, ligaments and blood vessels. Bone substitutes need to be strong and stiff, but other replacement parts need to be more elastic. In some cases, this is because they need to store and return energy. In the chapter *Higher, Faster, Stronger*, you measured some of the energy transfers in sporting activities, and energies of a few hundred joules were common (for example, in jumping and weight-lifting). In Activity 7, stretched UHMWPE's ability to store energy is not relevant to its use in hip joints, but it is important in some of its other uses (such as harbour fenders).

In *Higher, Faster, Stronger* you saw how to find elastic energy, ΔE_{el}, from a force–extension graph: it is equal to the area under the curve. If the sample obeys Hooke's law, then:

$$\Delta E_{el} = \frac{F\Delta x}{2} = \frac{1}{2}k(\Delta x)^2 \qquad (6)$$

where F is the magnitude of the force needed to produce an extension Δx (see Figure 3.19(a)). If the sample does not obey Hooke's law, then the area can be found by counting squares (Figure 3.19(b)).

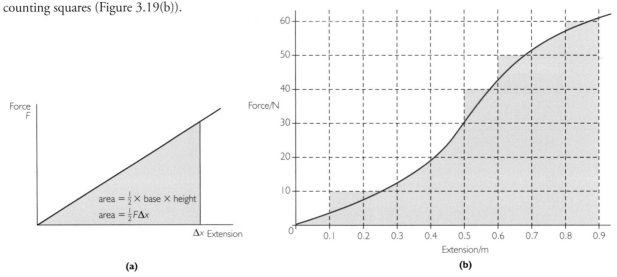

(a)   (b)

Figure 3.19 Elastic energy is found from the area under a force–extension graph either (a) by calculation or (b) by counting squares

What about stress–strain graphs? We have seen that the gradient of such a graph has a meaning (it is the Young modulus), but what about the area under the curve? Think of a material that obeys Hooke's law. At a given strain ε, the area under the stress–strain curve is $\sigma\varepsilon/2$ (Figure 3.20). Since (from Equation 1):

$$F = \sigma A$$

and (from Equation 3):

$$\Delta x = \varepsilon l$$

we can write Equation 6 as:

$$\Delta E_{el} = \frac{(\sigma A \times \varepsilon l)}{2} = Al \times \left(\frac{\sigma\varepsilon}{2}\right)$$

But Al is just the volume, V, of the sample, and so:

$$\text{area under stress–strain curve} = \left(\frac{\sigma\varepsilon}{2}\right) = \frac{\Delta E_{el}}{V} \qquad (7)$$

In other words, the area under the curve is equal to the energy stored per unit volume, or the **energy density**, U.

For a material that obeys Hooke's law, we can rewrite Equation 7 using the Young modulus, E (from Equation 5):

$$\sigma = \varepsilon E$$

and so:

$$U = \frac{\Delta E_{el}}{V} = \frac{E\varepsilon^2}{2} \qquad (8)$$

For a material that does not obey Hooke's law, Equation 8 does not apply, but the area under a stress–strain curve is still equal to the stored energy density.

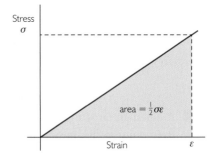

Figure 3.20 The area under a stress–strain curve for a material that obeys Hooke's law

Questions

15 Suppose you want a material that is capable of storing a large amount of elastic energy per unit volume (e.g. to make an artificial tendon). What particular mechanical properties would you look for?

16 (a) Look back at your answer to Question 13, and use your graph to find the energy density in steel when it breaks.

   (b) Given that the density of steel is about 8×10^3 kg m^{-3}, estimate the speed of a steel wire when it breaks.

2.5 The inside story

As you saw in Section 2.3, polymer materials are being developed to make substitute bones, and to make replacement ball-and-socket joints. These materials have mechanical properties (Young modulus, strength, smoothness) that enable them to match natural materials. Another recent development is artificial skin, which is particularly useful when treating people who have become badly burned or whose skin has become ulcerated. Artificial skin is much more elastic than artificial bone, to match natural skin. (You can demonstrate this elasticity by pinching a fold of skin on the back of your hand and noticing that it springs back. The elasticity of skin declines with age; ask an older person to repeat the 'pinch test' and compare observations.)

A skin substitute called Dermagraft, made by Advanced Tissue Science Inc. in the USA, combines a bioengineered human dermal (skin) layer with a synthetic polymer covering. On being transplanted, the polymer covering dissolves and the patient's own cells start to grow.

Why do polymers play such a key role in spare part materials? The clue lies in their small-scale structure.

Polymer structures

Figure 3.21 shows polymer molecules in a stretched and an unstretched polymer sample. Unstretched polymers have no regular structure and are said to be amorphous: the long-chain polymer molecules are intertwined and jumbled up (Figure 3.21(a)). A stretching force tends to uncoil the chains and straighten them into orderly lines (Figure 3.21(b)), and so they can produce regular diffraction patterns. When released, the molecules coil up again. When fully extended the chains are stiff because their

> **Study note**
>
> In the chapter *Digging Up the Past* you will see how diffraction studies can reveal the small-scale structure of materials.

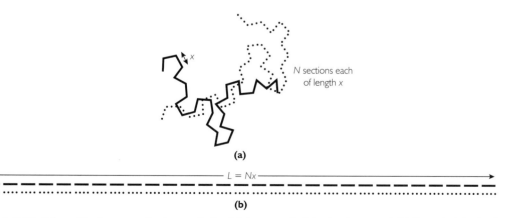

(a)

$$L = Nx$$

(b)

Figure 3.21 Schematic diagrams of polymer chains (a) in an unstretched polymer and (b) in a stretched polymer

interatomic bonds are then stretched directly. This explains why polymers have a Young modulus 10 000 times less than that of a metal but can be extended to perhaps ten times their original length (1000% strain).

Low-density low-molecular-weight polyethylene is similar to rubber with lots of jumbled-up polymer chains. In fact the very long molecules fold up as shown in Figure 3.22 and stack up in what is referred to as a lamella formation (lamella means 'little plate'). As can be seen, the vertical strands are parallel and regular and so resemble a crystalline structure.

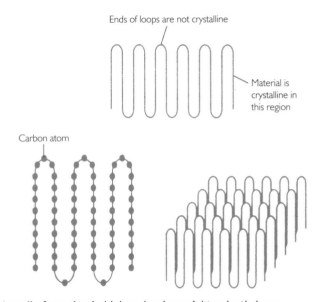

Figure 3.22 Lamella formation in high-molecular-weight polyethylene

When the material is unstretched, the lamellae are randomly orientated as shown in Figure 3.23. As the material is stretched, they start to align. This is a reversible effect, and so the material behaves elastically. It is quite stiff and so makes a good material for cups. Increased stress leads to a breakdown of the lamellae into fibrils (tiny fibres). The long polymer chains line up and then the material becomes strong as the carbon–carbon bonds are stretched.

Designer materials

Just as the Stone, Iron and Bronze Ages are so called because of the materials in common use at the time, perhaps the years around the late 20th and early 21st centuries will come to be called the 'Polymer Age'. Most of the man-made materials that we loosely call 'plastics' are in fact polymers. Polythene is one common example, and PVC (polyvinyl chloride) is another – you can probably think of others (having 'poly' in the name is a clue). Polymer materials such as PVC can be made relatively cheaply, can easily be shaped and coloured, and are also durable, hence their widespread use.

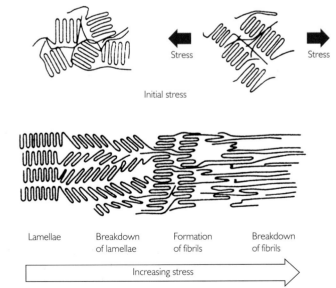

Figure 3.23 Stretching UHMWPE

Polyethylene and PVC are man-made polymers, but there are also many natural polymers found in living things. Plant fibres such as cotton are made up of polymers, as are animal fibres such as hair and wool. It is because they can imitate natural fibres that man-made polymers find a use in spare part surgery.

As you have seen, the mechanical properties of a polymer depend on the way its chain molecules arrange themselves, and this in turn depends on the length of the molecules and their chemical make-up. The disciplines of physics and chemistry come together in polymer science, enabling a vast range of polymers to be developed that have a very wide variety of properties.

2.6 Summing up Part 2

In this part of the chapter you have extended your knowledge of the physics of materials, and seen how a knowledge of the small-scale structure of a material can help explain its large-scale behaviour. This understanding can help materials scientists to develop materials with properties to suit a particular purpose.

Activity 8 Summing up Part 2

Look back through your work and make sure you know the meanings of all the key terms printed in bold. Then write each term on a slip of paper (e.g. a Post-it™ sticker) and arrange them on a large sheet of paper to make a concept map for this part of the chapter. Link the slips of paper by writing a few words or an equation or by sketching a diagram.

When completed, your concept map should provide you with a summary of what you have learned about large-scale and small-scale properties (particularly polymers) and

Question

Figure 3.24 shows a schematic stress–strain graph for a polymer material.

17 (a) How would a stress–strain graph for a metal differ from Figure 3.24?

(b) Suggest a physical interpretation for the area enclosed between the 'Loading' and 'Unloading' curves in Figure 3.24.

(c) Sketch the arrangement of molecules you would expect when the sample was at point A and at point C.

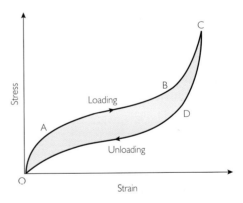

Figure 3.24 Stress–strain graph for a polymer

Further investigations

Explore the behaviour of various materials that appear 'bone-like', such as cuttle-fish bone (available from pet-shops), seaside rock and 'oasis' (used in flower-arranging). Try to measure the Young modulus and ultimate compressive strength. Observe the way they fracture. Comment on their resemblance (or otherwise) to real bone.

3 Recovery

3.1 Getting better

This chapter has touched on several areas of physics, and has built on much of the work that you have done in earlier chapters. Use the following activities to remind yourself of the work that you have done.

Activity 9 Getting better

Look back through your work on other chapters in this course, and look for links with this chapter. Annotate your notes to show where similar ideas come up in different situations, and extend them where you think it is relevant (for example, you could add some notes on stress and strain in the relevant sections of *Higher, Faster, Stronger* and *Good Enough to Eat*).

Activity 10 Bedside manners

Imagine that you are talking to someone who is to have spare part surgery, such as a knee or hip replacement, and they are a bit unsure about what is involved. Decide how you could help to inform and reassure them. Either talk to another student who is playing the role of the patient, or make a tape-recording that the patient could play on their own. Think carefully about what information would be helpful, how you could explain it in an appropriate way – and also think about what not to include.

3.2 Questions on the whole chapter

Questions

18   When new materials are developed, their physical properties are recorded so that users can decide whether a material is suitable for a particular purpose. The graphs in Figure 3.25 show the performances of two materials X and Y.

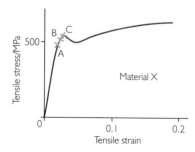
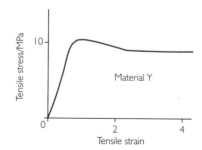

Figure 3.25 Graphs for Question 18

(a) Up to point A, the graph for material X is a straight line from the origin. Up to point B, the material returns to its original condition if the stress is removed. Beyond point C, the material continues to stretch while a constant stress is applied. What are the names of points A, B and C?

(b) Which of materials X and Y is the stiffer? Explain your answer.

19 If a video tape becomes stretched when in use, the playback will be affected. There needs to be some 'give' in the tape to prevent snapping, but not so much that the magnetic tape information gets badly distorted. A certain manufacturer requires that a tape of thickness 0.02 mm and width 1.3 cm is not to undergo a strain of more than 3.0×10^{-3} when a force of 1 N is applied to it.

When choosing material for the tape, what is the smallest allowable value for its Young modulus?

20 A student was attempting to measure the Young modulus for the sweet jelly used in an apple lace. He used a micrometer screw gauge to measure the diameter of the lace, and obtained five results, moving the gauge down the lace, and rotating it: 1.02, 0.99, 1.04, 1.03, 1.05 mm. He marked a length of 200 mm on the lace, and obtained the results in Table 3.3.

Load / g	Length / mm	Extension / mm	Strain	Load / N	Stress / Pa
0	200				
20	220				
40	240				
60	260				
80	280				
100	320				
120	380				
140	480				
160	snapped				

Table 3.3 Data for Question 20

You may find it simplest to use a spreadsheet programme to answer this question.

(a) Calculate the mean diameter of the lace, in m, and then the cross-sectional area in m².

(b) Copy and complete the table (use $g = 9.8$ N kg^{-1}).

(c) Plot a stress–strain graph.

(d) Measure the gradient of the linear part of the graph to obtain a value for the Young modulus of the apple jelly.

(e) (i) What was the uncertainty in the diameter of the lace?

(ii) What was the percentage uncertainty in the area?

(f) Assuming that lengths were measured to the nearest mm, what was the percentage uncertainty in the marked length?

(g) What was the percentage uncertainty on the extension for a load of 80 g?

(h) Assuming that the uncertainty in the masses used to stretch the lace was ± 1%, estimate the uncertainty in the Young modulus.

3.3 Achievements

Now you have studied this chapter you should be able to achieve the outcomes listed in Table 3.4.

Table 3.4 Achievements for the chapter *Spare Part Surgery*

Statement from examination specification	Section(s) in this chapter
22 obtain and draw force–extension, force–compression, and tensile/compressive stress–strain graphs. Identify the *limit of proportionality*, *elastic limit* and *yield point*	2.2
23 investigate, and use Hooke's law, $F = k\Delta x$, and know that it applies only to some materials	2.2
24 explain the meaning of, use and calculate *tensile/compressive stress*, *tensile/compressive strain*, *strength*, *breaking stress*, *stiffness* and *Young modulus*. Obtain the Young modulus for a material	2.2
27 calculate the elastic strain energy E_{el} in a deformed material sample, using the expression $E_{el} = \frac{1}{2}Fx$, and from the area under its force–extension graph	2.4

Answers

1   A shell

2   You might include: supports the body's weight; provides levers for movement; anchors the muscles; protects internal organs.

3   It would be very brittle (the fibres help it to become tough).

4   A hinge joint can move only in one plane, whereas a ball and socket joint can rotate and move in any plane.

5   Taking bone from another part of the body exposes the patient to two operations, leading to increased risk (and increased cost). Taking bone from another person may lead to infection.

6   Organic fibres (mainly collagen).

7   (a) Elastic, tough and durable

    (b) Stiff, tough, hard and durable

    (c) Stiff, tough, hard and durable

8   You need to estimate your weight and the area of your bones. If mass = 60 kg, then F = weight = $mg \approx 600$ N. If diameter of leg bone = 6 cm, then area of one bone $\approx \pi r^2 \approx$ 30 cm^2 = 30 × 10^{-6} m^2. When standing on both feet, total area A = 60 × 10^{-6} m^2.

$$\sigma = \frac{F}{A} \approx \frac{600 \text{ N}}{(60 \times 10^{-6} \text{ m}^2)} = 10^7 \text{ Pa}$$

$$= 10 \text{ MPa}.$$

Bones are commonly subject to much larger stresses e.g. when moving, or simply when standing on one leg.

9   (a) Using $v^2 = u^2 + 2as$, with $u = 0$, $a = 9.8$ m s^{-2}, $s = 1.5$ m (taking downwards as positive):

$$v^2 = 29.4 \text{ m}^2 \text{ s}^{-2}, v = 5.42 \text{ m s}^{-1}.$$

    (b) Magnitude of deceleration

$$a = \frac{\Delta v}{\Delta t} = 54.2 \text{ m s}^{-2}.$$

    (c) Magnitude of force
$$F = ma = 70 \text{ kg} \times 54.2 \text{ m s}^{-2} = 3.80 \times 10^3 \text{ N}.$$

(d) Total cross-sectional area = 60 cm^2

$$= 60 \times 10^{-4} \text{ m}^2.$$

$$\text{Stress} = \frac{3.80 \times 10^3 \text{ N}}{60 \times 10^{-4} \text{ m}^2}$$

$$= 6.3 \times 10^5 \text{ Pa}$$

$$= 0.63 \text{ MPa}.$$

10 Strain = 2.5 mm/400 mm = 6.3 × 10^{-3} = 0.63%. (Notice that you get the same answer if you express both lengths in cm, or both in m – it does not matter which units you use, provided you use the same for both lengths.)

11 Expressed as a decimal

$$\text{strain} = \frac{0.15}{100} = 1.5 \times 10^{-3}.$$

From Equation 3:

$$x = \varepsilon l = 0.15 \times 10^{-3} \times 50 \text{ cm}$$

$$= 7.5 \times 10^{-2} \text{ cm}$$

$$= 0.75 \text{ mm}.$$

If the original length were ten times longer, then the extension, too, would be ten times as great, i.e. 7.5 mm, to achieve the same strain.

12 (a) $A = 0.5 \times 10^{-6}$ m^2

    (b) $\sigma = \dfrac{F}{A} = \dfrac{75 \text{ N}}{0.5 \times 10^{-6} \text{ m}^2} = 1.5 \times 10^8$ Pa

    (c) $\varepsilon = \dfrac{6.0 \times 10^{-3} \text{ m}}{8.0 \text{ m}} = 7.5 \times 10^{-4}$

    (d) $E = \dfrac{\sigma}{\varepsilon} = \dfrac{1.5 \times 10^8 \text{ Pa}}{7.5 \times 10^{-4}} = 2.0 \times 10^{11}$ Pa

    (e) Steel has a much greater Young modulus than bone, so it would take most of the stress in a replacement joint and the natural bone would waste away.

    (f) Total area of cross section of 100 strands,

$$A = 100 \times 0.5 \times 10^{-6} \text{ m}^{-2}$$

$$= 5 \times 10^{-5} \text{ m}^2.$$

Weight of ten people,
$$F = mg = 850 \text{ kg} \times 9.8 \text{ N kg}^{-1}$$

$$= 8.33 \times 10^3 \text{ N}.$$

Stress due to weight of people,

$$\sigma = \frac{8.33 \times 10^3 \text{ N}}{5 \times 10^{-5} \text{ m}^2} = 1.67 \times 10^8 \text{ Pa.}$$

Strain $\varepsilon = \dfrac{\sigma}{E} = \dfrac{1.67 \times 10^8 \text{ Pa}}{2.0 \times 10^{11} \text{ Pa}} = 8.33 \times 10^{-4}.$

Extension $= \varepsilon l = 8.33 \times 10^{-4} \times 90 \text{ m} = 7.5 \times 10^{-2} \text{ m.}$

13 (a) See Table 3.5 and Figure 3.26.

Tension / 10^3 N	Extension / 10^{-3}m	Stress	Strain
0.00	0.00	0.00	0.00
35.0	0.10	2.69	1.53
38.0	0.20	2.92	3.08
36.5	0.40	2.81	6.15
36.0	0.70	2.77	10.8
37.0	1.00	2.85	15.4
40.0	1.40	3.08	21.5
45.0	2.10	3.46	32.3
50.0	3.10	3.85	47.7
55.0	5.10	4.23	78.5
57.5	7.20	4.42	111
60.0	9.10	4.62	154
60.0	13.1	4.62	202
57.5	14.7	4.42	226
55.0	15.2	4.23	234
52.5	15.8 (broke)	4.04	243

Table 3.5 The answer to Question 13(a)

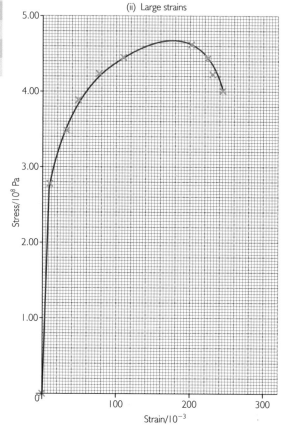

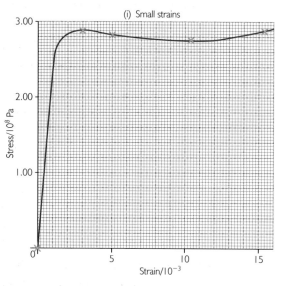

Figure 3.26 The answer to question 13(a)

(b) (i) Yield stress: about 2.6×10^8 Pa (where the graph in Figure 3.26 (i) starts to curve and the sample extends with no additional stress)

(ii) Ultimate tensile stress: about 4.7×10^8 Pa (from Figure 3.26 (ii))

(iii) Breaking strain $= 243 \times 10^{-3} = 0.243 \approx 24\%$

14 Area of string, $A = \pi(0.2 \times 10^{-3}\ \text{m})^2$
$$= 1.26 \times 10^{-7}\ \text{m}^2.$$

Strain produced by 8 turns,

$$\varepsilon = \frac{8 \times 4.0\ \text{mm}}{840\ \text{mm}}$$

$$= 3.8 \times 10^{-2}.$$

Stress, $\sigma = \varepsilon E = 3.8 \times 10^{-2} \times 3.0 \times 10^9$ Pa

$$= 1.14 \times 10^8\ \text{Pa}.$$

Tension $= \sigma A = 1.14 \times 10^8\ \text{Pa} \times 1.26 \times 10^{-7}\ \text{m}^2$

$$= 14.4\ \text{N}.$$

The Young modulus of the material must be high, and it must be able to achieve quite large strains without breaking (which means that its breaking stress must be high).

15 The material would need to have a large Young modulus (large E) and be capable of undergoing large strains (large ε) without failing.

16 (a) See Figure 3.27: each large square represents an energy density of 1×10^7 J m^{-3}, and each small square represents 4×10^5 J m^{-3}. From counting the squares, the area under the curve thus represents about 10.5×10^7 J m^{-3}.

(b) If we assume that all the stored elastic energy becomes kinetic energy when the wire breaks, and that all parts of the wire move at the same speed, and deal with a volume V of the wire, then we can write:

mass $m = \rho V$ (ρ is the density)

$E_{\text{el}} = UV$

kinetic energy $E_k = \frac{1}{2}(\rho V)v^2 = UV$

and so $\frac{1}{2}\rho v^2 = U$

$$v^2 = \frac{2U}{\rho}$$

$$= \frac{2 \times 10.5 \times 10^7\ \text{J m}^{-3}}{8 \times 10^3\ \text{kg m}^{-3}}$$

(Note that the units of v^2 are J kg^{-1} and that 1 J kg^{-1} = 1 m^2 s^{-2}.)

so $v = 1.6 \times 10^2$ m s^{-1}.

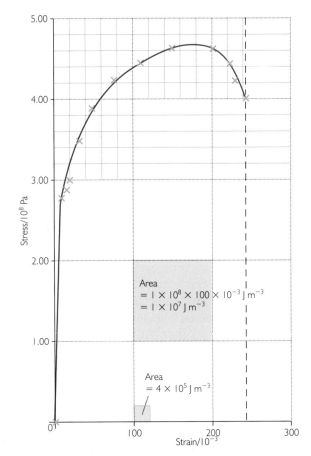

Figure 3.27 See the answer to Question 16(a)

This answer shows that the wire will move very fast when it breaks. As the kinetic energy will not, in practice, be evenly distributed, the moving end of the wire will in fact move at several hundred metres per second.

Maths reference

See Maths note 2.3
Derived units

17 (a) The graph for a metal would be much steeper (as its Young modulus would typically be a thousand times greater) and it would not reach such a large strain (polymers can be strained by serveral hundred percent, while metal break at strains of a few per cent.) The graph for the metal might curve over as it yielded (as in Figure 3.26 (ii)).

(b) The area represents the difference between energy transferred to the sample on stretching, and the energy recovered as it is unloaded; it represents the energy dissipated due to heating.

(c) For point A your sketches should resemble Figures 3.21(a) and for point C they should resemble Figures 3.21(b).

CHAPTER 4

The Sound of Music

Why a chapter called *The Sound of Music*?

The study of sound is surprisingly ancient. Its origin probably lies with Roman architects trying to control echoes in the amphitheatres used for plays and concerts at that time. Since those days, scientists have discovered how a sound is propagated, the speed at which it travels, and precisely what controls its pitch, its loudness and its quality. From its origins in folklore and religion, where it was perceived to be linked with divine and magical properties, music has also become an increasingly important part of many people's lives (Figure 4.1).

Over the years, the craft of constructing and developing musical instruments has become increasingly scientific. Science has improved the tuning of instruments and the quality and penetration of the sound they produce. New technologies, new materials and modern techniques of mass production have been incorporated into the construction processes. The advent of electronics has heralded a further revolution in music, throwing open the science of sound to the average interested person in the street who can now, with a minimum of effort, generate and combine sounds with infinite variety.

Figure 4.1a Mixer desk

Figure 4.1b Laser light show

Overview of physics principles and techniques

In this chapter you will build on ideas from GCSE about vibrations and waves, looking particularly at how these can be represented graphically. In part 1 you will see how waves combine by a process called superposition and how the waves produced by musical instruments are related to their physical properties. You will use computer software to explore and synthesise complex sounds.

Part 2 concentrates on the compact disc. You will discover how a compact disc stores sound and how the CD player retrieves it and how wave properties of the laser light are exploited in order to recover the information. In exploring the components of the CD player, you will learn how light is reflected and refracted. You will also see that, while a wave picture allows us to explain many properties of light, we need to introduce a different idea (that of photons) to explain how a laser works.

In the course of this chapter you will also be using and developing some key mathematical and ICT skills and techniques. Several activities are concerned with the generation and interpretation of graphs – the understanding of exactly what these graphs mean is crucial to the chapter. You will also use software, both as a data source and to analyse your own data.

In later chapters you will do further work on:

- travelling waves in *Technology in Space*, *Reach for the Stars* and *Build or Bust?*
- superposition, interference and standing waves in *Digging Up the Past* and *Build or Bust?*
- refraction and reflection in *Build or Bust?*
- photons and energy levels in *Technology in Space* and *Probing the Heart of Matter*
- signals in *Transport on Track* and *The Medium is the Message*.

1 Making sounds

1.1 Synthetic sounds

A person steps out on to a stage. Six steel wires are struck sharply. They vibrate only millimetres to either side, too fast to see. Simultaneously, a flood of distorted sound volleys outwards from speakers stacked house-high, and ten thousand people shift their attention forwards. A second person emerges, flicks a few switches and begins to finger a pattern on to rectangular plastic keys. A virtual orchestra of sounds and rhythms emerges from the electronics, skilfully fused together into a recognisable anthem – the band has begun and the audience, down to the very last person at the back of the vast stadium, begins to move to the pulsing of the sound waves washing over them.

This familiar scenario of a modern concert would have been inconceivable a century ago, when an orchestra was an orchestra consisting of dozens of individuals and their instruments. Nowadays an individual can be an orchestra: musical sounds can be recorded and replayed, time after time. And, most astonishing of all, a simple keyboard or computer can be programmed to reproduce the sounds of any instrument, or in fact any sound, at the press of a button by a process called synthesising. A typical electronic keyboard may contain the 'voices' of several hundred instruments and synthesised effects as well as pre-programmed percussive and instrumental accompaniments.

Activity 1 Synthetic sound – how realistic is it?

Listen carefully to two extracts of the same piece of music, one with musicians playing real instruments and one computer-synthesised version of the same piece.

Listen to synthesised versions of common sounds from children's interactive books.

Comment on the quality and realism of the synthesised sounds.

Getting from a real sound to a computer-synthesised version requires an understanding of the physical nature of sound. This is your aim in the first part of this chapter.

1.2 Oscillations

All sources of sound involve vibrations:

- A guitar string vibrates to produce a sound.
- The skin of a drum and a table top both vibrate when struck. (Can you suggest why the vibrations of the table top are smaller and die away faster?)
- Your vocal chords vibrate as you make sounds. Feel the vibration with your fingertips placed on the front of your neck as you hum a note.
- The prongs of a tuning fork vibrate. (You can see this if you touch the surface of water with a humming tuning fork.)

A motion that repeats itself over and over again, at regular time intervals, is called a **periodic oscillation**. Behaviour of this type is remarkably common (Figure 4.2). Oscillating periodic motion is often referred to as **harmonic motion** because of its relation to sound.

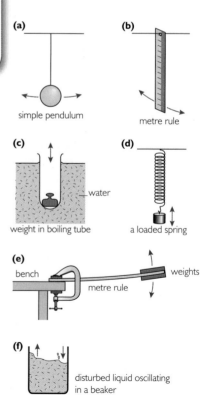

(a) simple pendulum

(b) metre rule

(c) weight in boiling tube / water

(d) a loaded spring

(e) bench / metre rule / weights

(f) disturbed liquid oscillating in a beaker

Figure 4.2 Oscillators

Describing and representing oscillations

The sequence of events that form one 'unit' of a periodic motion is called a **cycle**. In one cycle, the oscillating object moves to and fro, returning to its original position and direction, whereupon the cycle begins again. The time it takes for the system to complete one cycle is called its **time period**. The **frequency** of an oscillation is defined as the number of cycles executed per unit time. One cycle per second is called one hertz (Hz) – this is the SI unit of frequency. The frequency f and time period T are related:

$$T = \frac{1}{f} \qquad (1)$$

which can also be written:

$$f = \frac{1}{T} \qquad (1a)$$

As an oscillating object moves, the maximum **displacement** in either direction that it reaches from its **equilibrium position** is called the **amplitude**. The equilibrium position is where the object comes to rest after its oscillations die down.

> **Maths reference**
>
> Reciprocals
> See Maths note 3.3

Activity 2 Periodic oscillations

By measuring the time period and frequency investigate the motion of a variety of oscillating objects, such as those in Figure 4.2.

Displacement–time graphs

We often use displacement–time graphs to represent oscillations. Figure 4.3 shows such a graph for a metre rule which, when clamped to a bench with about 80 cm of its length projecting, oscillates with an amplitude of 30 mm and a frequency of 2.5 Hz.

A graph such as Figure 4.3 shows the 'shape' of an oscillation and is often referred to as a **waveform**. Waveforms like that in Figure 4.3 which have the same simple, smooth shape as graphs of $\sin\theta$ or $\cos\theta$ plotted against angle θ, are collectively described as **sinusoidal**.

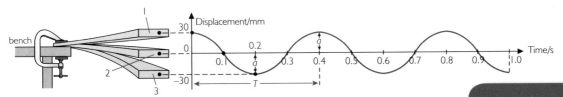

Figure 4.3 Displacement–time graph for a clamped metre rule

Sound waves produced by an oscillating object are detected when they set up oscillations with a similar waveform in a detector such as our ears or a microphone. The oscillations in a microphone can be displayed as a displacement–time graph using a cathode-ray oscilloscope (CRO) or on a computer screen. Their waveform closely matches that of the sound source.

> **Maths reference**
>
> Sine, cosine and tangent of an angle
> See Maths note 6.2
>
> Graphs of trigonometric functions
> See Maths note 6.3

Activity 3 Exploring waveforms with an oscilloscope

Use an oscilloscope or *PicoScope* to display a waveform produced by a sound source. Observe how the frequency and amplitude of the waveform are related to the pitch and loudness of the sound.

Activity 4 Exploring waveforms with *Audacity*

Use the sound analysis software *Audacity* to explore the waveforms of the sound from a variety of sources.

Phase

The term **phase** is used to describe the stage an oscillation has reached in its cycle. The two oscillations shown in Figure 4.4(a) are exactly in step. They are said to be **in phase**. Oscillations that reach their peaks and troughs at different times are said to have a **phase difference** between them. The phase of an oscillation is often expressed in terms of angles, drawing on the similarity between sinusoidal waveforms and graphs of sines and cosines.

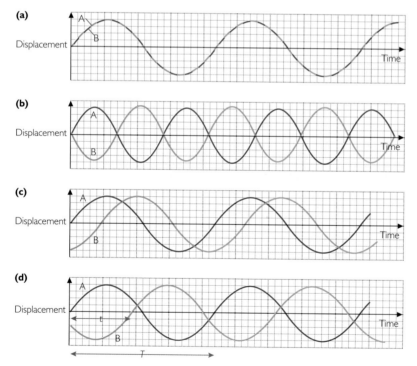

Figure 4.4 Graphs showing waveforms for oscillations that are (a) in phase, (b) in antiphase, (c) in quadrature, (d) the general case

When dealing with phase, angles can be expressed in degrees or, more commonly, in **radians**. One 'cycle' of a sine or cosine graph corresponds to one complete circle (360° or 2π radians), so oscillations that are exactly half a cycle out of step (Figure 4.4(b)) have a phase difference of 180° or π radians. Such oscillations are said to be **in antiphase**.

In Figure 4.4(c) oscillation A is one quarter of a cycle ahead of B. A leads B by a phase difference of 90° or $\frac{\pi}{2}$ radians – or, put another way, B leads A by $\frac{-\pi}{2}$. Oscillations with a phase difference of $\frac{\pi}{2}$ are said to be **in quadrature**. The general case is shown in Figure 4.4(d). A leads B by the fraction $\frac{t}{T}$ of a cycle. The phase difference is therefore $\left(\frac{t}{T}\right) \times 360°$ or $\left(\frac{t}{T}\right) \times 2\pi$ radians.

Maths reference

Degrees and radians
See Maths note 6.1

Questions

1 Figure 4.5 shows three waveforms drawn to the same scale. Write down expressions relating the frequencies of (a) A and B, and (b) A and C.

Study note

When π appears in a description of phase the units of radians are taken for granted and are sometimes omitted.

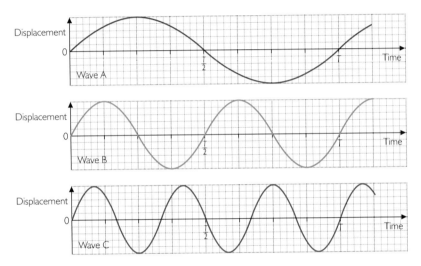

Figure 4.5 Diagrams for Question 1

2 On a sketch of a sinusoidal waveform (like Figure 4.6) use different colours to draw the following waveforms on the same axes:

(a) the same amplitude as the original and leading it by $\frac{\pi}{4}$ radians

(b) the same amplitude as the original and lagging behind it by 45°

(c) twice the amplitude of the original and leading it by $\frac{3\pi}{4}$ radians.

Figure 4.6 Waveform for Question 2

3 (a) Write down a general expression, in degrees and in radians, for a phase change that would produce a wave identical to the original.

(b) What, if any, is the difference between waves that differ from wave A in Figure 4.5 by each of the following:

(i) $\frac{\pi}{8}$ radians

(ii) $22.5°$

(iii) $\frac{17\pi}{8}$ radians

(iv) $-337.5°$?

4 Because the oscillations in Figure 4.5 have different frequencies, they will only be in phase at certain times. At what times are the following pairs in phase:

(i) A and B

(ii) A and C

(iii) B and C?

5 In the mid-1880s American photographer Eadweard Muybridge took several sequences of photographs showing horses in motion. (Figure 4.7). Treating the leg motion as a simple oscillation, state the phase relationships between the legs of a walking horse.

Figure 4.7 A walking horse

1.3 Travelling waves

Sound travels by spreading out in all directions from the vibrating source as a **wave** whose properties are closely linked to those of the oscillation. The waves transfer energy from the source to other places without any permanent transfer of matter – this is a property of all **travelling waves**.

It is sometimes useful to think of a travelling wave as a succession of **pulses** (a pulse is an oscillation that lasts for just one period). The behaviour of a travelling wave can be deduced by studying a single pulse. The leading edge of a single pulse is known as a **wavefront**.

There are two main classes of wave, both of which play a part in the production of sound from musical instruments, as you will see in Section 2. A **transverse** wave involves oscillations at right angles to the direction of wave propagation. In a **longitudinal** wave, the oscillations are along the direction of propagation.

Activity 5 Waves on a slinky

Using a slinky, generate and observe some transverse and longitudinal waves (Figure 4.8).

Mark one loop of the slinky and observe how it moves. Explore how to control the speed of the wave along the slinky. For example, try varying the tension in the slinky. Observe a single pulse as it reaches a fixed or a free end of the slinky.

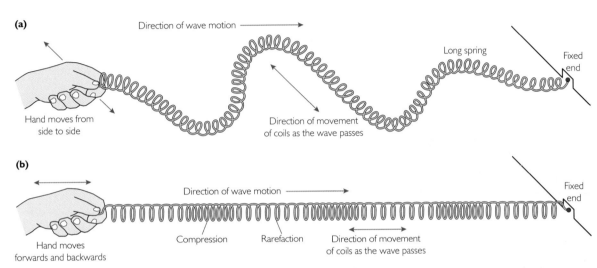

Figure 4.8 Using a slinky to generate (a) transverse and (b) longitudinal waves

In a water wave (Figure 4.9), the particles circulate rather than just moving up and down. This 'stretches' the surface and will create a normal wave shape at the surface if the amplitude is very small compared with the wavelength. Sharper crests form at larger amplitudes.

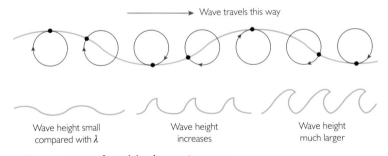

Figure 4.9 The movement of particles in a water wave

Question

6   Use some of the terms introduced so far to describe the following motions:

(a) a tablecloth shaken to get rid of crumbs

(b) a train of railway trucks shunted by a locomotive

(c) a 'Mexican wave' in a stadium

(d) the stop–go motion of traffic in a jam.

Sound waves

Sound waves are longitudinal, as can be demonstrated with a loudspeaker cone. As the cone vibrates, it first compresses the air next to it and then immediately allows it to spread out again (rarefy) before repeating the cycle. A series of **compressions** and **rarefactions** travels outwards from the cone. The movements produced by loud sounds can make a candle flame flicker (Figure 4.10). Sound cannot travel in a vacuum, since the waves need a material that can be compressed or stretched.

Figure 4.10 Demonstrating longitudinal oscillations associated with sound waves

In a human ear (Figure 4.11) the ear drum (a small membrane) is made to vibrate by the air compressions and rarefactions. These small vibrations are then amplified via a mechanical linkage of tiny bones. These larger pressure variations are detected by nerve endings within the inner ear.

A piezoelectric microphone (Figure 4.12) works in a very similar way to the ear. Variations of pressure caused by a sound wave exert tiny stresses on a piezoelectric crystal – a material that generates a voltage across it when stressed. This voltage can be detected as a small electrical signal.

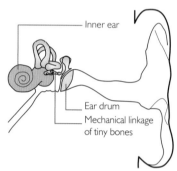

Figure 4.11 A human ear

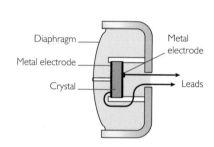

Figure 4.12 A piezoelectric microphone

Activity 6 Seeing the pressure change

Use an oscilloscope or *PicoScope* to display the voltage from a microphone. The in–out motion of the microphone diaphragm is displayed as an up–down motion of the spot in the screen. Clap your hands by the microphone to produce a voltage spike on the screen.

1.4 Graphs of travelling waves

There are several ways in which we can use graphs to represent travelling waves. Since transverse waves are perhaps easier to visualise, we will deal with them first and then apply the same ideas to longitudinal waves.

Transverse waves

Activity 7 Freezing a travelling wave

Generate a travelling transverse wave on a rubber cord, rope or chain. If you do this in a darkened room, a strobe light can be used to 'freeze' the motion if the frequency of the strobe and the wave are suitably adjusted.

Safety note

Strobe lighting can cause fits in epilepsy sufferers, especially at frequencies below about 20 Hz. Do not take part in this activity if you think you might be affected.

Displacement–position graphs

The 'snapshot' that you should have seen in Activity 6 looks something like Figure 4.13, a displacement–position (or displacement–distance) graph. We can use such a diagram to define the **wavelength** λ as the distance between adjacent crests or, more generally, the distance between two places where the oscillations are in phase.

The motion of each particle in Figure 4.13 is slightly out of phase with that of each of its neighbouring particles, and so the wave pattern travels, as shown in Figure 4.14

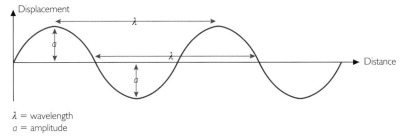

λ = wavelength
a = amplitude

Figure 4.13 A displacement–position graph of a travelling transverse wave

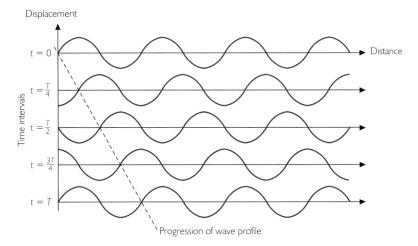

Figure 4.14 Successive 'snapshots' of a travelling wave

The wave equation

The wavelength λ and the frequency *f* of the wave are connected via the **wave equation**

$$v = f\lambda \tag{2}$$

where *v* is the speed of propagation of the wave. Note that equation (2) applies to *all* types of wave.

Question

7 The human ear is sensitive to sounds between about 20 Hz and 20 kHz. (The ability to hear high frequencies diminishes with age. It can also be greatly reduced by prolonged exposure to loud sounds.) Taking the speed of sound in air as 340 m s^{-1}, calculate the corresponding range of wavelengths.

Displacement–time graphs

We can represent the wave by drawing a displacement–time graph for just one particle in the material through which the wave is travelling, as we have already done for oscillating objects. You will have noticed that displacement–time and displacement–position graphs have very similar shapes. To be sure of distinguishing between these two sorts of graphs, make sure you *always* label the axes.

Questions

8 Figure 4.15 shows a wave travelling in the positive x direction, away from the origin.

  (a) What are the wave's

  (i) wavelength

  (ii) time period?

  (b) Sketch a displacement–time graph for the particle marked A.

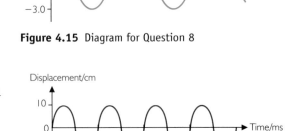

Figure 4.15 Diagram for Question 8

9 A travelling wave whose speed is 300 m s^{-1} has the displacement–time graph shown in Figure 4.16. Sketch a displacement–position graph for this same wave.

Longitudinal waves

Figure 4.17 shows a schematic diagram of the compressions and rarefactions in a sound wave. As you will see, there are various ways in which we can represent such a wave on a graph.

Figure 4.16 Diagram for Question 9

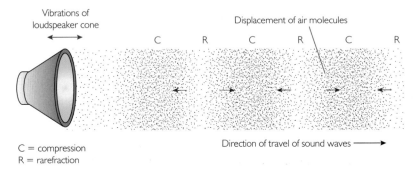

C = compression
R = rarefraction

Figure 4.17 A 'snapshot' of a sound wave

Displacement–position and pressure–position graphs

One of the ways to turn Figure 4.17 into a graph is to plot displacement against position, and to define the amplitude as the maximum longitudinal displacement caused by the wave.

Activity 8 Graphs for a longitudinal wave

Figure 4.18 shows a row of 20 undisplaced particles, and below it the same particles displaced by the passage of a longitudinal wave. By tracing the motion of each particle, generate graphs of displacement versus position.

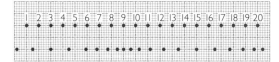

Figure 4.18 Diagram for Activity 8

Figure 4.19(a) shows the displacement–position graph for the wave of Figure 4.17. This graph looks 'transverse' even though it represents a longitudinal wave. The displacement is measured *along* the direction of wave motion (along the *x*-direction) but is plotted up the *y*-axis of the graph.

However, another way to turn Figure 4.17 into a graph is to plot pressure against position. This is shown in Figure 4.19(b). As you should have found in Activity 8, the two graphs are not in phase. Positions of maximum compression or rarefaction correspond to zero displacement, and positions of maximum displacement correspond to 'normal' pressure.

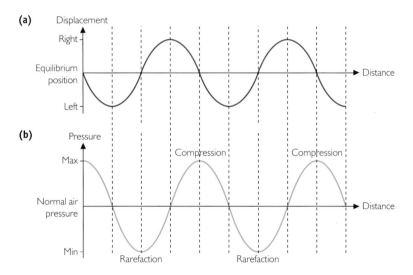

Figure 4.19 Displacement–position and pressure–position graphs for a longitudinal wave

Displacement–time and pressure–time graphs

We can also use displacement–time or pressure–time graphs to represent longitudinal wave motion in exactly the same way as transverse waves. If you look back at Questions 8 and 9, you will see that there is nothing in either question to indicate the type of wave they were dealing with – they apply equally to either type.

1.5 Superposition and standing waves

You have seen how an oscillating object can produce a sound wave. But what controls the frequency (the pitch) of the sound? How can a musical instrument be tuned to give a desired note? And why do notes of the same pitch sound different when played on different instruments? What determines the 'quality' of a sound?

Superposition

The answers to all the questions above involve **superposition** – the combination of two or more waves. When two or more waves arrive at the same place at the same time, the resultant displacement is equal to the sum of those due to the individual waves. The effect is most noticeable if the waves combined are **coherent**, meaning that they are of the same frequency (or wavelength) and with a constant phase relationship. Figure 4.20 shows the result of combining two waves of the same frequency and phase but different amplitude.

The displacement–time graphs in Figure 4.20 show the effect of the resultant wave on just one particle: the first two graphs are simply added together to produce the resultant displacement–time graph. If the two superposing waves are both travelling along 'on top of one another' we can simply add their displacement–distance graphs to produce a 'snapshot' picture of the resultant wave. For sound waves, we could plot graphs using pressure rather than displacement: these too can simply be added together.

Activity 9 Superposition

Using the University of Salford's *Sounds Amazing* software, investigate what happens when two simple sinusoidal waves superpose.

In Activity 9 you should have seen examples of two special cases of superposition. If two waves combine so that they are always reinforcing one another, this is called **constructive superposition**; and if they are always cancelling one another, this is called **destructive superposition**.

Questions

10 Sketch graphs similar to Figure 4.20 to show the resultant wave in each of the following cases:

(a) coherent waves of identical amplitude superposed in phase

(b) coherent waves of identical amplitude superposed with a phase difference of 90° or $\frac{\pi}{2}$ radians

(c) coherent waves of identical amplitude superposed with a phase difference of 180° or π radians.

11 The driver of the jet-propelled car 'Thrust' that broke the world land speed record in 1997 sat directly between two high-powered jet engines. He would have been deafened but for a technique called 'active sound suppression', which involves generating sounds in the driver's headphones to cancel out the engine noise. What can you say about the sounds that would need to be generated?

12 What conditions must two waves satisfy if they are to undergo

(a) maximum constructive superposition

(b) maximum destructive superposition?

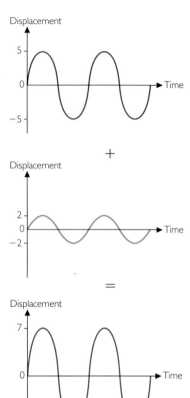

Figure 4.20 Superposition of two waves

Standing waves

Interesting things can happen when waves of a single frequency (or wavelength) travel back and forth 'on top of one another', as might happen if waves are reflected back and forth. Figure 4.21 shows the superposition of a 'red' wave travelling from left to right with a 'green' wave of the same frequency travelling in the opposite direction.

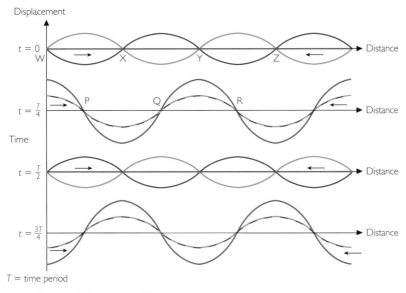

Figure 4.21 Superposition of travelling waves

The waves could be …

- transverse, with the y-axis representing transverse particle displacement
- longitudinal, with the y-axis representing particle displacement along the direction of travel
- longitudinal, with the y-axis representing changes in pressure.

Superposition of the 'red' and 'green' waves gives the 'blue' resultant waves, which doesn't travel in either direction, but merely remains where it is and changes profile as shown in Figure 4.22. The wave goes through the sequence 1, 2, 3, 4, 5, 4, 3, 2, 1 in one cycle. This type of wave is called a **standing wave** (or in some books, a **stationary wave**). A point on the standing waves such as P, Q or R where the value plotted on the y-axis is always zero is called a **node** (there is **no** change). A point such as W, X, Y or Z where the amplitude reaches a maximum is called an **antinode**.

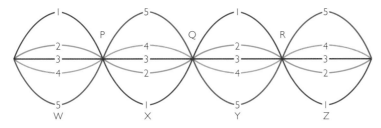

Figure 4.22 A standing wave

Activity 10   Standing waves

Generate and study some standing waves of various types. Examples might include: water waves in a bowl; waves on a cord or a slinky; sound waves in a tube; waves on a stretched wire.

Questions

13   What is the distance on the standing waves between two adjacent nodes? (Express your answer in terms of the wavelength of the travelling waves.)

14   What is the phase relationship of all the oscillating particles between a pair of adjacent nodes? How does that compare with the phase of wave particles between the pair of nodes either side?

1.6 Summing up Part 1

So far in this chapter you have learned how to describe oscillations and waves using words and mathematical expressions.

Activity 11   Describing waves

Look back through this chapter and your notes and make sure you understand the meaning of all the terms highlighted in bold. Then for each of the following, write a sentence or two and sketch a labelled diagram to explain its meaning:

- wavelength
- wavefront
- antiphase
- longitudinal wave

- coherent
- superposition
- antinode.

2 Musical notes

In this part of the chapter, you will apply ideas about waves to two sorts of musical instrument and see how they produce notes. You will also see why the same note sounds different when played on different instruments.

2.1 Wind instruments

Getting notes from a wind instrument (such as a saxophone or a flute) can be surprisingly difficult for a person trying it for the first time. The process is, however straightforward to describe at a basic level.

Activity 12 Playing a wind instrument

Read the passage below and answer Questions 15 to 20 that follow. Answers to many of the questions can be found in the passage and in earlier sections of this unit. Skim read the passage first to get an idea of what is in it and then read the questions, so that on your second, more careful, reading of the article you will be able to pick out the important points.

Wind instruments (Figure 4.23) rely on the column of air molecules inside the tube of the instrument being made to oscillate as a standing wave. This can be done in several ways. Trumpets and other brass instruments are played with lip vibrations causing the air to oscillate; oboes and clarinets rely on a flexible reed vibrating as it is blown; recorders have a notched air vent at the top which protrudes into the air column slightly, setting up eddies in the air which generate molecular oscillations.

Figure 4.23 Wind instruments

The standing wave of air vibrations makes the instrument and the air around it vibrate at the same frequency, generating a sound wave that travels outwards in all directions.

There are two basic categories of wind instrument:

- *open tubes – a flute or a recorder behaves like this type of tube, with both ends open*
- *closed tubes – a clarinet behaves like this type of tube, with one end open and one closed; the closed end is the reed end.*

The standing waves are set up because sound waves are reflected from either a closed or an open end. The wavelengths of the standing waves are governed by the following end conditions:

- *At a closed end, air molecules are not free to move so their amplitude of vibration is zero (a displacement node) but the changes in pressure are maximum.*
- *At an open end, the amplitude of air molecule vibrations is a maximum (a displacement antinode) but the pressure does not change (a pressure node).*

These conditions mean that a given tube can only sound certain precisely fixed frequencies that are known as its **resonant frequencies**. Figure 4.24 shows the longest standing wave that can be set up in an open tube, and Figure 4.25 shows the longest standing wave that can be set up in a closed tube of the same length. The lowest frequency that can be produced from a given tube is called its **fundamental frequency**.

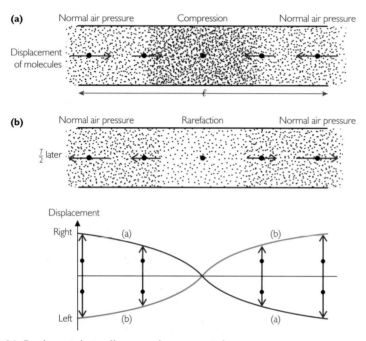

Figure 4.24 Fundamental standing wave in an open tube

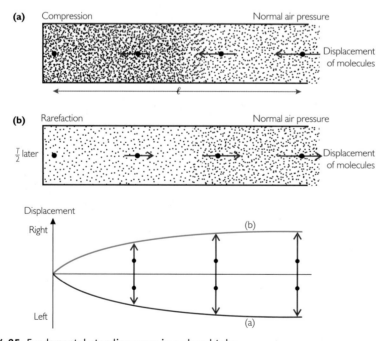

Figure 4.25 Fundamental standing wave in a closed tube

When an instrument is played, the pitch is controlled by opening or closing the holes along the tube. At an open hole, the air molecules can move freely but the pressure does not change. So opening or closing the holes effectively alters the length of the tube in which a standing wave can be set up.

Questions

15 Explain why the air molecule displacement graphs in the diagrams in Figures 4.24 and 4.25 have both positive and negative values.

16 Sketch graphs showing how air pressure will vary with distance along each tube in Figures 4.24 and 4.25.

17 What fraction of a wavelength λ corresponds to the tube length l for

(a) an open tube fundamental

(b) a closed tube fundamental?

18 (a) From you answer to Question 17, work out an expression for the fundamental frequency of an open tube in terms of its length, l, and the speed of sound in air, v.

(b) Repeat (a) for a closed tube.

19 What properties of a recorder do you think determine the fundamental frequency of the instrument?

20 When playing a recorder you cover up different numbers of holes to get different frequencies. In Figure 4.26 a black circle indicates a covered hole. On a sketch of Figure 4.26 mark the effective length of the tube in each case. Which of the two cases shown should give the lowest pitch (frequency) of note?

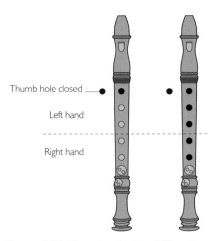

Figure 4.26 Fingering for two different notes on a recorder

Activity 13 Can your recorder tell you the speed of sound?

Your answers to Question 18 involved a relationship between the frequency of a note, the length of the tube and the speed of sound in air. Using this relationship, and with the help of *Audacity*, measure the speed of sound using a recorder.

2.2 Stringed instruments

Stringed instruments such as guitars or violins (Figure 4.27) produce a transverse standing wave on a string held between two fixed supports. The particles in the string are normally set into oscillation by plucking or bowing, but piano strings are struck by felted hammers.

The standing wave is generated by the superposition of the waves travelling along the string and being reflected at the support. A phase change of 180° (π radians) takes place at the reflection (Figure 4.28(a)) so the incident and reflected waves always superpose destructively at that point and there will be a displacement node at both supports (Figure 4.28(b)).

Figure 4.27 Stringed instruments

(a)

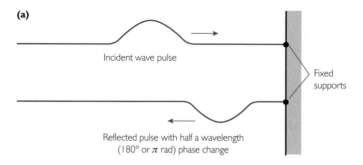

Incident wave pulse

Fixed supports

Reflected pulse with half a wavelength
(180° or π rad) phase change

(b) These diagrams show a single full wave being reflected.
(I) = incident part of wave. (R) = reflected part.
Snapshots (i) (ii) & (iii) are half a time period apart.

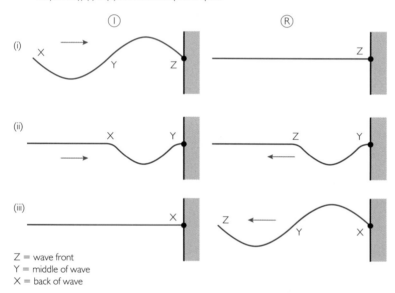

(I)   (R)

(i)

(ii)

(iii)

Z = wave front
Y = middle of wave
X = back of wave

Figure 4.28 Reflection at a fixed boundary (a) produces a phase change which (b) ensures that there is a node at the boundary

Study note

You can see the phase change if you send a single pulse along a slinky that is fixed at one end. You might have seen this in Activity 5.

Question

21 (a) Sketch a diagram showing the lowest-frequency standing wave that can be produced on a string with two fixed ends.

(b) Write down

(i) a relationship between the wavelength λ of this wave and the length l of the string

(ii) a relationship between the length l of the string and the frequency f and speed v of this wave.

(c) When we refer to 'the wavelength' and 'the speed' in part (b), do we mean the wavelength and speed of the transverse waves travelling along the string, or the wavelength and speed of the resulting sound waves in air?

The sound box of a stringed instrument

As the string oscillates as a standing wave, it generates a succession of compressions and rarefactions in the air, which travel outwards as a sound wave (Figure 4.29).

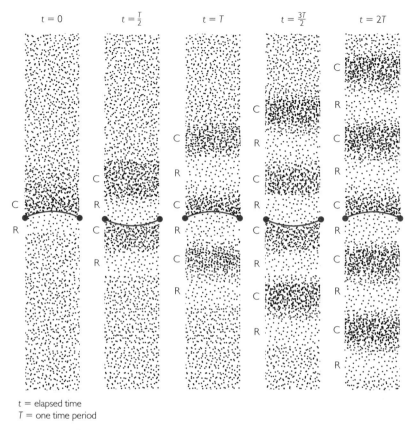

t = elapsed time
T = one time period

Figure 4.29 An oscillating string generates a sound wave

If the string were the only thing vibrating, then the sound would be barely audible because the string itself affects very little air. To make the sound louder, the string is attached to a sound box, which resonates with the string and sets a greater mass of air in motion, so the sound is louder.

Figure 4.30 Testing the sound box of a violin

Question

22 Classical guitars have a very obvious sound box to 'resonate' the sound. How do electric guitars achieve the same effect?

23 Figure 4.30 shows part of a violin under construction, being tested by driving it at two of its resonant frequencies, with tea leaves on the surface of the wood. Explain what is happening to the tea leaves.

Notes from a stringed instrument

Even if you are only vaguely familiar with a guitar (or any other stringed instrument), you will know that there are several ways in which the player can control the pitch (frequency) of the note.

> ### Activity 14 What affects the note produced by a string?
>
> Use a guitar or a sonometer (Figure 4.31) to explore one factor that affects the pitch of its note. Try to deduce a mathematical relationship between frequency and the factor you are investigating.

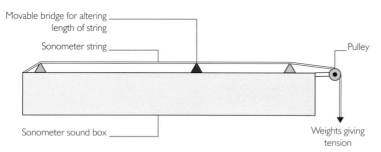

Movable bridge for altering length of string

Sonometer string

Pulley

Sonometer sound box

Weights giving tension

Figure 4.31 A sonometer

In Question 21, you saw that the lowest-frequency standing wave that could fit on to a stretched string of length l has a wavelength λ where

$$\lambda = 2l$$

This wavelength is related to the frequency f of oscillation, by the wave equation:

$$v = f\lambda \qquad (2)$$

where v is the *speed of the transverse waves travelling along the string*. We can therefore write:

$$f = \frac{v}{\lambda} = \frac{v}{2l} \qquad (3)$$

In Activity 14, you should have found that f is inversely proportional to the length l of the string. The shorter the string, the higher the note – as described by Equation 3.

You should also have found that the frequency does indeed depend on the tension and the mass per unit length of the string. Both of these factors affect the speed of transverse waves travelling along the string. In Activity 5 you will have seen that you can control the speed of waves along a slinky: the greater the tension, the greater the speed. The speed also depends on the mass per unit length of the string (or spring): the heavier the string, the lower the speed. The speed v is in fact related to the tension T and mass per unit length μ by the following equation:

$$v = \sqrt{\frac{T}{\mu}} \qquad (4)$$

Combining Equations 3 and 4 gives an expression for the frequency of the fundamental standing wave on a string:

$$f = \frac{1}{2l} \times \sqrt{\frac{T}{\mu}} \qquad (5)$$

Questions

24 Explain which one of the following pairs of similar instruments (Figure 4.32) should produce notes with a higher range of frequencies:

(a) a bass guitar and a banjo

(b) a piccolo and a flute

(c) a side drum and a timp.

Figure 4.32 Pairs of similar instruments

25 Explain in words how Equation 5 accounts for the following:

(a) Before being played, stringed instruments are tuned by twisting a series of screws or pegs.

(b) Violinists and guitar players press their fingers against the strings while playing.

(c) In a piano, the strings that sound the high notes are much thinner than the strings that sound the low notes.

2.3 Complex sounds

Most people can recognise when a note produced in two different ways has the same frequency. Vocalists, for example, often need to hear a note from a piano to adjust their voice to the right key. They can hear the note and produce a note at the same frequency. But why do these two notes sound different? And why does a trumpet or a guitar, or a didgeridoo for that matter, sound different again?

Speech tells us the answer

A waveform of human speech (Figure 4.33) gives a clue. At first sight, the wave looks complex and 'untidy', but you can see that it actually has a clear repetitive pattern or **periodicity**. You can pick out a basic period of about 10 ms, corresponding to a frequency of about 100 Hz. But the waveform clearly does not correspond to a simple sinusoidal wave. In Activity 9 and Question 10, you saw that the superposition of sinusoidal waves with different frequencies can produce waveforms with different shapes. The waveform in Figure 4.33 takes this to extremes – it can be reproduced by the superposition of a very large number of sinusoidal waves.

The frequencies and amplitudes of sinusoidal waves needed to make up a given sound can be found with an instrument called a **spectrum analyser**, which analyses (splits up) the sound into its component frequencies – rather as a prism splits light into its different frequencies or colours. The spectrum analyser produces a **sound spectrum**, which is a plot showing the frequencies and amplitudes of these component sinusoidal waves. The spectrum of the human speech (Figure 4.33) has a very large number of components, dominated by frequencies of a few hundred hertz.

Activity 15 A note about notes

Use *Audacity* to explore the waveforms and frequency spectra of notes that sound similar in frequency and yet are different in quality. Look for waveforms that have similar periods (similar 'repeat distances' on the screen).

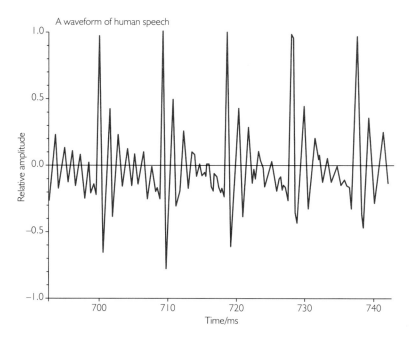

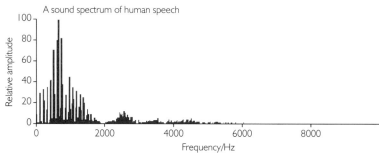

Figure 4.33 A waveform of human speech and its sound spectrum

Sounds from different instruments or other sources may have similar
pitch due to similar periodicities in their waveforms, but they may have
very different tones or qualities due to the presence of other components.
Each individual sound will have its own characteristic frequency spectrum.
Because the patterns in speech sounds are recognisable, spoken words can be
converted into digital code as a particular word. This technology is used in
software that allows you to 'speak' the instruction into an instrument
(Figure 4.34).

Figure 4.34

Harmonics and overtones

A dictionary may define musical 'harmony' as 'a pleasing combination of concordant
sounds'. Exactly what property makes one combination of sounds 'pleasing' to the ear,
and another not, fascinated ancient scholars such as Pythagoras. Early experiments
found patterns and mathematical links between lengths of strings, for example, which
produced concordant sounds when plucked.

You have already seen that, if a string or air column is made to vibrate in its
fundamental mode, it will emit a sound corresponding to the longest standing wave
that will fit within the boundaries. But it is also possible to produce higher-frequency
standing waves, and sounds from these higher-frequency standing waves are called

overtones or harmonics. When a musical instrument is played, the air column or string is does not vibrate with the simple 'tidy' vibration corresponding to the fundamental frequency. Rather, the vibration is a superposition of the fundamental frequency with many harmonics at smaller amplitudes. The resultant complex vibration has the same pitch as the pure fundamental vibration, but the sound quality is different. The characteristic quality of sound from an instrument depends on the combination of harmonic frequencies that are present, and these in turn depend on the shape of the instrument as well as its size.

2.4 Summing up Part 2

You should now be well versed in the science of sounds. You should be aware of what makes sounds different and how complex waveforms can be treated as combinations of simple sinusoidal waves. Activity 16 uses these ideas in the **synthesis** of familiar sounds. Synthesis means 'bringing together'. Activity 16 is about synthesis of complex sounds from simple components, and it also involves bringing together what you have learned in this part of the chapter.

The idea is simple: if, by analysing a sound spectrum, you can pick out the main frequencies present in their correct proportions, then by reproducing those frequencies simultaneously you should get a reasonable copy of the sound. The more of the constituent frequencies you can reproduce, the better the copy will be. This method of reproducing sounds is often used by electronic devices and less sophisticated keyboards.

Activity 16   Synthesising sounds

Use *Audacity* to explore the sound spectra of some synthesised sounds and compare them with the spectra of the same sounds produced naturally.

As a final word, it is worth mentioning sampling – the method by which better keyboards achieve much more realistic musical instrument copies.

The main disadvantage of reproducing a sound as in Activity 16 is that the constituent frequencies and their relative proportions may change considerably during the duration of the sound. Sampling gets round this by recording digitally the make-up of a sound at specific stages during the sound.

We can also think of this as making a detailed digital copy of the shape of the sound's waveform. You can improve a sample in several ways: Firstly by increasing the sampling rate, i.e. how many times you record information about the sound per second. Secondly, you can increase the number of digital bits – pieces of digital code – that represent each tiny section of the sound. You will meet the idea of digital signals again later in this chapter.

Questions

26 A guitar string is 0.5 m long, has a mass per unit length of 3.75×10^{-4} kg m^{-1} and is held under a tension of 15 N.

(a) What is the frequency of its fundamental vibration?

(b) If the guitar player wishes to produce a note of frequency 400 Hz from this string (without re-tuning), how far along the string should the finger be placed?

27 You may have witnessed a person breathing in helium gas and then speaking – the voice sounds much higher in pitch. Suggest a physical explanation for this effect.

28 For the waveform shown in Figure 4.35, determine its period and (fundamental) frequency.

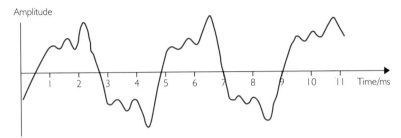

Figure 4.35 Waveform for Question 28

Further investigations

If you have the opportunity, here are some things that you might like to investigate:

- the frequency spectrum of a clarinet (or other instrument) as it plays through its complete range
- the frequency spectrum of notes from a plucked elastic band
- sounds produced from simple percussion instruments, e.g. a block of wood suspended and struck with a hammer
- the quality of sounds synthesised by a cheap keyboard (the cheaper the better!).

---

3 Sound sense

3.1 A word in your ear

Noise is perhaps best defined as unwanted sound. Our environment is becoming increasingly noisy, whether it is neighbours in an adjoining house or the person with a MP3 player sitting next to us on the bus. Road traffic and machinery can also produce continuous background sounds that may become very intrusive. We all make noise, but what is pleasurable noise to one person can be intolerable to another and destroy their quality of life. There is also evidence that listening to loud sounds (whether wanted or not) can damage the ears permanently.

The amplitude of a sound wave can be measured as a pressure, which has SI units of Pa. The human ear is sufficiently sensitive to notice a pressure change of 2×10^{-5} Pa (normal air pressure is about 10^5 Pa). This corresponds to air molecules moving by less than 10^{-10} m, that is less than the diameter of an air molecule. The greater the amplitude of the sound wave, the greater the intensity of the sound.

Intensity is not quite the same as our perception of the loudness of the sound, as our ears are more sensitive to some frequencies than others. Most people will perceive sounds in the range of a few kHz as louder than other frequencies, even though they may be receiving the same intensity of sound at their ears. Sound loudness is measured on the decibel scale (dB). This is not a linear scale and an increase of 10 dB will be perceived as a doubling of the loudness of the sound.

Activity 17 How loud does it sound?

Connect a signal generator to a loudspeaker and monitor the amplitude via an oscilloscope or *PicoScope*. Turn the signal through a range of frequencies and, with the amplitude constant, make judgements as to the range of frequencies over which you perceive the sound to be loudest.

The dangers of noise pollution

Different types of people react to noise in different ways as it gets louder (Figure 4.36). Prolonged exposure to high levels of noise or exposure to peaks of noise of a very high level can cause hearing damage. There is a statutory requirement for employers to protect their workers from such exposure under the Noise at Work Regulations.

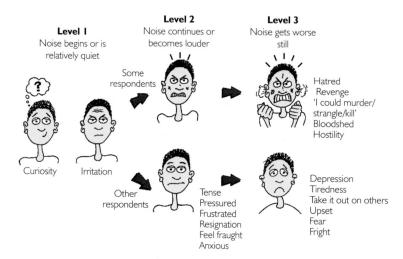

Figure 4.36 Different responses to noise

After exposure to loud noises you may have experienced a dullness in your hearing due to fatigue of the sensitive hair cells in your inner ear. Prolonged exposure leads to permanent hearing loss due to atrophy of the sensitive hair cell. You may experience a ringing in your ears (tinnitus) which is again due to temporary damage done to your ears though the precise cause of the ringing sensation is not yet understood by scientists.

However noise need not be loud to disturb activities such as conversation, relaxation or sleep and can cause emotional effects such as annoyance and tension. Research into people's psychological response to environmental noise has influenced legislation, regulations, codes of practice and guidance from government departments which provide the framework for the control of environmental noise affecting people in their homes.

Psychoacoustic researchers claim to have found evidence of community noise exposure resulting in hypertension (high blood pressure), heart attack, low birth weight babies and psychiatric problems. The most convincing research results suggest that aircraft noise can reduce scholastic achievement.

Activity 18 Noise survey

Using a sound-pressure-level meter, undertake an Environmental Pollution Survey measuring the sound around a motorway, ring road, factory, building site or similar location in which you can determine the way in which sounds 'drop off' with distance and the effects of mounds, barriers, fences or trees etc. in absorbing the sound.

You could supplement your measurements by interviewing people who live or work in the locations you have surveyed.

Consider sharing your data with other school via e-mail. This is particularly useful if you do not have a suitable site for your own investigation nearby.

Discuss ethical arguments for and against introducing laws to limit environmental noise.

3.2 Active noise control

Prevention, they say, is better than cure. Although it's not always possible to prevent noise entirely, the best way forward is to devise ways of reducing the effects of noise through sound insulation and through other means of control. When noise is a problem within a confined space (inside a building or a vehicle) then some ingenious engineering can reduce the problem using the superposition of waves.

Keeping noise low at the ear makes it possible to enjoy music without raising the volume unnecessarily. Ear defenders are useful in places where continuous engine noise may be a problem, but unusual noises may need to be heard, such as a tractor cab or airline cabin.

Noise cancelling headphones reduce unwanted ambient sounds by means of **active noise control**. Essentially, this involves generating an 'anti-noise' sound wave with the same amplitude as the sound wave arriving at the microphone, but in antiphase with it. This results in destructive interference, which cancels out the noise within the enclosed volume of the headphone.

Several components are required to achieve this effect, as shown in Figure 4.37.

Microphone

A microphone placed inside the ear cup 'listens' to external sounds that cannot be blocked passively.

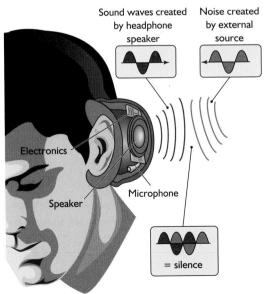

Figure 4.37 Active noise control

Noise-cancelling circuitry

Electronics, also placed in the ear cup, sense the input from the microphone and generate a 'fingerprint' of the noise, noting the frequency and amplitude of the incoming wave. Then they create a new wave that is 180 degrees out of phase with the waves associated with the noise.

Speaker

The 'anti-sound' created by the noise-cancelling circuitry is fed into the headphones' speakers along with the normal audio. The anti-sound erases the noise by destructive interference, but does not affect the desired sound waves in the normal audio system.

Battery

The term 'active' refers to the fact that there must be some energy input to system to produce the noise-cancelling effect. The source of that energy is a rechargeable battery.

An animation of how these headphones work can be found at the University of Salford webside. See (www.shaplinks.co.uk) for details of the URL.

Noise cancelling headphones typically only cancel the lower-frequency portions of the noise; they depend upon traditional noise suppression techniques (such as their earcups) to prevent higher-frequency noise from reaching the interior of the headphone. They work well for sounds that are continuous, such as the hum of a refrigerator, but are ineffective against speech or other rapidly changing audio signals. Noise cancelling headphones are able to provide an additional reduction in noise that means about 70 per cent of ambient noise is effectively blocked, making such headphones ideal for airline and train travel, open office environments or any other location with a high level of background noise.

Activity 19 Active noise control

It is possible to simulate (partially) an ANC system using two speakers connected to the same signal generator.

Place the speakers a couple of metres apart in the lab and walk across the room in front of them. Notice the regular variation of sound intensity as you move. Explain what you hear.

Place speakers in a box with a listening port where one would expect to find destructive interference. Note the effect of changing frequency on the production of destructive interference.

3.3 Summing up Part 3

In this part of the chapter you have seen how sound pollution can affect the environment, and how physics can be used to combat the problem in some circumstances.

Activity 20 Summing up Part 3

Write a short explanation of noise-cancelling headphones. Include the following terms in your account:

- phase
- amplitude
- superposition
- microphone.

Suggest a reason why these headphones may be less successful at blocking out higher frequency, rapidly changing signals.

Activity 21 Neighbourhood noise

Either:

design an information leaflet persuading people to be 'good neighbours' and explaining how sound can often transmit to unexpected places and how problems might be avoided by effective reduction of sound transmission.

Or:

prepare a presentation which defends the location of a motorway, ring road or factory on the basis of proposed sound reduction methods to be put in place.

4 Compact discs

Sometimes we want a permanent record of some favourite music, a famous voice or the strange sounds of whales. If so, it is more than likely that the permanent storage of that recording will be on a compact disc (CD, Figure 4.38), as this is one of the most popular formats for sounds. Digital Video Discs (DVDs), Blu-ray disc (BD) or High Definition DVDs (HD-DVDs) all work in a similar way to CDs. In this part of the chapter you will explore many aspects of the physics required to manufacture a CD, DVD, Blu-ray disc or HD-DVD player.

Figure 4.38 A CD player

4.1 What's on the discs?

Compact discs store information in a way quite unlike the previous generations of discs, as we now discover.

Before the CD

Thomas Edison was the first to record sound; he used wax cylinders. Another American, the German-born Emile Berliner, invented the flat disc-type record in Philadelphia in May 1888. In the early part of the 20th century the records were played back rotating at 78 revolutions per minute, so were called 78s. Later, a way was devised to store more music on vinyl discs that rotated more slowly – hence the name long-playing record or LP. In all these types of storage, vibrations caused by sound waves are used to drive a tool that cuts a wavy groove into the surface of rotating record. The shape of the groove replicates the waveform of the sound. This type of recording is known as **analogue** recording, as the stored information is an analogue, or copy, of the original sound. The sound is retrieved using a stylus, which is a flexible arm with a fine point that rests in the groove. As the record rotates the stylus moves with the wobbles in groove. These movements are converted to electrical signals which are amplified and produce the sound in the speakers.

Activity 22 Comparing discs

Examine the surfaces of a CD and an LP (Figure 4.39) and their playback systems (just the visible exterior parts). Use any magnification available. Briefly describe the appearance of each surface and comment on how many grooves (if any!) you can see. Which surface is 'read' during play?

Listen to two extracts of music, one from an LP and one from a CD played through the same amplifier. Comment on any difference in the quality of sound from the two systems you heard.

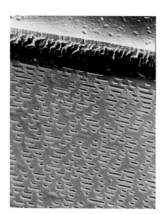
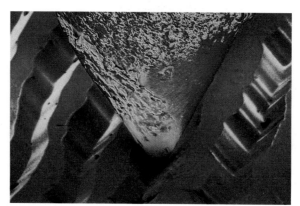

Figure 4.39 Magnified views of (a) a CD and (b) an LP surface

Activity 23 CDs – the sound of science

Read the article below and answer Questions 29 to 36 that follow. Answers to many of the questions can be found in the article. If you skim read the article first, you will get an idea of what is in it. Then read the questions so that on your second, more careful, reading of the article you will be able to pick out the important points. Have a dictionary at hand to assist with any unfamiliar words.

The sound of science

Compact discs offer better sound quality than vinyl discs, with up to an hour's uninterrupted playing time and, most important perhaps, they are more durable. A compact disc will still play back perfectly even with a 2 mm hole drilled through the playing area, and handling a compact disc present no problems as they are immune to a few scratches, dirt and grease. There are no grooves in the surface of the compact disc, and the CD player has no stylus. The audio information or signal recorded onto the disc is processed and stored there in a fundamentally different way, which is where computer electronics come in.

Storage and retrieval

Old music systems like vinyl and tape were analogue storage and retrieval systems. In analogue systems the sound was recorded as a wobbly groove or varying magnetic field similar in form to the sound wave to be reproduced. CDs use **digital** recording – a completely different approach in which the sound is stored as **binary codes** of 0s and 1s. For the compact disc, Philips developed an elegant optical scanning system, using a low-powered laser to 'read' tiny bumps in a reflective metal surface (Figures 4.40 and 4.41).

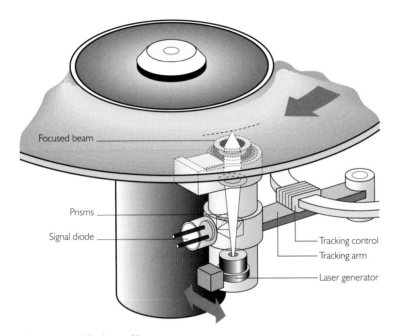

Figure 4.40 Playing a CD

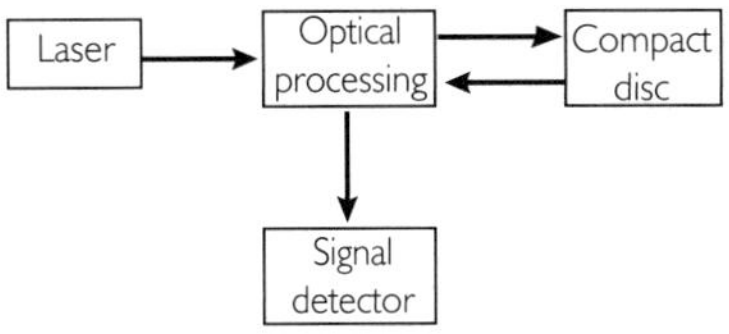

Figure 4.41 Block diagram of a CD player system

Metallic layer

Sandwiched by a protective plastic coating 1.6 mm thick, the silver metallic layer in the compact disc is etched with a spiral track of bumps of literally microscopic proportions. The pitch of the spiral is in fact 1.6 μm, which makes the 'micro grooves' of a conventional LP look quite big. When light, in fact a beam of highly concentrated laser light, is focused on this pattern, a reflected light signal is received from the flat surface, which is compared with the light reflected from the bumps. In the player's optics, both reflected rays are passed to a photodiode, where superposition produces a series of electrical 'ones' and 'zeros'. Here the fundamental benefit of working with digits can be realised – it does not matter exactly how much light is received, because anything above a predetermined level is read as a 'one' and everything below that level as a 'zero'. Because the playback system only has to identify these two conditions, fidelity to the original information is theoretically perfect. The laser is accurately focused on the reflective metal layer so dust particles or marks on the plastic coating are ignored.

Questions

29 What is meant by the statement that CDs are durable?

30 (a) Does the CD system of sound recording use a digital or an analogue method?

(b) In a couple of sentences, explain the different between the two methods.

31 What do you think the author means by 'optical scanning?' What is the equivalent for an LP?

32 The distance between one turn of the spiral track and the next is called its 'pitch'.

(a) Write down the size of the pitch of a CD expressed in standard form.

(b) How broad can the light beam be if it is to distinguish between one set of bumps and another?

Maths reference

Standard form
See Maths note 1.2

33 The article mentions two materials. The compact disc has a 'reflective metallic layer' and a 'protective plastic coating'.

(a) What property, other than protective, must the plastic coating have?

(b) What is important about the metallic layer?

(c) Which reflection is more important, that from the metal or that from the plastic?

34 Sketch diagrams to show

(a) what is meant by a focused beam and a parallel beam

(b) how it is possible for an accurately focused beam to ignore dust particles and marks on the plastic surface.

35 Reflection is mentioned many times. Jot down what you understand by the term.

36 The article includes a labelled diagram. What are: a signal diode; a tracking arm, and a prism?

4.2 Optical scanning

The digital information is coded on a CD as a series of small raised areas created as the playing disc is pressed from a master disc in which pits have been etched (Figure 4.42).

The pits are made, using a binary code, from audio measurements made 44 100 times a second. Each binary number represents the sound signal at one instant in time converted to a sequence of bumps or no bumps (rather like Braille) on a spiral track. Using this system we only have to read binary 1 or 0. For an optical reader this means detecting whether a light signal is 'on' or 'off'.

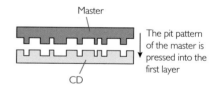

Master

The pit pattern of the master is pressed into the first layer

CD

Figure 4.42 Pressing a CD from a master disc

Superposition

Somehow a series of bumps on a disc have to turn the light on and off as it scans over them. It is all done by waves – in particular the property of waves known as **superposition**.

The following discussion, with Questions 37 to 39 and Activity 24, shows how superposition of coherent waves plays a key part in reading a CD.

Study note

Superposition is discussed in Section 1.5 of this chapter.

Questions

37 This question is about light waves A, B and C.

- Wave A has frequency f and amplitude a.
- Wave B has frequency f and amplitude $1.5a$, and is in phase with A.
- Wave C has frequency f and amplitude a, and is $180°$ out of phase with A.

State what an observer would see in each of the following situations. Give the frequency and amplitude of the resulting wave in each case and sketch graphs to illustrate your answers.

(a) (i) A is combined with B to give wave P.
    (ii) A is combined with C to give wave Q.

(b) Wave B becomes $180°$ out of phase with A before combining with it to give wave R.

(c) Wave C becomes in phase with A before superposing with it to give wave S.

38 For digital use the signal detector requires two states – the light is on or off.

(a) Look at your answers to Question 37 and decide which results are equivalent to 'on' and which are to 'off'.

(b) What are the requirements for two waves to produce total cancellation?

In practice an electronic circuit can be calibrated to recognise 'on' signals for anything above a predetermined level of brightness, and if the light level falls below that the circuit records 'off'. In our example the conditions for cancellation could then be set so that the two waves had to have the same period, be out of phase and their amplitudes only need be similar, not identical.

Question

39 If 'on' is taken to mean 'has an amplitude greater than a' and all other brightness levels are deemed 'off', reconsider your answer to Question 38(a).

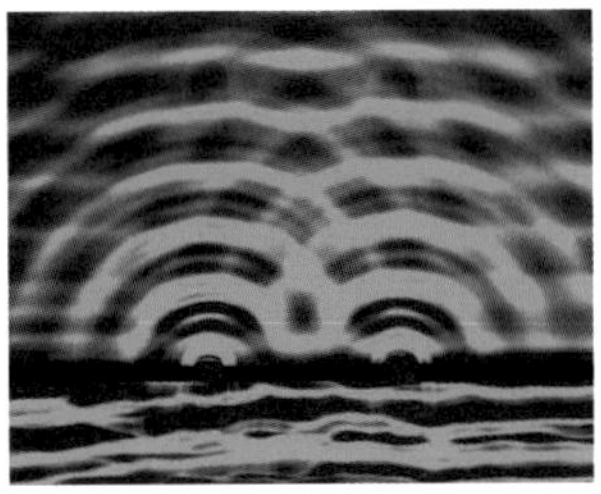

Coherent waves

To get a steady signal from the superposition of two waves, they need to be **coherent**, otherwise the signal keeps changing (see Question 37). Light waves from two separate sources will be incoherent because the light is emitted in short random bursts each lasting about a nanosecond. One way to be sure of getting coherent waves is to use a single beam of light. But how could a phase difference be caused across a single beam? It can be caused in several ways. Two parts of the wave could start together in phase, travel different distances and meet up again (Figure 4.43). Or part of the wave could be **reflected** and meet up with the unreflected part (Figure 4.44)

Figure 4.43 Superposition of waves that have travelled different paths

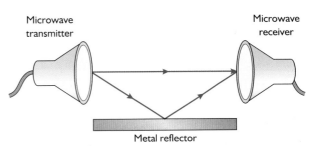

Figure 4.44 Superposition of reflected waves

Superposition in a CD player

In a CD player the laser light source produces a single beam with a diameter equal to twice that of a raised bump. In the absence of a bump, all the light is reflected from the background surface with no phase difference across it – so it produces constructive superposition i.e. the signal diode detects that the light is 'on' (Figure 4.45).

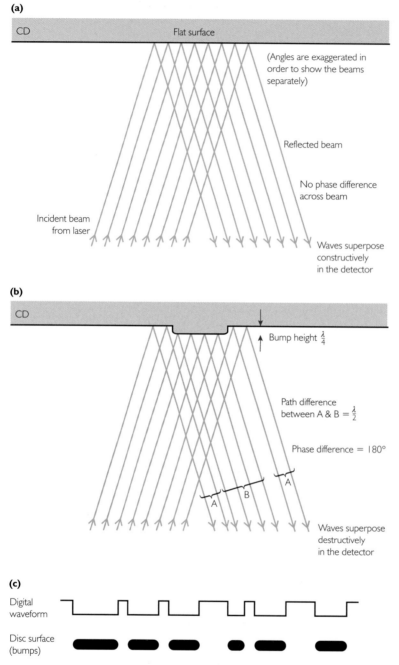

Figure 4.45 Superposition of waves reflected from a CD

In the presence of a bump, half the light will go to the background surface and back and get out of phase with the part reflected from the shorter route via the top of the bump. If the **path difference** between the two reflected beams is carefully arranged, the *phase difference* between them will give rise to destructive interference when they come together in the detector. The light is recorded as 'off' by the signal diode even if the cancellation is not absolutely complete. The light beam moving over the spiral track gives a sequence of on–off signals, which is then processed to reconstruct the original sound.

In the case of the CD, one part of the light beam can destructively interfere with the other part because of a difference in the distance the both travel. This will only work if the laser light is **monochromatic** (i.e. of a single frequency) and the incident and reflected beams are coherent.

Activity 24 Model CD

Use the apparatus shown in Figure 4.46 to illustrate reading a CD. The 500 nm wavelength laser light used in a CD player is far too fine for laboratory use, so this apparatus uses electromagnetic waves with a wavelength of 3 cm to model the action. The size of the bumps is scaled up by the same factor as the wavelength.

Describe how this model works, using the terms *path difference* and *phase difference*.

Figure 4.46 A large-scale model of a CD and scanner

Questions

40 (a) If two beams of light are to interfere destructively, what is the smallest possible difference in the paths travelled?

(b) If two beams of light interfere constructively, what are the possible differences in the paths travelled?

(c) If the wavelength of the light is 500 nm, how high must the bumps be on a CD be to produce destructive interference?

41 You have seen that a path difference of half a wavelength is equivalent to a phase difference of half a cycle (180 degrees or π radians). Use this to deduce a general expression relating path difference to phase difference.

4.3 The coating on the disc

One of the greatest selling points for CDs is that the surface is protected by a transparent coating – no spills, scratches, etc., can alter the sound quality. But strange things can happen to light as it enters and leaves a transparent material (i.e. crosses a boundary between different materials). What are the consequences for the optical scanning system? This section explores a behaviour, known as **refraction**, and shows its consequences for the CD player. But first some terminology.

Figure 4.47 shows are ray of light crossing a boundary, labelled with the conventional terms. Note the convention of measuring all the angles with the **normal** and not with the surface.

Question

42 (a) There are two materials in Figure 4.47. If they are air and glass, which is which? What helped you to decide?

(b) If you interchanged the materials, what would be different?

(c) If a ray is incident along the normal, what happens to it in the second material?

(d) Now imagine the materials are altered so that the ray of light is refracted less. How would the angle, r, alter?

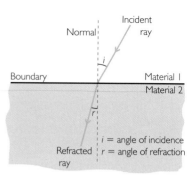

Figure 4.47 A light ray crossing a boundary

Light changes its direction in this way because it travels at different speeds in different materials. The speed is greatest in a vacuum (almost the same as in air) and smaller in all other materials. It is only when light meets a surface at an angle that the effect of this becomes apparent. We get a similar experience when a car swerves into the kerb because the inside wheels meet a large puddle of water. The slowing effect turns the vehicle.

Snell's law of refraction

It is not enough to be able to say which way the ray will 'turn' as it crosses a boundary. We need to be able to predict the exact path from a given angle of incidence. Figure 4.48 shows waves crossing a boundary.

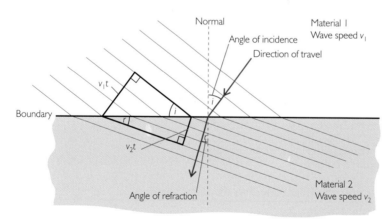

Figure 4.48 Waves crossing a boundary

In a time t, the waves in material 1 travel a distance $v_1 t$ while the wave in material 2 travel a distance $v_2 t$. By drawing right-angled triangles as shown in Figure 4.48, we can write down some useful relationships. The triangles share a hypotenuse, so:

$$\frac{v_1 t}{\sin i} = \frac{v_2 t}{\sin r} \qquad (6)$$

> **Maths reference**
>
> See Maths Note 6.2
> Sine of an angle

Cancelling t, and multiplying by $\sin r$ and by $\sin i$ we can write the relationship known as **Snell's law**:

$$\frac{\sin i}{\sin r} = \frac{v_1}{v_2} \qquad (7)$$

For any given materials, the ratio $\frac{v_2}{v_1}$ is constant, and is known as the **refractive index**, and given the symbol $_1\mu_2$.

$$_1\mu_2 = \frac{\sin i}{\sin r} = \frac{v_1}{v_2} \qquad (8)$$

Strictly, we should always talk of the refractive index *between* two materials and use the labels 1 and 2 before and after the μ. But when dealing with light we often talk of the refractive index *of* a material, taking it for granted that the other material is air (or a vacuum) and dropping the labels.

> **Study note**
>
> In some books you will find the symbol n rather than μ.

Activity 25 Measuring refractive index

By tracing rays of light through a rectangular block of transparent material and measuring the angles at the interfaces (Figure 4.49), use Snell's law to calculate the refractive index of that material.

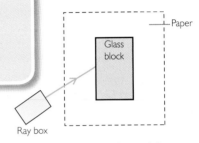

Figure 4.49 Diagram for Activity 25

Questions

43 A ray of light travels from air into another transparent material. Use Snell's law to determine the missing values in Table 4.1.

Angle of incidence, i	Angle of refraction, r	Refractive index, μ
40°	A	1.33
B	30°	1.47
64°	36°	C

Table 4.1 Data for Question 43

44 (a) Extend Table 4.2 by adding your own experimental values of refractive index from Activity 25.

(b) Given that light travels at 3.00×10^8 m s^{-1} in air, what is the speed of light in each material in Table 4.2?

Material	Refractive index between air and material
glass	1.47
water	1.33
polystyrene	1.60

Table 4.2 Data for Question 44

> **Study note**
>
> There are many different types of glass and the refractive index of each type depends on its composition. Table 4.2 refers to just one type.

Changing the wavelength

Refraction means that we have to look again at the value of the wavelength of the light used in a CD player. Earlier the height of a bump was set at 125 nm based on a wavelength of 500 nm. But this is inside the plastic coating. Will the light have the same wavelength in the air?

Have another look at Figure 4.48. Notice that when the waves slow down they get closer together. Frequency remains constant. Starting from the link between speed v, frequency f, and wavelength λ of a wave:

$$v = f\lambda \qquad\qquad \text{(Equation 2)}$$

we can answer the question as follows:

Worked example

Q If λ = 500 nm in plastic with refractive index 1.55 what is the wavelength in air?

A Using the labels a = air, p = plastic:

$$_a\mu_p = 1.55 = \frac{v_a}{v_p}$$

and

$$\lambda_p = 500 \text{ nm.}$$

We can write:

$$f_a = \frac{v_a}{\lambda_a} \text{ and } f_p = \frac{v_p}{\lambda_p}.$$

Frequency f does not change and so:

$$f_a = f_p = f$$

and therefore:

$$\frac{v_a}{\lambda_a} = \frac{v_p}{\lambda_p}$$

which we can rearrange to give:

$$\frac{\lambda_a}{\lambda_p} = \frac{v_a}{v_p} = {_a\mu_p}$$

$$\lambda_a = 1.55\lambda_p$$

$$\lambda_a = 1.55 \times 500 \text{ nm} = 775 \text{ nm.}$$

This wavelength is much longer in air than it was inside the plastic coating of the CD. This alters the specification for the laser in a CD player.

From the worked example above, we get an important general result that extends Equation 8:

$$_1\mu_2 = \frac{v_1}{v_2} = \frac{\lambda_1}{\lambda_2} \qquad\qquad (9)$$

4.4 A focused beam

The bumps on a CD are so small and so close together that they can only be detected by a very narrow beam of light. As you have seen in Section 4.1, the diameter of this beam must be about 1.6 μm across. In this section you will see how a lens is used to narrow the laser beam down to a small spot just at the surface of the CD.

Converging lenses

Any lens that is fatter at its centre than at its edge will converge (bring together) a beam of light (Figure 4.50). Such lenses are known as **converging** or **convex** lenses. In Figure 4.50(d) the lens is reducing the divergence of the beam i.e. converging it. Figures 50(a) and (c) show beams converging through a point.

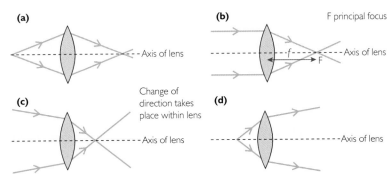

Figure 4.50 Converging a beam of light

Figure 4.50(b) shows a parallel beam converging through a point F known as the **principal focus** of the lens. (There is one on either side of the lens.) The distance from the lens to this point is called the **focal length**, *f*, of the lens.

A lens is used to focus the laser beam onto the metallic surface of the CD. By accurately focussing the laser on the metallic surface of the CD scratches and dirt are ignored because they are out of focus (Figure 4.51).

Some DVDs and BDs have two layers of information on the same side of the disc. The laser can be focussed on the top or the bottom layer using a lens (see Figure 4.52), which doubles the amount of information that can be stored on the disc.

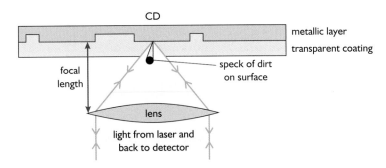

Figure 4.51 A speck of dirt does not affect the way a CD is read

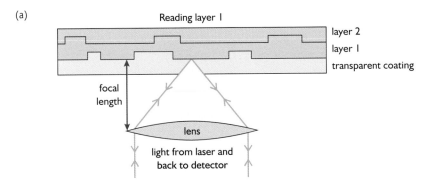

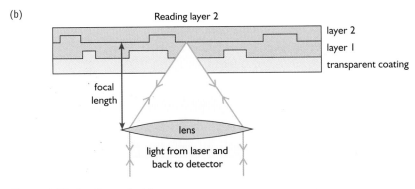

Figure 4.52 Reading a dual layer

Questions

45 Figure 4.53 shows two rays about to enter a magnified section of a lens. Copy the diagram and continue the paths to show how Snell's law of refraction predicts convergence.

46 In the scanning system of a CD player, a diverging beam passes through two strong converging lenses: the first makes the beam parallel and the second brings it to a point on the playing surface (see Figure 4.40 in Section 4.1). If the second lens is 2 mm from the surface, what must be its focal length?

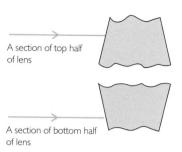

A section of top half of lens

A section of bottom half of lens

Figure 4.53 Light rays approaching a lens

4.5 Summing up Part 4

In this part of the chapter you have seen how information is encoded on CDs, and you have used ideas about superposition from earlier in the chapter to explain how the information can be read. You have also seen how refraction of light plays a part in the way CDs are read – you will do more on refraction in the next part of the chapter.

Activity 26 Summing up Part 4

Check back through your work on this part of the chapter and make sure you understand the meaning of all the terms printed in bold.

Question

47 Blu-ray discs (BDs) and HD-DVDs use a light with a wavelength in air of 405 nm.

(a) If the plastic surface coating has a refractive index of 1.5 what is the wavelength of the light in the coating of a BD or HD-DVD?

(b) How high are the bumps on a BD or HD-DVD?

5 CD players

In this part of the chapter, attention turns to the way the laser beam is manipulated in a CD player, and to the production of the beam itself.

5.1 Splitting the beam

So far we have seen how a focused beam shines on the prepared surface of a compact disc and is reflected from it, full of coded information. It's not much help if the light reflected from the CD then gets mixed up with the incident light from the laser. Somehow it must retrace a different route to reach a detector and be decoded. There are many different ways the beam can be split and different players use different methods. Activity 27 shows how the beam can be diverted into the detector.

Activity 27 Ray tracing on the way out

Using blocks of various shapes (as in Figure 4.54), look at light as it comes out of a block and see how a single beam can be split into two parts.

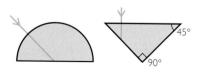

Figure 4.54 Block shapes for Activity 27

In Figure 4.55 light emerges from a dense material into a less dense one, for example from a Perspex block into air. It is incident on a boundary where both reflection and refraction are possible. Your sketches from Activity 27 will show that reflection is always possible but refraction not always. The largest possible internal angle of incidence that will allow light to emerge is called the **critical angle**, *C*. If the light cannot refract out of the block we say that there is **total internal reflection** (sometimes abbreviated to TIR).

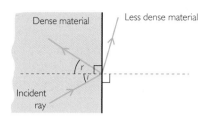

Figure 4.55 Light emerging from a block

Activity 28 Critical angle

Decide how you could apply the laws of refraction to light emerging from the block to derive a relationship between critical angle and refractive index.

Look back at your results for Activity 27 and find the size of the critical angle for one of the blocks. Use this value to determine the refractive index for the material of the block.

Questions

48 Light is incident at 42° at the inner surface of a transparent material. For each of the materials listed in Table 4.3, decide whether it will be split.

Material	Refractive index
glass fibre	1.55
Perspex	1.48
water	1.33
optical (flint) glass	1.61

Table 4.3 Data for Question 48

49 The silvering on the back of a mirror can deteriorate with time, so instruments often use a prism as a reflector in preference to a conventional mirror. Draw diagrams to show how a prism with angles 45°, 45°, 90° can be used

(a) to reflect a beam totally through 90°

(b) to reflect a beam through 180°.

5.2 Polarisation

Many CD players use a device called a 'polarising beam splitter' to direct the reflected light from the CD into the detector. This relies on a property of waves called **polarisation**.

You may already be familiar with **polarised light** through the use of polarising sunglasses, which reduce the intensity of light passing through them and are particularly effective at reducing glare from reflected surfaces. If you take two pieces of Polaroid from such sunglasses, and place one in front of the other and rotate it, you will notice that the brightness of the light changes with angle. At certain positions, the light is blocked completely; this arrangement is referred as 'crossed Polaroids'. See Figure 4.56.

Figure 4.56 View through crossed and un-crossed Polaroids

Polaroid was first produced by George Wheelwright III, in America, in 1938. It is made by first stretching a piece of plastic to align its long chain molecules and then dipping it into iodine solution so that iodine atoms become attached to the chains and line up along them. Polaroid is a trade name; more correctly and generally, such a piece of plastic should be called a **polarising filter**.

To explain how polarising filters work we need to think about the behaviour of light waves.

Light waves, and all other electromagnetic waves, are *transverse* – that is as you saw in Section 1.4, they involve oscillations at right-angles to their direction of travel. What's actually oscillating is an electric and a magnetic field at right angles to each other and to the direction of travel, as shown in Figure 4.57.

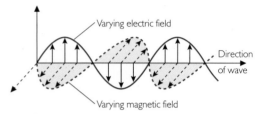

Figure 4.57 An electromagnetic wave

> ### Study note
>
> An electric field is a region in which a charged particle experiences a force. The direction of the electric field is defined as the direction of the force on a positive charge. A magnetic field is a region in which a magnet (e.g. a compass needle) experiences a force; its direction is defined as that in which the north-seeking pole of a compass needle would point.

Polarised waves

We can model the behaviour of light using transverse waves on a rope. If you shake the end of a rope up and down, you produce a travelling wave where the oscillations take place only in a vertical plane; the wave is said to be **polarised** or, more correctly, **plane polarised**. If you shake the rope in all transverse directions, you still produce a transverse travelling wave, but now the oscillations are no longer confined to a plane, and the wave is **unpolarised**. If the rope passes through a narrow slit as in Figure 4.58, only those oscillations parallel to the slit can get through, so the slit turns unpolarised into polarised waves – it acts as a polarising filter.

> ### Study note
>
> A Polaroid filter absorbs radiation that is polarised in one plane and allows the unimpeded passage of radiation that is polarised at right angle to that plane. The mechanism is not simply a matter of allowing waves through a slit, though the slit model illustrates the principle for waves on a rope.

If polarised waves encounter a second polarising filter parallel to the first, they can travel through unimpeded, but if the slit is at 90° to the first, then it completely blocks their passage and the slits are behaving like crossed Polaroids. With the second filter at an intermediate angle, waves still emerge through it; their amplitude is reduced and plane of polarisation is parallel to the second slit.

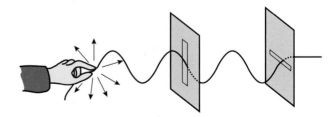

Figure 4.58 Demonstrating polarised waves on a rope

If you repeat this exercise with *longitudinal* waves (e.g. on a Slinky), then the slits have no effect on the passage of the waves – they always get through the slits. Longitudinal waves cannot be plane polarised. The fact that electromagnetic waves can be plane polarised in fact shows that they *must* be transverse.

Rotating the plane of polarisation

Light is blocked by crossed Polaroids because the two filters are aligned at 90° to one another. However, there are some materials that rotate the plane of polarisation of light. Sugar solution is one such **optically active** material; others include turpentine and many plastics, particularly when stretched. If you place an optically active material between two crossed Polaroids, then some light can emerge, and you need to rotate one of the Polaroids to produce extinction.

Activity 29 Exploring polarisation

Use a rope and a Slinky to demonstrate the polarisation of transverse waves, and to show that longitudinal waves cannot be plane polarised. Use two polarising filters to observe the behaviour of some optically active materials. Look through a single polarising filter at light reflected at a shallow angle from a polished surface (e.g. a bench top) and hence explain why Polaroid sunglasses are particularly good at reducing glare.

The rotation of the plane of polarisation can be measured with a polarimeter (Figure 4.59). In its simplest form, such an instrument for studying liquids consists of two polarising filters, one fixed and one that can be rotated against a protractor scale, with a tube of liquid placed between them.

Activity 30 Polarimetry

Use a polarimeter to measure the rotation of the plane of polarisation for some sugar solutions.

Figure 4.59 A polarimeter

Polarising beam splitter

In a polarising beam splitter, light from the laser is polarised by the polarising prism. Vertically polarised light passes through the polarising prism towards the CD. The light passes through a device called a 'quarter wave plate' which rotates the plane of polarisation of the light by 90° so that the reflected light is now horizontally polarised. The reflected beam is totally internally reflected into the detector because the materials in the prism have different refractive indices for different polarisations of light. See Figure 4.60

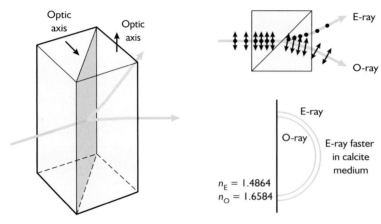

Figure 4.60 Polarising beam splitter

Further investigations

In Activity 30 you probably noticed colours produced by optically active materials between crossed Polaroids. You might like to devise a way to explore how the rotation of the plane of polarisation depends on wavelength. Or you might look into variations with temperature.

Engineers sometimes make plastic models of structures (such as bridges) and observe them through crossed Polaroids; the colour and intensity of light that emerges changes according to the load on the structure, indicating where the structure is under greatest stress. You might think of devising a way to explore this phenomenon.

Question

50 The specific rotation of a material that rotates the plane of polarisation is defined as:

$$\text{specific rotation} = \frac{\theta}{cL}$$

where θ is the rotation in degrees, c is the concentration in g ml^{-1} for a substance in solution or its density in g ml^{-1} for a pure substance, and L is the length of the light path (or depth of substance) in decimetres. (1 dm = 10^{-1} m.)

(a) What is the specific rotation of a sugar solution which produced a rotation of the plane of polarisation of 33° with a concentration of 0.5 g ml^{-1} and a light path of 1 dm?

(b) Some materials will, for a 1 dm light path, rotate the plane of polarisation by many hundreds of degrees. Since it is not feasible to measure more than 360° directly, suggest how you might calculate their specific rotation.

5.3 Laser

In Section 4.2 we said that the light beam in a CD player had to be monochromatic (single coloured) – i.e. have a single frequency. The way to achieve this is to use a laser. In order to explain what is special about laser light, we will first look at light from sources that are not monochromatic.

Coloured light

You will probably have seen all the colours that make up white light separated out by a prism and displayed as a spectrum. This shows us what is there but doesn't explain how it got there. To understand about the lasers used in CD players, we need to develop a theory about light production.

In Activity 31 you will observe the spectra from various light sources. To display the spectra you need a **spectrum analyser** (a device that separates out the different frequencies present). This could be a prism, but a filter called a diffraction grating allows you to spread the colours over a wider range of angles so that they are more easily distinguished.

Activity 31 Observing spectra

Use a prism or a diffraction grating to observe the colours that make up the light from various sources. See Figure 4.61.

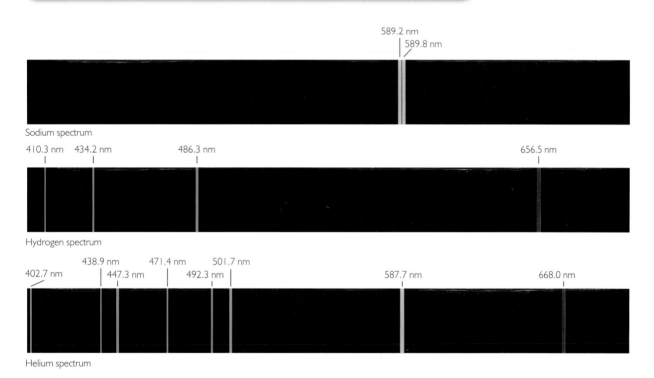

Figure 4.61 Light from various sources observed through a diffraction grating

Modelling light

In our first encounters with light it was enough simply to say that we need it in order to see. Then, perhaps through physics lessons, you learn that it involves energy transfer, and then you observe the phenomenon of superposition, which convinces you that light has wave properties. This helps to explain colour and refraction – different colours are attributed to different wavelengths, and refraction is linked to the wave speed. So, is it enough to say that light is a wave? Well, no. Once we get to the point of having to explain what is happening as light is created or absorbed, we need another way of looking at this thing called light; we need another **model**.

The energy emitted by a light source is the result of millions of individual events. Each event radiates a packet of energy, which means light also has to be modelled as a particle. A light particle is called a **photon**. These two models, wave and particle, describe different aspects of the phenomenon of light. Which one we use depends on the particular situation.

Both models can be used to describe colour. According to the wave model, colour depends on frequency: violet light has a higher frequency than red light (about twice as high). But according to the particle model, colour depends on energy: a violet photon transfers more energy than a red photon (about twice as much). The two models are related via the Planck equation:

$$E = hf \qquad\qquad (10)$$

Where E is the energy of the photon and f is the frequency of the wave. The constant h is known as the Planck constant: $h = 6.63 \times 10^{-34}$ J s.

> **Maths reference**
>
> Units
> See Maths note 2.2

Questions

$c = 3.00 \times 10^8$ m s^{-1} $h = 6.63 \times 10^{-34}$ J s.

51 (a) Copy and complete Table 4.4 so that it contains data relating to the wave and photon models of light.

Colour	Wavelength λ / nm	Frequency f / Hz	Photon energy E / J
infrared	775		
red	656		
green	486		
blue	434		
purple	410		
ultraviolet	389		

Table 4.4 Data for Question 51

(b) How do the values of photon energy help explain warnings about sunburn?

52 Power is the rate at which energy is transferred. A power of 1 W corresponds to 1 J s^{-1}. A laser for a CD player has a power of 0.2 mW and a wavelength of 775 nm.

(a) What is the frequency of the light?

(b) At what rate does the laser emit photons?

(c) What difference would it makes to your answers if the laser had a smaller wavelength but the same power?

Atomic line spectra

In Activity 31 you will have seen some **line spectra** – light that contains only a few distinct colours separated by gaps, rather than a continuous range. This light was emitted by atoms in an electrical discharge tube. Each element has its own distinctive set of colours.

The photon model can help explain the origin of atomic line spectra. In a discharge tube, energy is transferred from the electrical supply to the atoms in the gas, giving their electrons additional energy. The electrons lose this extra energy by emitting light – each photon given out corresponds to a single electron losing energy. When electrons

> **Maths reference**
>
> Manipulating powers on a calculator
> See Maths note 1.4

are bound in an atom, they can only have certain energies, which are known as electronic energy levels, and the photons they emit correspond to electrons making transitions between these levels. By analysing the energies of photons given out by excited atoms, it is possible to work out their **energy levels**.

As hydrogen is the simplest element (just one electron), its line spectrum is the one most readily worked out. Some of the calculated energy levels for hydrogen are shown in Figure 4.62, with arrows indicating some of the possible energy transitions.

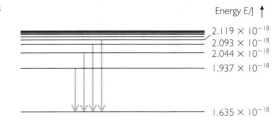

Questions

53 Calculate the energy transitions corresponding to the four arrows in Figure 4.62, and match them up with four of the lines listed in your answer to question 51.

54 Without doing any calculations, what can you say about the photon emitted when an electron makes a transition to the lowest energy level in Figure 4.62 (called the ground state)?

Wave–particle duality

If light can be represented as a wave *and* as a stream of particles, what exactly is light and what do you have to do about it? Both models are 'right' in that they each explain some aspects of the behaviour of light, and we have to treat light as behaving *like* waves part of the time and *like* particles at others – notice that we have not said that light *is* either of these!

Figure 4.62 An energy level diagram for a hydrogen atom

Questions

55 Given a choice between the wave and photon models, say which you think best explains each of the following:

(a) a lens refracts a beam of light

(b) a light source produces a line spectrum

(c) a ray of light is reflected.

56 'A photon is the means by which the energy carried by the waves is ultimately delivered and takes effect.'

(a) What does the speaker of this quotation mean about energy being delivered?

(b) How does this short statement indicate when to use the wave and particle models?

> **Study note**
>
> You will return to the question of wave–particle duality in *Technology in Space*.

Laser light

A **laser** emits light of just one frequency. In other words, it has a special sort of line spectrum, corresponding to just one energy transition. Laser light has two other unusual properties, both of which are useful in the CD player:

- the light is emitted in a narrow beam rather than coming out in all directions
- the emission is coherent – different atoms emit light in phase with one another, rather than randomly.

The name laser stands for **l**ight **a**mplification by **s**timulated **e**mission of **r**adiation. This phrase relates to the way that a laser produces light, which is in turn related to the light's special properties.

A directed beam of coherent monochromatic light is produced if excited atoms are stimulated to emit a photon in a chosen direction. An identical photon will do this. One photon enters the excited atom and two leave (the light is 'amplified'). The photons travel in the same direction, have the same wavelength and are in phase with one another (see Figure 4.63).

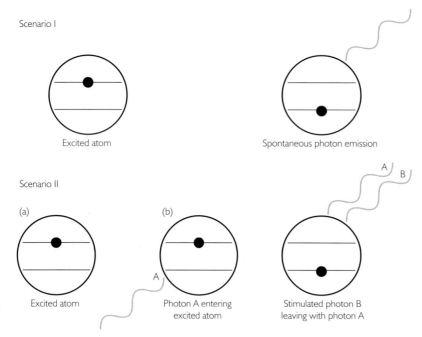

Figure 4.63 (a) spontaneous and (b) stimulated emission of radiation

The first consideration is how to get the atoms excited. The energy from an electric current or discharge through the material works well. Almost immediately photons will be spontaneously emitted. Mirrors are used to make sure some of these photons are kept travelling up and down the space, stimulating others to 'get in step'. A small hole in one mirror allows a small proportion to escape as a laser beam.

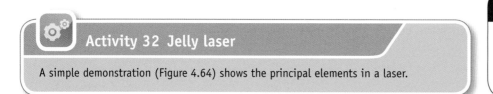

Activity 32 Jelly laser

A simple demonstration (Figure 4.64) shows the principal elements in a laser.

Safety note ⚠

Do not allow light from the flash gun or the laser to shine directly into anyone's eye.

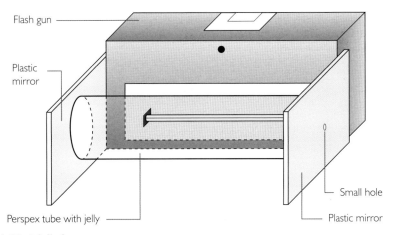

Figure 4.64 A jelly laser

Many types of material can be made to 'lase'. Sometimes a gas is used; but to get a compact, robust portable laser, solid materials are used. The laser for a CD player must be extremely small. To get an intense monochromatic light from a small component, the answer is to use a semiconductor material – most commonly gallium arsenide. The disadvantage of a semiconductor laser is that the beam spread is about 10°, hence our attention to focusing in Section 4.4.

Detecting the signal

Our journey around the working of a CD player is almost complete. There is just space for a word or two about the light detector. The digital signal is going to need electronic decoding and processing before the final sound is heard. All this requires a transfer of energy from a light to an electrical signal. A suitable component is a photodiode (Figure 4.65). When a photon hits it surface, the energy is absorbed by an electron, which moves, i.e. it contributes to an electric current.

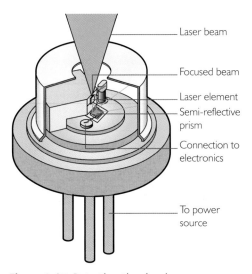

Figure 4.65 Detecting the signal

5.4 Summing up Part 5

This part of the chapter and Part 4 have covered many of the bits and pieces contained within a compact disc player. To summarise what you have learnt, it is a good idea to bring all the parts together. Activities 33 to 35 and the questions that follow are designed to help you do this.

Activity 33 Summing up Part 5

Skim through Part 5 of this chapter and make sure that your notes include a clear definition or explanation of each of the terms printed in bold type. Look back at the article about CDs *The Sound of Science* in Section 4.1 and see how it relates to what you have been learning.

Activity 34 How it works

Imagine you are helping to put together a page of the 'How it works' type for a newspaper or magazine. The page is on the CD player. Decide who your imaginary readership might be. The page is to consist of a large diagram (similar to Figure 4.66) showing the path of light through it, with plenty of explanatory labels that give a brief outline of the physics behind each of the components on your diagram.

A moment or two spent planning the final appearance of your layout can save you time in the end.

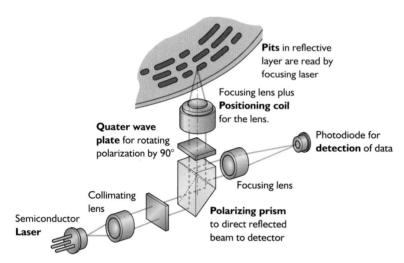

Figure 4.66 Diagram for Activity 34

Activity 35 A model CD player

Set up a large-scale model of the optical system of a CD player, using lenses and blocks like those in Activities 25, 27 and 28, that could be displayed at an open day or parents' evening. Prepare a brief explanation of your model (either written or spoken) suitable for GCSE students or other visitors.

Further investigations

In this part of the chapter, you have measured the refractive index of some materials using a beam of white light. You have also seen that white light consists of a mixture of colours and that colour is related to the frequency (and wavelength) of light. For most materials, refractive index depends on wavelength – which is why a prism can act as a spectrum analyser.

If you have an opportunity, you could measure the refractive index for different colours of light and compare various transparent materials to see how good they are at dispersing light (i.e. separating it into components of different wavelength). You could relate your findings to information about materials that are chosen for particular purposes because of their dispersive properties.

Question

57 Bicycle reflectors are made from sheets of small plastic prisms that reflect light (e.g. from car headlamps) back towards the source (Figure 4.67(a)).

(a) Using the terms *critical angle*, *total internal reflection* and *refractive index*, explain what happens to the light in Figure 4.67(a) after it enters the prism.

(b) Figure 4.67(b) shows a light ray entering the same prism at a different angle.

    (i) Using values from Figure 4.67, find the refractive index of the plastic.

    (ii) By calculating the critical angle, decide whether this prism would act as a good reflector when light enters as shown in Figure 4.67(b).

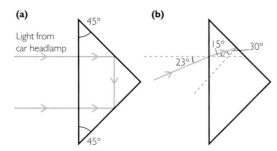

Figure 4.67 Using a prism as a reflector (a) 'square-on' and (b) when light enters at an angle

6 Encore

In this chapter you have studied several aspects of the behaviour of sound and light. This concluding section is intended to help you to look back over the whole chapter and consolidate your knowledge and understanding.

6.1 Waves

In studying this unit you have learned some fundamental pieces of physics which all relate to waves. You have been studying sound waves in air, waves on stretched strings, and light waves. Much of what you have learned about waves in this chapter can be applied to other natural and man-made phenomena that can be described and explained in terms of waves – earthquakes, microwave ovens and starlight are just three examples in addition to those you have met in this chapter. But there are also some important differences between the various types of waves.

Activity 36 Waves

Table 4.5 lists properties of waves that you have studied in this chapter. Copy the table. Use ticks and crosses to show which properties apply to which type of wave, and which are common to waves of all types. If you made a large copy of this table you could add brief notes in each box – as has been done in the second row.

| Property | Type of wave | | | |
	sound waves	waves on string	light waves	all
obeys wave equation $v = f\lambda$				
speed depends on material	✓	✓ $v = \sqrt{\dfrac{T}{\mu}}$	✓ ref. index $\mu = \dfrac{V_1}{V_2} = \dfrac{\sin i}{\sin r}$	✓
can travel in a vacuum				
transverse				
longitudinal				
undergoes reflection				
phase change of π when reflected at 'hard' boundary				
undergoes superposition				

Table 4.5 Summary of wave properties for Activity 36

6.2 Questions on the whole chapter

58 In the manufacture of microchips, sizes and positions have to be measured very precisely. This may be done using a laser, as shown schematically in Figure 4.68.

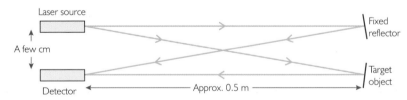

Figure 4.68 Measuring small changes in position

The detector measures the intensity of light produced by the superposition of the two light beams. Small changes in the intensity indicate small changes in position of the 'target' object. Using light of wavelength 600 nm, changes in position as small as 0.15 nm can be measured.

(a) Figure 4.69 shows two waves arriving at the detector in phase. Suppose the 'target' is moved away from the source through 150 nm. How much further, approximately, does the light beam now have to travel via the target to reach the detector?

Via reflector

Via target

Resultant signal

Figure 4.69 Waves arriving at a detector

(b) Draw a diagram to show the two waves that now arrive at the detector, and the resultant signal.

59 Ruari and Rachel are talking about different kinds of waves. Ruari is certain that both sound and light from a spaceship travel through outer space so that you can hear and see the engine. Rachel says that you only get sound effects in *Star Wars* films. What do you think?

60 This question is about the vivid colours seen in the wings of many moths, butterflies and birds. These so-called iridescent colours are produced by superposition when light is partially reflected and partially refracted by one or more thin layers of cuticle (scaly material) and then recombines as shown in Figure 4.70.

Figure 4.71 shows the light paths through a single cuticle layer. If light of a particular wavelength undergoes constructive superposition, when beams 2 and 3 recombine, then the cuticle layer appears to 'shine' with light of that colour.

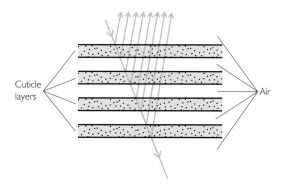

Cuticle layers

Air

Figure 4.70 Reflection and refraction by thin layers

Suppose that light is incident on the cuticle at an angle of $\theta = 0°$. In order for constructive superposition to occur, the wavelength of the light in air λ_a, and the cuticle thickness, d, must be related by the expression:

$$2d = \frac{\left(n + \frac{1}{2}\right)\lambda_a}{\mu}$$

where n is any whole number (0, 1, 2, …) and μ is the refractive index of the cuticle. The following questions show how this expression comes about.

(a) (i) Explain what is meant by *constructive superposition*.

(ii) How must the *phases* of beams 2 and 3 be related if constructive superposition is to take place when they recombine?

(b) Light reflected at A, the upper surface of the cuticle, *undergoes a phase change of 180° or π radians*. Explain what is meant by this phrase in italics.

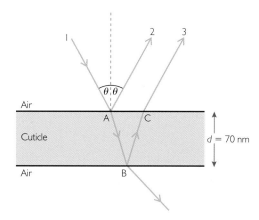

Figure 4.71 Light paths through a single cuticle layer

(c) In order for beams 2 and 3 to superpose constructively when $\theta = 0°$, what must be the relationship between the wavelength, λ, and the thickness, d?

(d) The expression in (c) must involve the wavelength of light in the cuticle. If the cuticle has a refractive index of μ, show how the wavelength in the cuticle, λ_c is related to the wavelength, λ_a, of the same light in air.

(e) When $n = 0$, the expression given earlier in the question is satisfied by one particular colour of visible light. The cuticles in a Urania moth wing have a refractive index $\mu = 2.45$. Given that visible light has a wavelength range from about 400 nm (violet) to 700 nm (red), what will be the colour of Urania moth wing when viewed at $\theta = 0°$?

61 In 1809 Étienne-Louis Malus (1775–1812) discovered that light can be partially or completely polarised by reflection from a shiny surface.

(a) Explain how you could use polarising filter to demonstrate that light has become polarised on reflection.

(b) David Brewster (1781–1868) discovered that totally polarised light can be obtained on reflection when the refractive index μ of the reflecting material equals the tangent of the angle of incidence of the light, as shown in Figure 4.72. (The angle at which this happens is known as the Brewster angle.)

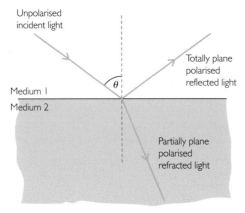

Figure 4.72 Obtaining polarised light by reflection

  (i) If the refractive index from air to glass is 1.5, what must be the angle of incidence in order to produce totally plane polarised light on reflection?

  (ii) What would be the angle of reflection of this polarised light from the glass surface?

(c) When light is polarised by reflection from glass, only about 8% of the light is reflected; the rest enters the glass. Suggest and explain how, using glass reflective surfaces *only*, you might boost this percentage.

6.3 Achievements

Now you have studied this chapter you should be able to achieve the outcomes listed in Table 4.6.

Table 4.6 Achievements for the chapter *The Sound of Music*

	Statement from examination specification	Section(s) in this chapter
28	understand and use the terms *amplitude, frequency, period, speed* and *wavelength*	1.2, 1.3, 2.1
29	identify the different regions of the *electromagnetic spectrum* and describe some of their applications	4.2 (and see SPC + DIG)
30	use the wave equation $v = f\lambda$	1.4, 2.1, 4.3
31	recall that a sound wave is a longitudinal wave which can be described in terms of the displacement of molecules	1.4
32	use graphs to represent *transverse* and *longitudinal* waves, including standing waves	1.4, 1.5
33	explain and use the concepts of *wavefront, coherence, path difference, superposition* and *phase*	1.3, 1.5, 2.1, 3.1, 4.2
34	recognise and use the relationship between *phase difference* and *path difference*	4.2
35	explain what is meant by a *standing (stationary) wave*, investigate how such a wave is formed, and identify *nodes* and *antinodes*	1.5, 2.1, 2.2
36	recognise and use the expression for *refractive index* $_1\mu_2 = \sin i/\sin r = v_1/v_2$, determine refractive index for a material in the laboratory, and predict whether *total internal reflection* will occur at an interface using critical angle	4.3, 5.1
37	investigate and explain how to measure refractive index	4.3
38	discuss situations that require the accurate determination of refractive index	4.3
39	investigate and explain what is meant by *plane polarised light*	5.2
40	investigate and explain how to measure the rotation of the plane of polarisation	5.2
44	recall that, in general, waves are transmitted and reflected at an interface between media	2.1, 2.2, 5.1
45	explain how different media affect the transmission/reflection of waves travelling from one medium to another	5.1
63	explain how the behaviour of light can be described in terms of waves and *photons*	5.3 (and see SPC)
68	explain atomic line spectra in terms of transitions between discrete *energy levels*	5.3

Answers

1   (a) $f_A = \dfrac{f_B}{2}$ because the period of A is twice that of B.

(b) $f_A = \dfrac{f_C}{3}$ because A has three times the period of C.

2   See Figure 4.73.

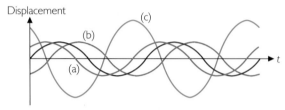

Figure 4.73 The answers to Question 2

3   (a) The phase difference must correspond to a whole number of cycles i.e. by $360n°$ or by $2\pi n$ radians, where n is a whole number.

(b) They are all the same.

$\dfrac{\pi}{8}$ radians = 22.5° so (i) and (ii) are the same.

$\dfrac{17\pi}{8} = 2\pi + \dfrac{\pi}{8}$ so (iii) is the same as (i) and (ii).

$-337.5° = 22.5° - 360°$ so (iv) is also the same as (i), (ii) and (iii).

4   (i) A and B are in phase at times 0 and T.

(ii) A and C are in phase at times 0, $\dfrac{T}{2}$ and T.

(iii) B and C are in phase at times 0 and T.

5   The order in which the feet are placed on the ground is front right, rear left, front left, rear right, front right … So the two front legs move in antiphase with one another, as do the two rear legs. Each front leg is a quarter of a cycle (90°) behind the rear leg on the same side.

6   (a) The motion of the cloth approximates to a travelling transverse wave pulse, with an amplitude and frequency that depends on how it is shaken.

(b) The motion of the trucks approximates to a travelling longitudinal wave pulse.

(c) The Mexican wave is a travelling transverse wave of roughly constant amplitude and speed.

(d) If the traffic is confined to one lane its motion approximates to a series of longitudinal pulses. If lane changing is possible, then there is also some transverse motion.

7   Using Equation 2, wavelength $\lambda = \dfrac{v}{f}$

$$= \dfrac{340 \text{ m s}^{-1}}{20 \text{Hz}}$$

$$= \dfrac{340 \text{ m s}^{-1}}{20 \text{ s}^{-1}}$$

$$= 17 \text{ m}.$$

Similarly, when $f = 20$ kHz, $\lambda = 17 \times 10^{-3}$ m

$$= 17 \text{ mm}.$$

8   (a) (i) $\lambda = 0.8$ m

(ii) $f = \dfrac{v}{\lambda}$

$$= \dfrac{5.0 \text{ m s}^{-1}}{0.8 \text{ m}}$$

$$= 6.25 \text{ s}^{-1} (= 6.25 \text{ Hz}).$$

$$T = \dfrac{1}{f} = \dfrac{1}{6.25 \text{ s}^{-1}} = 0.16 \text{ s}.$$

(b) See Figure 4.74.

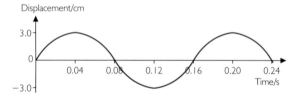

Figure 4.74 The answer to Question 8(b)

9   See Figure 4.75. The period T is 5 ms (5×10^{-3} s), so the wavelength can be found using Equations 1 and 2:

$$\lambda = \dfrac{v}{f} = v \times T$$

$$= 300 \text{ m s}^{-1} \times 0.005 \text{ s}$$

$$= 1.5 \text{ m}.$$

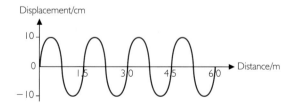

Figure 4.75 The answer to Question 9

10 See Figure 4.76.

(a)

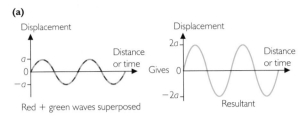

Red + green waves superposed — Gives — Resultant

(b)

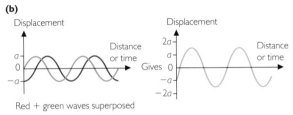

Red + green waves superposed — Gives

(c)

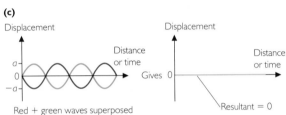

Red + green waves superposed — Gives — Resultant = 0

Figure 4.76 The answers to Question 10

11 The sound must have the same amplitude as the engine noise, but be in antiphase with it.

12 In both case the waves must have the same amplitude and frequency. For (a) they must be in phase and for (b) they must be in antiphase.

13 Distance between nodes = $\frac{\lambda}{2}$.

14 All particles between a given pair of nodes oscillate in phase. They are in antiphase with particles immediately beyond each of those nodes.

15 Sign indicates the direction of displacement, and the molecules oscillate first to one side of their equilibrium position and then to the other.

16 See Figure 4.77.

17 (a) Open tube: there is a displacement antinode at each end, so $l = \frac{\lambda}{2}$.

   (b) Closed tube: the length of the tube is the distance from node to antinode i.e. $l = \frac{\lambda}{4}$.

18 (a) From Equation 2, $f = \frac{v}{\lambda}$. For an open tube, $\lambda = 2l$ (see Question 17) so $f = \frac{v}{2l}$.

   (b) For a closed tube, $\lambda = 4l$ so $f = \frac{v}{4l}$.

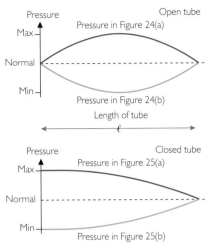

Figure 4.77 The answers to Question 16

19 The fundamental frequency depends primarily on the length of the air column from the tip of the mouthpiece to the nearest open hole. (The width of the tube might also affect the frequency, as would the overall pattern of uncovered holes.)

20 Case (b) gives the lower note because it has a longer air column than (a) – it has a longer uninterrupted row of covered holes.

21 (a) There is a node at each end and an antinode in the middle. See Figure 4.78.

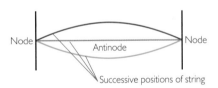

Figure 4.78 The answer to Question 21(a)

   (b) $l = \frac{\lambda}{2}$ so $f = \frac{v}{2l}$. (Compare with the answers to Question 17 about waves in an open tube.)

   (c) We are referring to the transverse waves on the string, because it is these waves that superpose to form the standing waves that define the fundamental frequency of the string's vibration. (The sound waves that are generated in the air have the same frequency as these vibrations, but travel at a different speed and so have a different wavelength.)

22 In an electric guitar, the vibrations are amplified electronically.

23 As the wood vibrates, the tea leaves are set in motion and settle at places where the wood is not moving i.e. at the nodes. The pattern of tea leaves therefore reveals a two-dimensional standing wave (such standing waves are sometimes called Chladni vibrations after the 19th century scientist who first investigated them).

24 (a) Banjo, (b) piccolo, (c) side drum. In each case, the smaller instrument produces the higher-pitched notes, because the standing waves that correspond to its fundamental vibration will be shorter and hence of higher frequency. (We can only compare similar instruments in this way, because there are other factors that affect the speed of the waves set up within the instrument and hence the frequency of the fundamental vibration.)

25 (a) Twisting the screws adjusts the tension in the strings so that they produce the desired notes. The greater the tension, the higher the frequency of vibration of a given string, and so the higher the pitch of its note.

    (b) Placing a finger on the string shortens the length that is free to vibrate. The shorter the string, the higher the frequency of its vibration, and so the higher the note it produces.

    (c) Reducing the mass per unit length of a string increased the frequency of its vibration and hence raises the pitch of its notes. (If all the strings of a piano were the same thickness, they would have to have a very large range of lengths, and the high-note strings would have to be held under much greater tension than the low-note strings. Using strings of different thickness reduces the extremes of length and tension that are needed, making the instrument easier to construct and use.)

26 (a) Substituting values into Equation 5

$$f = \frac{1}{1\ \text{m}} \times \sqrt{\frac{15\ \text{N}}{3.75 \times 10^{-4}\ \text{kg m}^{-1}}}$$

$$= \sqrt{(4 \times 10^4)}\ \text{s}^{-1} = 200\ \text{Hz}.$$

    (b) The frequency is to be doubled, therefore the length of vibrating string must be halved, so the finger must be placed 0.25 m from the end.

27 You can treat the 'voice box' as a hollow cavity, in which standing waves are set up whose wavelengths depend on the size of the cavity. In helium, the frequency of a standing wave of given wavelength is higher than when the person breathes air, indicating that sound travels faster in helium. (This is indeed the case. Sound speed in a gas is governed by the speeds of the molecules. Helium molecules are lighter than the nitrogen and oxygen molecules that make up air, and at a given temperature they move faster.)

28 Time period $T \approx 4.3$ ms, so $f \approx \dfrac{1}{0.0043\ \text{s}} = 233$ Hz.

29 A CD does not easily wear out. Scratches (or small holes), dirt and grease do not affect playback, and the stylus does not wear down the disc surface.

30 (a) Digital

    (b) An analogue method replicates the waveform as a continuous signal, either electrical or mechanical (a stylus oscillating along a record groove). A digital system measures a series of sound samples and converts each value into a binary number.

31 An optical system is one that uses light. If you look up scanning in a large dictionary you get many choices. Our context is an electronic one so 'to move a beam of light in a predetermined pattern over a surface to obtain information' is probably what the author means. In other words, a laser beam will be moved along a spiral path from the centre to the edge of a CD. A stylus or pick-up is used with LPs.

32 (a) 1.6×10^{-6} m. (This could also be written 1.6 µm, as 1 µm = 10^{-6} m.)

    (b) In a perfect world the beam could have a diameter of 1.6×10^{-6} m but in practice it must be smaller to allow for wobbles and imperfections in the focusing of the light.

33 (a) The plastic coating must be transparent so that the light can reach the metallic layer.

    (b) The information is stored in the metallic layer as a spiral track of bumps.

    (c) The important reflection takes place at the metal surface to collect the information codes in its bumps.

34 The beam does not focus at the surface of the plastic but lower down on the metal layer. At the surface it is wide enough so that the amount of light blocked by a speck of dust is not critical (see Figure 4.79).

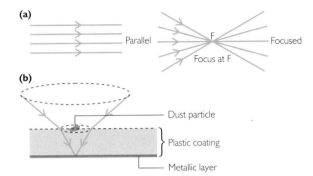

Figure 4.79 Diagram for the answer to Question 34

35 You should have drawn a diagram similar to Figure 4.80 illustrating which angles are equal when light rays change direction at a reflecting boundary. Reflection is not a property of light alone. Sounds, light and balls reflect from surfaces, often with some loss of energy.

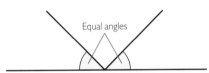

Figure 4.80 Diagram for the answer to Question 35

36 The diode is a photodiode which absorbs light to produce an electrical signal. The tracking components control the movement of the laser light so that it keeps on the spiral track as the disc turns. A prism is a block of transparent material – glass or similar – with a triangular cross-section.

37 All resultant waves have period T, frequency f.

  (a) (i) P amplitude 2.5a, brighter than A or B individually.

     (ii) Q amplitude zero, totally dark.

  (b) R amplitude 0.5a, fainter than either A or B.

  (c) S amplitude 2a, brighter than either individually.

38 (a) Combination Q is 'off'. All the others are 'on'.

  (b) They must have exactly the same frequency and amplitude and be 180° out of phase.

39 Now both Q and R count as 'off'.

40 (a) The smallest path difference between the two beams is half a wavelength. This gives a 180° phase difference.

  (b) It must be zero or a whole number of wavelengths, i.e. $n\lambda$ where n is a whole number.

  (c) 125 nm high since a half wavelength difference is achieved by an outward and return journey to make the 250 nm (half wavelength) difference.

41 If the path difference x between two superposing waves of wavelength λ is $k\lambda$, the phase difference will be k cycles, or $2\pi k$ radians, or $360k$ degrees. In terms of x, the phase difference is $\dfrac{2\pi x}{\lambda}$ radians or $\dfrac{360x}{\lambda}$ degrees.

42 (a) Material 1 is air, material 2 is glass. The turn towards the normal indicates that the second material is the denser (i greater than r).

  (b) The ray would turn away from the normal as it emerged (r is greater than i). The path is the exact reverse of the one in the diagram. This is an example of the so-called 'reversibility' of light. If you plot the path of light going one way through a block it will trace the identical path if the direction is reversed.

  (c) It continues in the same direction.

  (d) It gets larger.

43 A: $\sin 40° = 0.643$

      $\dfrac{0.643}{\sin r} = 1.33$

      $1.33 \sin r = 0.643$

      $\sin r = \dfrac{0.643}{1.33}$

      $\sin^{-1} 0.483 = 28.9° \approx 29°$

   B = 47.3°

   C = 1.53

44 (a) Depends on your own results

(b) $\mu = \dfrac{\text{speed in air}}{\text{speed in material}}$ (from Equation 8) so:

speed in material $= \dfrac{\text{speed in air}}{\mu}$

speed in glass $= \dfrac{3.00 \times 10^8 \text{ m s}^{-1}}{1.47}$

$= 2.04 \times 10^8 \text{ m s}^{-1}$.

Similarly:

speed in water $= 2.26 \times 10^8 \text{ m s}^{-1}$

speed in polystyrene $= 1.88 \times 10^8 \text{ m s}^{-1}$.

45 See Figure 4.81.

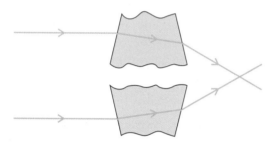

Figure 4.81 The answer to Question 45

46 The focal length $f = 2$ mm $= 2 \times 10^{-3}$ m.

47 (a) From Equation 9, with subscripts a and p denoting air and plastic,

$\lambda_p = \dfrac{\lambda_a}{{}_a\mu_p}$

$= \dfrac{405 \text{ nm}}{1.5}$

$= 270$ nm.

(b) Bump height $= \dfrac{\lambda_p}{4} = 67.5$ nm.

48 In Activity 25 you should have derived the relationship $\sin C = \dfrac{1}{\mu}$.

glass fibre: $\sin C = \dfrac{1}{1.55} = 0.645$

$C = \sin^{-1} 0.645 = 40.2°$.

Similarly for Perspex, $C = 42.5°$; water, $C = 48.8°$; flint glass, $C = 38.4°$.

TIR will occur when $42° > C$, which is the case for both types of glass. For water and Perspex, internal reflection will not be total and so the beam will split.

49 See Figure 4.82.

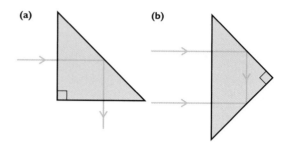

Figure 4.82 The answers to Question 49

50 (a) Specific rotation $= \dfrac{\theta}{cL}$

$= \dfrac{33°}{(0.5 \text{ g ml}^{-1} \times 1 \text{ dm})}$

$= 66°$.

Note: the specific rotation is, by convention, expressed in degrees alone although it may appear more sensible to have used ° ml g^{-1} dm^{-1}.

(b) Measure the rotation angle for a much smaller light path and multiply by a suitable factor, (say 1 cm and multiply the rotation angle by 10). For even greater rotations light paths of millimetres could be used.

51 (a) See Table 4.7.

Colour	Wavelength λ / nm	Frequency f / Hz	Photon energy E / J
infrared	775	3.87×10^{14}	2.57×10^{-19}
red	656	4.57×10^{14}	3.03×10^{-19}
green	486	6.17×10^{14}	4.09×10^{-19}
blue	434	6.91×10^{14}	4.58×10^{-19}
violet	410	7.32×10^{14}	4.84×10^{-19}
ultraviolet	389	7.71×10^{14}	5.11×10^{-19}

Table 4.7 The answers to Question 51(a)

(b) Photons of ultraviolet radiation have more energy than those of visible or infrared radiation, so they can cause ionization and do more damage. Labels on barrier creams refer to their ability to absorb ultraviolet radiation.

52 (a) $f = \dfrac{c}{\lambda}$

$= \dfrac{3.00 \times 10^8 \text{ m s}^{-1}}{775 \times 10^{-9} \text{ m}}$

$= 3.87 \times 10^{14}$ Hz.

(b) Energy of single photon

$E = hf$

$= 6.63 \times 10^{-34} \text{ J s} \times 3.87 \times 10^{14} \text{ Hz}$

$= 2.56 \times 10^{-19}$ J.

$\left\{ \begin{array}{c} \text{Number of} \\ \text{photons} \\ \text{emitted per} \\ \text{second} \end{array} \right\} = \left\{ \dfrac{\text{total energy emitted}}{\text{per second}}{\text{energy of each photon}} \right\}$

$= \dfrac{0.2 \times 10^{-3} \text{ J s}^{-1}}{2.56 \times 10^{-19} \text{ J}}$

$= 7.80 \times 10^{14} \text{ s}^{-1}.$

Note that we have included the units at each step, and this gives sensible units for the final answer i.e. a number per second.

(c) The photon energy would be greater, so there would need to be fewer photons per second in order to deliver the same power.

53 Going from energy level 3 to level 2, the energy lost by the electron is

$(1.94 - 1.63) \times 10^{-18} \text{ J} = 3.02 \times 10^{-19} \text{ J}.$

From Table 4.7, this corresponds to the energy of a photon of red light (give or take a slight difference in the third figure due to rounding). Similarly, the transitions from level 4 to 2, 5 to 2 and 6 to 2 correspond, respectively, to the photon energies for the green, blue and violet light that you calculated for Table 4.7.

54 Even the smallest possible transition to the ground state (from level 2 to level 1) involves an electron losing 1.635×10^{-18} J, which is larger than any of the photon energies corresponding to visible light, so this radiation lies in the ultraviolet (or even shorter wavelength) part of the spectrum.

55 (a) Refraction can best be explained using a wave model. A stream of particles will change direction when crossing a boundary, but when their speed is reduced they veer away from the normal, unlike what is observed for light.

(b) Line spectra can only be explained using the particle model.

(c) Many aspects of reflection can be explained using either model, but not the phase change – so the wave model is better.

56 (a) The speaker seems to mean that when a beam of light meets a surface it is absorbed in packets as photons.

(b) When light is emitted or absorbed it seems to behave as a particle, but on its travels it is behaving as a wave.

57 (a) Light undergoes total internal reflection at back surfaces. The angle between the light ray and the normal to the surface (45°) must therefore be greater than the critical angle C. Critical angle is related to the air–Perspex refractive index μ i.e. $\sin C = \dfrac{1}{\mu}$.

(b) (i) Using angles measured from Figure 4.67(b),

$\mu = \dfrac{\sin 23°}{\sin 15°} = 1.5$

(ii) $\sin C = \dfrac{1}{1.5}$ so $C \approx 41°$.

The angle between ray and normal is less than 41° so some light will escape through the back of prism i.e. it will not be a very good reflector.

CHAPTER 5

Technology in space

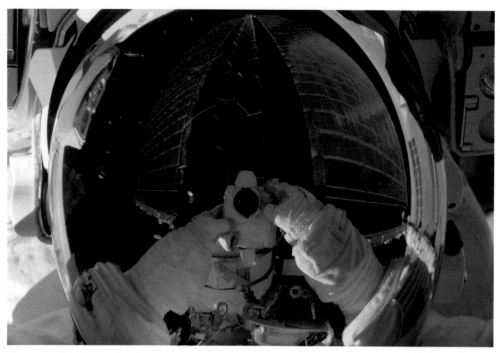

Figure 5.1(a) Space Shuttle Mission STS 117 unfurling the solar arrays for the International Space Station, as reflected in astronaut Suni Williams' visor, June 2007

Why a chapter called *Technology in Space*?

Since the mid 20th century, space technology has played an increasingly important part in our lives, both through satellites orbiting the Earth and through space probes that travel to distant parts of the solar system.

Communications satellites (Figure 5.1) transmit radio signals for telephones and television. Meteorological satellites monitor the atmosphere and the Earth's surface, using remote sensing to provide data for weather forecasting and adding to our understanding of long-term environmental changes. Astronomical satellites provide clearer views of the cosmos than is possible from the ground. Satellites also play a part in locating oil and other mineral deposits and in navigation.

The design, building and operation of a satellite or space probe involves the understanding and application of many areas of physics. In this chapter you will first see how a self-contained renewable electrical power supply can be designed using solar cells, and then how changing temperatures affect the properties of electrical components as a spacecraft moves between sunlight and shadow. Attention then turns to remote sensing, and you will see how reflected signals can provide information about the location and movement of distant objects.

In the second year of this course, you will study two more chapters connected with space and communications. *The Medium is the Message* is about modern telecommunications, and *Reach for the Stars* is about the stars and how we study them.

Overview of physics principles and techniques

In this chapter you will begin by using solar cells and revisiting some ideas about current, voltage and resistance in electric circuits. In part 2 you will meet the idea of internal resistance and you will learn how resistors and power supplies can be combined in circuits to perform particular functions. Part 3 is about the physics behind solar cells, which involves the photon model of light. In part 4, you will see how solar radiation input can be measured and then how a simple model of a conductor can be used to explain changes in its resistance with temperature. Finally in part 5 you will see how pulse-echo techniques and the Doppler effect are used in remote sensing.

In the course of the chapter, you will also be using and developing some key mathematical and ICT skills and techniques – in particular, you will be using algebra and graphs and using a spreadsheet.

Many of the principles and techniques that you meet in this chapter will be picked up and developed further in later chapters. This approach of introducing, revisiting and building on ideas as and when they are relevant to a particular situation is a key feature of this course.

In this chapter you will extend your knowledge of:

- properties of waves from *The Sound of Music*
- the photon model of light from *The Sound of Music*
- properties of materials from *Good Enough to Eat* and *Spare Part Surgery*.

In other chapters you will do further work on:

- DC electric circuits in *Digging Up the Past* and *Transport on Track*
- explanations of why materials behave in certain ways in *Digging Up the Past* and *The Medium is the Message*
- radiation in *The Medium is the Message* and *Reach for the Stars*.

Figure 5.1(b) Solar panels on Skylab

1 Satellites in space

1.1 A space engineer

Jeremy Curtis is an engineer and Business Development Manager for Space Science at the Rutherford Appleton Laboratory (RAL) in Oxfordshire. The RAL provides research and technology development, space test facilities, instrument and mission design, and studies of science and technology requirements for new missions. Much of the Department's work is in collaboration with UK university research groups and a range of institutes around the world, mostly with European Space Agency (ESA) and NASA missions, but also other countries and organisations including Australia, Japan, Morocco, Pakistan, Russia and the European Union (see Figure 5.2).

Jeremy says 'I trained as a mechanical engineer, but I find space engineering exciting because I have to work with all sorts of experts such as astronomers, physicists, designers, programmers and technicians working around the world.' He was sponsored by RAL during his university degree and then spent several years on designs for a very large proton synchrotron (a machine for accelerating protons to very high energies) before moving over to space instrument design. In the following passage, he describes some aspects of space engineering.

Figure 5.2 Jeremy Curtis (centre) working on part of a satellite-borne telescope before it is tested at RAL

Why satellites?

Getting spacecraft into orbit is a very expensive activity with typical launch costs generally measured in tens of thousands of pounds per kilogram. So what makes it worth the bother? There are three key reasons. First, a satellite is at a good vantage point for studying the Earth's surface and atmosphere – just think how many aircraft would be needed to photograph the whole Earth, or how many ships to monitor the temperatures of the oceans. Second, if we want to study most of the radiation coming from distant parts of the universe we have to get above the atmosphere. The Earth's atmosphere absorbs nearly everything that tries to get through it – from X-rays to ultraviolet and from infrared to millimetre waves. Only visible light and radio waves can get through. In fact, even visible light suffers – convection in the Earth's atmosphere makes stars seem to jump about or twinkle, blurring telescope images, so a telescope in space produces sharper images than is possible on Earth.

Finally and not least, a communications satellite can beam TV pictures across the globe and link telephone users on different continents.

The problem with space

Once you've gone through the huge trouble and expense of launching your satellite, a new set of problems confronts you in space.

First, a typical spacecraft may need several kilowatts of power – but where do you plug in? The only convenient renewable source of power is the Sun, so most spacecraft are equipped with panels of solar cells. You can see these on the Infrared Space Observatory (ISO) (Figure 5.3). Unlike on Earth, there is no worry about what to do on cloudy days, but batteries are still needed for periods when the

Figure 5.3 The Infrared Space Observatory was launched in 1995 by the European Space Agency (ESA) and carries several instruments designed and built in the UK

satellite is in the Earth's shadow (usually up to an hour or two per orbit) and the satellite has to be continually steered to keep the panels pointing at the Sun.

So now we have our spacecraft floating in orbit and pointing the same face to the Sun all the time. Although the solar cells provide partial shade from sunlight this surface starts to heat up, and with no air to convect the heat away the temperature can rise dramatically. To add to the difficulties, the other side of the spacecraft faces cold space (at about 3 K or −270 °C) and so begins to cool down. Unchecked, this would distort the structure, wreck the electronics and decompose the materials that make up the spacecraft. So most surfaces of the spacecraft are covered in 'space blanket' – multilayer insulation made of metallised plastic which reflects the radiation away and insulates the spacecraft. This is the crinkly shiny material you can see in the photo (Figure 5.4).

Figure 5.4 Installing a radiator on the outside of a JET-X, an X-ray telescope launched in 1999

1.2 Studying with satellites

The UoSAT satellites are small, relatively low-cost spacecraft whose purpose is to test and evaluate new systems and space technology and to enable students and amateur scientists to study the near-Earth environment. They are designed and built by the University of Surrey Spacecraft Engineering Research Unit. Figure 5.5 shows UoSAT 2, also known as Oscar 11. Its sensors record the local magnetic field, providing information about solar and geomagnetic disturbances and their effects on radio communications at various frequencies. Instruments on board also measure some sixty items relating to the satellite's operation. These include: the temperatures of its faces, its batteries and other electronic devices; the current provided by its solar arrays, and the battery voltages. It can also receive, store and transmit messages to simple radio receivers anywhere in the world. UoSAT's orbit takes it over both poles at a height of about 650 km about the surface, and the spinning of the Earth allows us to receive data six times a day.

Each UoSAT spacecraft is designed to last for several years. Even small spacecraft such as these need electricity to run all the on-board systems, from the computer that controls it all, to the radio transmitters and receivers that send and receive data to and from ground stations on the Earth's surface.

UoSATs are small, each with a mass of typically 50 kg and about 0.5 m across. For comparison, JET-X (Figure 5.4) is about 540 kg in mass and about 4.5 m long. Communications satellites are larger still, with masses of typically 2 to 5 tonnes. At the top end of the scale is the International Space Station (ISS) (Figure 5.6) – a cooperative venture between 13 nations, including the United Kingdom. Construction and testing started in 1995 and completion is due in 2010. The completed station will have a mass of about 470 tonnes, measure 110 m from tip to tip of its solar arrays, and have pressurised living and working space for its crew of six almost equal to the passenger space on two 747 jet airliners. It will have a power demand of over 110 kW.

Figure 5.5 UoSAT 2 (the frame at the top of the photo is about 0.5 m across)

Figure 5.6 The partially completed International Space Station in 2007. Notice how the solar panels are orientated to capture maximum sunlight

Activity 1 Finding out about spacecraft

Use the Internet to find out about the history of space flight. How far have we advanced in the last fifty years? Make a time line and mark on the major advances. You might like to start at the NASA website or Astronaughtix, details of which are provided at www.shaplinks.co.uk.

Activity 2 Data from space

Examine some satellite data and see what you can deduce about conditions on board. Depending on the equipment available to you, you might use data stored on a disk, accessed via the Internet, or direct from a satellite.

Look particularly for information about current and voltages in power supplies and about the temperature at various locations in the spacecraft.

Print out a sample of records showing how these measurements vary with time. You will be learning more about these aspects of satellite design and operation as you study this chapter.

1.3 Spacecraft power systems

Figure 5.7 shows the three main elements in a spacecraft power system. The *primary source* involves the use of a fuel to produce electrical power. Primary sources include fuel cells in which a chemical reaction between hydrogen and oxygen produces electricity (with drinking water as a useful by-product), and radioisotope thermoelectric generators (RTGs) in which a radioactive decay process produces heating in a thermoelectric module that generates electricity. In spacecraft, the most common primary source is the photovoltaic cell, powered by the solar radiation: here the initial fuel is protons in the Sun, which undergo nuclear fusion.

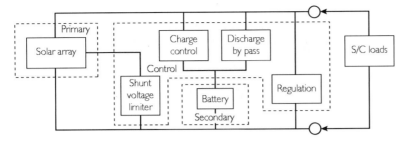

Figure 5.7 Schematic diagram of a spacecraft power system

The *secondary source* is the energy storage system – usually a set of batteries. Sometimes regenerative fuel cells are used in which power from solar arrays electrolyses water to produce hydrogen and oxygen gases during the 'charge' cycle followed by hydrogen and oxygen recombining to make water during the 'discharge' cycle.

An electronic *power control and distribution system* controls and adjusts the voltage and current inputs and outputs, often using primary and secondary sources together to boost the overall output power.

There are other systems available and these are shown in Figure 5.8. Information on these can be found via the (manual link to Heavens Above Site, which give times when ISS can be seen in the sky) Internet using the NASA Spacelink page, or the Heavens Above Site, details of which an be found at www.shaplinks.co.uk.

Question

1 Using Figure 5.8, decide which would be the most suitable power source(s) for a spacecraft needing:

(a) 1 kW power output for just one week

(b) 10 kW for five years.

The most common primary source used in satellites is the

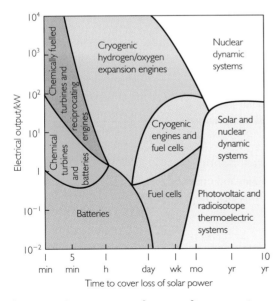

Figure 5.8 Power outputs of spacecraft power systems

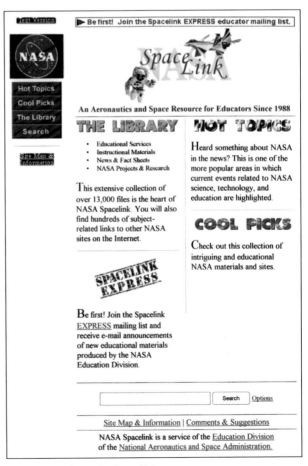

Figure 5.9 The NASA Spacelink page

photovoltaic cell or solar cell. Hundreds of many thousands of such cells are connected together to make up solar arrays. In Figure 5.5 and 5.6 you can see the arrays of solar cells on UoSAT 2 and the ISS.

Solar cells have one important characteristic: they only generate electricity when illuminated. Orbiting satellites undergo between 90 and 5500 eclipses, moving into the Earth's shadow, each year (Figure 5.10). The former is typical of a geostationary telecommunications satellite, the latter of a satellite in a low orbit like UoSAT 2. The ISS has sixteen 30 minute periods of shadow a day. The secondary power supply is therefore vital, because during eclipses electrical power has to be supplied by batteries. There are also occasions when the batteries are needed to provide power in addition to that of the solar panels.

The spacecraft's solar panels are used to recharge its batteries when it emerges into

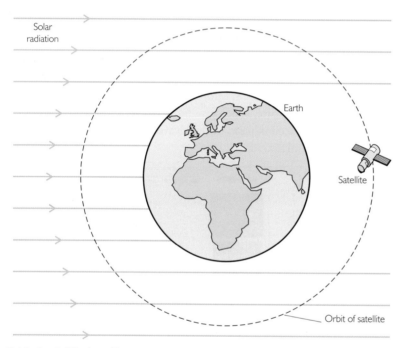

Figure 5.10 A satellite in eclipse

sunlight. To do this they must produce a high enough voltage – higher than the battery's own voltage. (A charger for a 12 V car battery provides about 13 V.) The power system must therefore be carefully designed to ensure that the solar panels can charge the batteries and that the batteries can operate the electrical equipment on board.

So what voltage does a solar cell provide? How does this voltage vary with the brightness of the light? How can we connect up solar cells in order to charge batteries and operate equipment? These are questions that you will be exploring in Part 2 of this chapter.

> **Study note**
>
> A geostationary satellite is one whose orbit keeps it always above the same spot on the Earth's surface. Another term used is geosynchronous satellite.

2 Solar cells and electric circuits

In this part of the chapter you will learn about solar power supplies and electric circuits in order to see what is involved in designing a power supply for a spacecraft. The things that you learn about circuits can be applied to electrical systems operated by other types of power supply – not just to those with solar cells.

2.1 Solar cells

Activity 3 A first look at solar cells

In this short activity, your task is to measure the voltage produced by a solar cell under various conditions of illumination.

Connect a voltmeter across just one of the solar cells as shown in Figure 5.11. Observe how the voltage across the solar cell changes as you vary the separation between the cell and the lamp. Write a sentence summarising what you have observed.

Figure 5.11 A solar cell connected to a voltmeter and illuminated

You should have found that even the highest voltage generated was quite small, so, to charge up the on-board batteries, large numbers of solar cells must be connected together.

Activity 4 Solar array voltage

Have a look at some satellite data and see how the solar array voltage varies with time. Think how these variations might be explained in terms of the illumination of the satellite.

How does a solar cell work?

A solar cell (technically, a **photovoltaic cell**) is an electrical power supply. Incoming radiation provides energy that is transferred to an electric circuit via the motion of charged particles. Figure 5.12 shows a schematic diagram of a photovoltaic cell. The

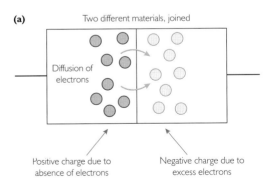

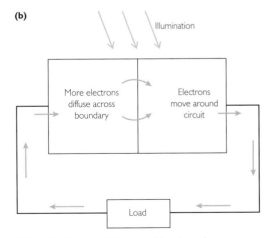

Figure 5.12 Schematic diagrams of a photovoltaic cell (a) in the absence of illumination and (b) when illuminated

199

two materials are designed so that electrons spontaneously drift from one to the other, giving one a negative charge and the other a positive charge until drifting is halted by the build up of electric charge. When radiation is absorbed by the cell, some of the electrons gain enough energy to move freely. Some move back across the boundary but most move around the external circuit and are replaced by more electrons drifting across the boundary.

Study note

In Part 3 of this chapter you will learn more about the physics behind solar cells.

Current and charge, voltage and energy

The continuous flow of charged particles constitutes an electric **current**. Current is defined as the rate of flow of charge past a point in the circuit:

current (I) = charge (ΔQ) ÷ time interval (Δt)   (1)

The SI unit of charge is the coulomb (C) and the unit of current is the ampere or amp (A):

$1\,A = 1\,C\,s^{-1}$

If an amount of charge ΔQ flows past a point in a time interval Δt, then (adapting Equation 1) the current I can be written:

$$I = \frac{\Delta Q}{\Delta t}$$ (1a)

Maths reference

Index notation
See Maths note 1.1

Index notation and units
See Maths note 2.2

Charge is not created or destroyed (it is **conserved**) and there is no build up of charge anywhere in a circuit, so the rate at which charge flows towards any point in the circuit must be the same as the rate at which it flows away. So in a circuit where all components are joined in **series** (Figure 5.13(a)) the current is the same throughout, and where components are joined in **parallel** (Figure 5.13(b)) the sum of currents flowing into a junction is equal to the sum of currents flowing out of it:

$I = I_1 + I_2 + I_3$ (2)

Maths reference

The delta symbol
See Maths note 0.2

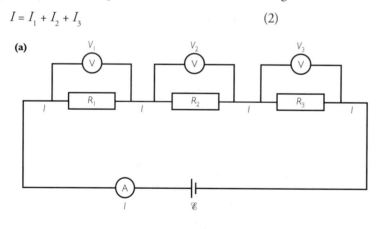

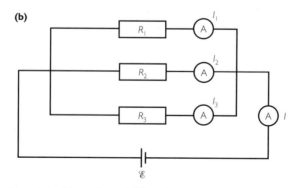

Figure 5.13 (a) Resistors joined in series and (b) resistors joined in parallel

Energy transfers in electric circuits are often expressed in terms of energy per unit charge. The SI unit of energy is the joule (J) and the unit for energy per unit charge is the volt (V): $1V = 1 J C^{-1}$. Energy per unit charge is more usually called **potential difference** (pd) or **voltage** and given the symbol V. Using the symbol W for energy:

$$W = QV \qquad\qquad (3)$$

The **emf** of a power supply is a measure of the total energy that it supplies to each coulomb of charge. Energy cannot be created or destroyed (it is conserved) so the energy transferred to the components in a circuit must be equal to the energy transferred from the power supply. In the time that it takes for one coulomb to pass any point in the circuit, the energy transferred by the power supply is numerically equal to its emf $\mathscr{E}$, and the energy transferred in each resistor is equal to the pd, V, across it. So the sum of all the pds across all the components in a series circuit (Figure 5.13(a)) is equal to the emf of the supply:

$$\mathscr{E} = V_1 + V_2 + V_3 \qquad\qquad (4)$$

And when components are connected in parallel (Figure 5.13(b)) each has the same potential difference across it:

$$V_1 = V_2 = V_3 = \mathscr{E} \qquad\qquad (5)$$

> **Study note**
>
> Emf stands for electromotive force, but that's rather a misleading term since emf is a measure of energy not a measure of force.

Question

2　A torch battery (chemical cell) has an emf $\mathscr{E} = 1.5V$. When it is connected to a bulb, there is a current of 0.5 A in the circuit.

(a) How much charge passed each point in the circuit during a time interval $\Delta t = 2.0s$?

(b) Use delta notation to write an expression relating the energy ΔW transferred by the cell to a small amount of charge ΔQ.

(c) How much energy does the cell transfer to this amount of charge?

In parts (a) and (c), make sure you show how to manipulate the units as well as the numbers.

Voltages by design

In designing power supplies for spacecraft, it is not sensible for each to be a one-off. Rather, systems are designed using agreed standards so that items of equipment can be used 'off the shelf' rather than each having to be purpose built. In science and engineering the term 'standard' is used to refer to a particular design specification.

In the early days of space exploration the spacecraft of the USA and the former Soviet Union (USSR) were very different from each other. However, when it was considered a good idea to be able to dock spacecraft with each other (Figure 5.14), the two countries agreed to have identical docking ports and laid down standards, or specifications, for them.

Figure 5.14 The first international docking of spacecraft, in 1975. The Apollo Soyuz test docking system is shown attached to the final Apollo spacecraft

Large organisations such as NASA (National Aeronautics and Space Agency), Energia (the Russian space agency) and ESA (European Space Agency), which are responsible for the design and construction of spacecraft, lay down standards for equipment design. Spacecraft have a lot of electrical equipment on board and these require specific voltages to operate correctly. To ensure that the required voltages are available, the designers specify a type of circuit, known as a **bus** (Figure 5.15), into which equipment can be connected. It is much like the ring main that connects the main sockets in houses, or the wiring harness of a car. In Europe the bus provides voltages of 28 V or 50 V. NASA in the USA has buses supplying voltages in the range 21 V to 35 V.

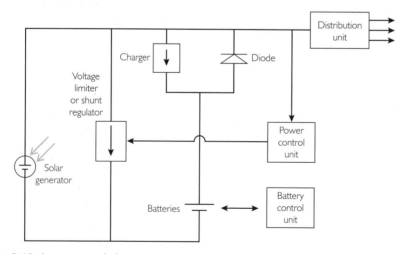

Figure 5.15 A power supply bus

Activity 5 Joining solar cells

The bus on an ESA spacecraft needs a voltage far greater than that of a single solar cell. Your task for this activity is to see how solar cells can be joined together to provide high enough voltages. Notice that the circuit symbol for a solar cell (Figure 5.16) is like a dry cell with arrows indicating illumination.

Questions

3 If a single cell, under specified lighting conditions, generated a voltage of 0.5 V, what voltage would be produced for each of the circuits shown in Figure 5.16?

4 What is the smallest number of solar cells that would be needed to supply 28 V if a single cell had an output of 0.5 V?

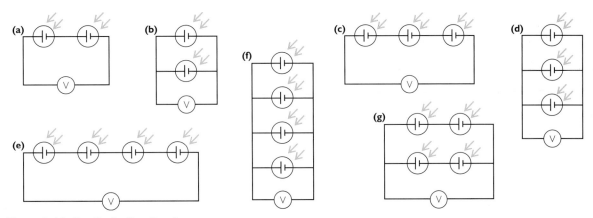

Figure 5.16 Circuits for Question 3

2.2 Cells and circuits

You have seen that instruments on board satellites are designed to use a standard voltage, and you have seen how to join cells in series and in parallel to produce different voltages. However, there is more to designing a satellite's electrical system than just arranging cells to provide a standard voltage and then connecting up the instruments. The following demonstration illustrates the problem.

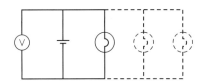

Figure 5.17 Connecting a load to a power supply

Activity 6 Connecting a load to a power supply

The voltmeter in Figure 5.17 measures the **terminal potential difference** – the pd between the terminals of the power supply. The purpose of this activity is to show what happens to the voltmeter reading and/or to the brightness of the lamps when first one lamp is connected and then more lamps are added in parallel.

Study note

The general term 'load' could mean a lamp, motor or heater or something more complex such as a measuring instrument or a radio transmitter.

Activity 6 raises some important questions that need to be answered in order to design any electrical system:

- Why does the terminal potential difference change when the supply is connected in a circuit?
- How is that change related to the external load?
- What happens when different loads, or combinations of loads, are connected to the supply?

The answers to all these questions relate to electrical resistance, which is discussed in some detail below.

Electrical resistance

The current in an electrical device connected to a power supply (Figure 5.18) depends on the potential difference (voltage) applied between its terminals and on its own internal properties. The more easily charge can flow within it, the lower its **resistance** and the greater the current will be for a given voltage. Resistance is defined by the resistance equation:

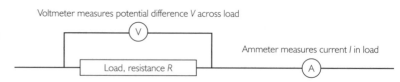

Voltmeter measures potential difference V across load

Ammeter measures current I in load

Load, resistance R

Figure 5.18 Defining resistance

resistance (R) = potential difference (V) ÷ current (I) (6)

$$R = \frac{V}{I} \qquad (6a)$$

The SI unit of resistance is the ohm (Ω); $1\ \Omega = 1\ \text{V A}^{-1}$.

A device (or a material) whose resistance remains constant when measured under constant physical conditions (e.g. constant temperature) over a wide range of voltages is said to obey **Ohm's law** and is often called an **ohmic** device, an ohmic conductor or an ohmic material. Another way to describe ohmic behaviour is to say that potential difference is directly proportional to current under constant physical conditions, i.e. the graph of pd against current is a straight line through the origin (Figure 5.19).

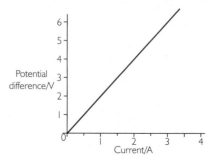

Figure 5.19 A current–voltage graph for an ohmic conductor

Questions

5   What is the resistance of the conductor in Figure 5.19?

6   Which of the graphs in Figure 5.20 show ohmic behaviour?

7   How can you compare the resistances of two ohmic conductors just by looking at their current–voltage graphs plotted on the same axes (without doing any calculations)?

8   For each of the graphs in Figure 5.20 that does not show ohmic behaviour, describe in words what happens to the resistance as the current is increased.

> **Maths reference**
>
> Graphs and proportionality
> See Maths note 5.1

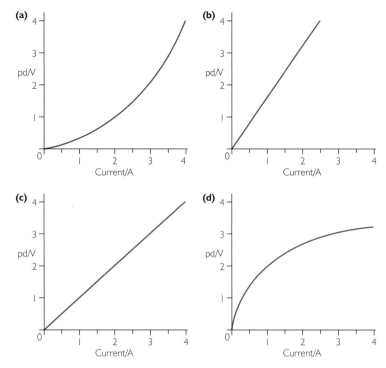

Figure 5.20 Current–voltage graphs for various conductors

Deciding whether a sample obeys Ohm's law

How can you tell, from a set of experimental measurements, whether a conductor obeys Ohm's law? The simple way is to plot a graph of voltage and current and see whether you can join the points with a straight line through the origin. Since experimental measurements on an ohmic conductor will rarely lie exactly in a straight line, you need to take account of **experimental uncertainty**: plot **error bars** on the points and use them to draw **error boxes**. If you can draw a straight line through the origin that passes through all the error boxes, then you can say that the conductor obeys Ohm's law *within the limits of experimental uncertainty*.

Figure 5.21(a) shows some voltage and current measurements plotted without error bars. In this particular example, the uncertainty in the voltage was $\Delta V = \pm 0.5$ V, and that in the current was $\Delta I = \pm 0.25$ A. When these (rather large) uncertainties are plotted in Figure 5.21(b), you see that the material does seem to obey Ohm's law within the limits of the experimental uncertainty; but if you only had Figure 5.21(a) it would be impossible to say.

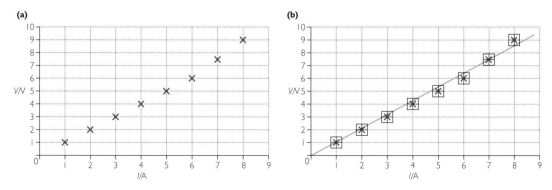

Figure 5.21 A plot of some experimental measurements of current sond voltage (a) without error boxes and (b) with error boxes

Question

9 By plotting a graph, decide whether the measurements in Table 5.1 obey Ohm's law within the limits of the experimental uncertainty.

Current (I/A)	Uncertainty $\Delta I = \pm 0.1$ A	Potential difference (V/V)	Uncertainty $\Delta V = \pm 0.1$ V
0.5		0.8	
1.0		1.4	
1.5		1.8	
2.0		2.0	
2.5		2.2	

Table 5.1 Data for Question 9

Combinations of resistors

To answer the questions raised in the points listed below Activity 6, which relate to the designing of a circuit for a particular purpose, we need to think what happens when there is more than one resistor in a circuit. Resistors can essentially be combined in two ways. Figure 5.22 shows some resistors connected in **series** and in **parallel**.

Any combination of resistors can be replaced by a single resistor without changing the currents and potential differences in the rest of the circuit. You can show, using Equations 4 and 6, that the net resistance of a *series* of resistors is equal to the sum of their separate resistances:

$$R = R_1 + R_2 + R_3 \qquad\qquad (7)$$

You can also show, using Equations 2, 5 and 6 that, to find the net resistance R of several resistors *in parallel*, you need to add their reciprocals:

$$\frac{1}{R} = \frac{1}{R_1} + \frac{1}{R_2} + \frac{1}{R_3} \qquad\qquad (8)$$

The net resistance of several resistors in parallel is always less than the smallest individual resistance – use that to check that your answer is reasonable.

Maths reference

Reciprocals
See Maths note 3.3

Activity 7 Resistors

Use circuit boards or the *Crocodile Clips* circuit simulator to review your knowledge of resistance.

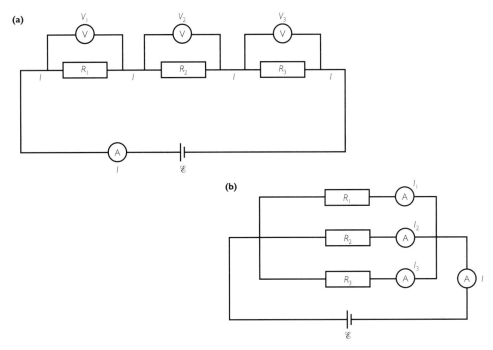

Figure 5.22 Resistors joined (a) in series and (b) in parallel

Question

10 Calculate the net resistance of each of the arrangements shown in Figure 5.23. Show your working and include units at each step. (When there is a mixture of series and parallel arrangements, you will need to break down the calculations into stages. Begin by looking for groups of resistors that you can easily replace by a single resistor.)

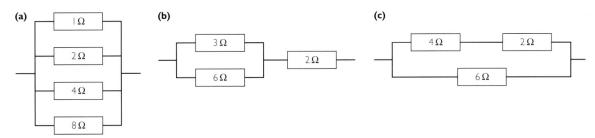

Figure 5.23 Combinations of resistors for Question 10

Internal resistance

The drop in terminal potential difference in Activity 6 can be explained using the idea of **internal resistance**. Any power supply (a photovoltaic cell, a chemical cell, a dynamo) has some electrical resistance due to the materials from which it is made. This resistance cannot be removed from the power supply, but it is normally shown in circuit diagrams as a resistance r in series with the power supply of emf $\mathscr{E}$ (Figure 5.24). Often a dotted line is drawn round $\mathscr{E}$ and r to indicate that they are inseparable.

When the power supply is connected to an external load R, there is a current I throughout the circuit – in r as well as in R. Some energy is transferred in the

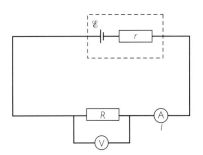

Figure 5.24 Internal resistance

internal resistance of the power supply (it gets warm) as well as in the external load, so the potential difference across the external load (the terminal pd) is less than the full emf of the power supply.

We can apply the resistance equation (Equation 6) to the whole circuit:

$$\mathscr{E} = IR + Ir \qquad (9)$$

and to the external load:

$$V = IR \qquad \text{(Equation 6)}$$

so:

$$V = \mathscr{E} - Ir \qquad (9a)$$

The quantity Ir is sometimes called the **lost volts**.

In your GCSE course, you probably treated the terminal pd of a power supply as being fixed. You have now seen that that is not the case. If you are designing a circuit that uses a real power supply rather than an ideal one, you need to take account of the internal resistance.

Activity 8 How do real power supplies behave?

Your task is to explore the behaviour of a power supply that has some internal resistance, using a voltmeter, ammeter and various combinations of resistors.

Predict what will happen to the terminal pd and to the current as the external load is varied, and test your prediction.

2.3 Getting the most from your power supply

Space engineers have to design spacecraft power systems that operate as effectively as possible. The following passage discusses what this means.

The output from a power supply depends on what it's connected to, and when we're designing an electrical system for a satellites with a limited energy input (from the Sun), we want to make sure it operates as effectively as possible. This raises questions about what we mean by 'output' and 'operate effectively'. I think we can say straight away that simply looking at the current (Figure 5.25) is not a sensible way to measure performance.

*The current is greatest when there is a **short circuit** between the terminals – a connection of low (essentially zero) resistance. The current is certainly large, but all that happens is that the power supply heats up because of its own internal resistance. Not only is that not useful but overheated power supplies have a nasty habit of catching fire or blowing up.*

*The other extreme – going for the greatest possible terminal pd – is equally useless, though less hazardous. The terminal pd is greatest when the supply is on **open circuit** – meaning that there's an extremely large (essentially infinite) resistance between the terminals – in other words, they are not connected via a circuit at all!*

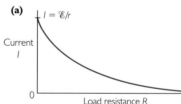

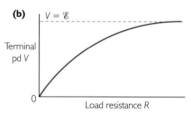

Figure 5.25 (a) Terminal pd and (b) current versus load resistance for a power supply with some internal resistance

No, to get the best performance out of a power supply, we have to look at maximising the energy transfer to an external circuit, and that involves a trade-off between having a large current on the one hand, and a large voltage on the other. It turns out that there is a fairly straightforward way of designing circuits so that they do get the best possible energy transfer from a power supply. You'll come across the terms 'maximum power' and 'impedance matching' in this context – understand those, and you've got it cracked.

Power in electric circuits

Power is the rate at which energy is transferred:

$$\text{power } P = \text{energy transferred} \div \text{time taken} \qquad (10)$$

Power has SI units of joules per second ($J\ s^{-1}$) or watts (W); $1\ W = 1\ J\ s^{-1}$. Using ΔW to represent energy transfer, Equation 10 can be written as:

$$P = \frac{\Delta W}{\Delta t} \qquad (10a)$$

The symbols ΔE, ΔW, ΔQ are all used to represent an energy transfer and you are likely to see all three if you consult other books. ΔW is sometimes used only for electrical energy and ΔQ for thermal energy – but that is not a hard and fast rule.

In an electrical circuit, power is related to current and to potential difference. The current I is the number of coulombs per second flowing past a point in a circuit, and the potential difference V across a load is the number of joules each coulomb transfers to the load. So the rate of energy transfer in the load, the power P, is given by:

$$P = IV \qquad (11)$$

> ### Worked example
>
> For example if $I = 2\ A$ ($= 2\ C\ s^{-1}$) and $V = 3\ V$ ($= 3\ J\ C^{-1}$), in each second 2 coulombs flow into the load, and each coulomb transfers 3 joules to the load. So in each second, 6 joules are transferred:
>
> $$P = 2\ C\ s^{-1} \times 3\ J\ C^{-1} = 6\ J\ s^{-1} = 6\ W$$

Maths reference

Index notation and units
See Maths note 2.2

Algebra and elimination
See Maths note 3.4

If the load has resistance R, we can use the resistance equation (Equation 6) to eliminate either V or I:

$$P = I^2 R \qquad (12)$$

or

$$P = \frac{V^2}{R} \qquad (13)$$

Questions

11 A domestic kettle is marked 230 V, 2000 W. When connected to a 230 volt supply, what is the current in the element?

12 What is the resistance of a light bulb designed to have a power of 60 W when connected to the 230 volt mains supply?

Activity 9 Maximising the power

How can we get maximum power transfer to the external load in a circuit? This activity shows several ways to tackle this question.

Patterns

Calculate the power transferred to various loads by a given power supply, and look for patterns. Try using 'ideal' values that have been made up to give round numbers, and try your own experimental measurements from Activity 8. This may be done on a spreadsheet as shown in Figure 5.26, or you could use a calculator (which would take much longer).

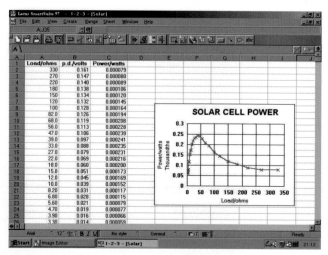

Figure 5.26 Using a spreadsheet to calculate power

Algebra

Starting from $\mathscr{E} = IR + Ir$ and $P = VI$, derive an expression for the power in the load resistance R. Eliminate V and I, and end up with an expression that involves only P, $\mathscr{E}$, R and r.

The condition for P to have a maximum value depends on the relationship between the external load R and the internal resistance r. To make this relationship clearer, write $R = fr$ and cancel as many common factors as you can. You should get an expression that involves just P, $\mathscr{E}$, r and f:

$$P = \frac{\mathscr{E}f}{(1 + f)^2 r} \tag{14}$$

For a particular power supply $\mathscr{E}$ and r remain fixed. Try calculating $f/(1 + f)^2$ for various values of f (0, 0.5, 1, 1.5, 2, …) and see if you can deduce the value of f that gives the greatest value of P. Compare your result with what you found using the spreadsheet.

Impedance matching

The introduction to this section mentioned the term **impedance matching**. Impedance is a more general term than resistance that related current to potential difference – there are some types of electrical device where the relationship between current and pd is more complicated than that described by the resistance equation, particularly when they are connected to an alternating, rather than a direct, supply. However, when dealing with a steady current and pd, impedance can be treated as meaning the same thing as resistance.

Impedance of solar cells

We have been treating the internal resistance (or impedance) of solar cells as though it had a fixed value for any given cell. In fact, the impedance of a solar cell can vary according to conditions, and this needs to be taken into account when designing a power system for a satellite.

Activity 10 Currents and voltages in satellites

Have a look at some satellite data and see how the battery current and voltage change with time. Think how you might explain any variations in these measurements.

Further investigations

When you used solar cells, you ensured (we hope!) that the level of illumination remained the same all the time. If you have an opportunity to carry out more detailed investigative work, you might like to explore what happens to the power output and/or internal resistance of a solar cell as the level of illumination changes. You could try using different light sources, or tilting a cell so that it intercepts different amounts of incoming radiation.

Level of illumination is one factor that could affect the internal resistance of a solar cell. You might be able to suggest other factors and explore them experimentally.

Maximum power – maximum efficiency?

In our work we always try to be efficient and, by that we mean that we make good use of our time and efforts. We also know instinctively what is meant when someone says something is 100% efficient – we know that it cannot be improved. But what does efficiency mean in the context of power supplies?

The **efficiency** of any device or system that transfers energy (such as an electric circuit) has a precisely defined meaning:

$$\text{efficiency} = \frac{\text{energy usefully transferred}}{\text{total energy transferred}} \qquad (15)$$

or:

$$\text{efficiency} = \frac{\text{power usefully transferred}}{\text{total power transferred}} \qquad (16)$$

Efficiency is often expressed as a percentage. For example, if a system is 20% efficient, then 20% (one fifth) of the energy supplied by the power source is transferred usefully in the external load, while the remaining 80% is wasted in heating the power supply and the surroundings.

Maths reference

Fractions and percentages
See Maths note 3.1

Activity 11 Maximum power – maximum efficiency?

In an electrical system, is the condition for maximum efficiency the same as that for maximum power? Spend a few minutes discussing the meaning of these two terms, making sure you can distinguish between them. Then try to decide whether the conditions are the same for both. Think what happens to the power output and the efficiency when the load resistance is much greater than the internal resistance, or much smaller, and when the two are equal. You might find it helpful to invent some numerical examples to illustrate what happens.

Designing an electrical system

In Activity 11, you should have convinced yourself that efficiency is greatest when the external load is made as large as possible, which is not the same as the condition for maximum power output. When designing an electrical system, should you go for maximum power or for maximum efficiency? If you want to get as much as possible from a small source of power, you aim for maximum power transfer and match the external load to the internal resistance. This is what is generally done in power systems for spacecraft. A circuit where an amplifier delivers power to a loudspeaker is another example where matched impedances are used to achieve maximum output power (speakers are generally labelled with their impedance). However, if the powers involved are large, it is better to aim for a higher efficiency and to reduce the amount of energy wasted due to the internal resistance of the supply. For example, electric vehicles are designed so that the internal resistance of the power supply is as small as possible and the external load resistance is much larger in comparison.

Question

13 Calculate the output power and the efficiency of the circuit in Figure 5.27. Explain why this arrangement would not be a sensible way to use the power supply.

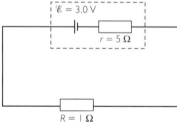

Figure 5.27 Circuit for Question 13

2.4 Summing up Part 2

So far in this chapter, you have reviewed and extended your knowledge of DC electric circuits, learned how cells and resistors can be combined, and learned about the conditions under which a power supply transfers maximum power to a load.

Activity 12 is designed to help you review your progress, and Questions 14 and 15 and Activity 13 are designed for you to reinforce and put into practice what you have been learning.

Activity 12 Summing up Part 2

Spend a few minutes checking through your notes – use the following exercises to help you do this.

Skim through Part 2 of this chapter and make sure that your notes include a clear definition or explanation of each of the terms printed in bold type.

Write a short paragraph, illustrated by a circuit diagram, to explain what is meant by impedance matching and why it is important in designing a power system for a satellite. Include *at least five* of the terms that are printed in bold type in Part 2.

Questions

14 A single 2.0 V cell in a lead–acid battery has an internal resistance, dependent on its state of charge and temperature, of around 0.005 Ω.

(a) If a battery is made up of six cells in series what is:

(i) its total internal resistance

(ii) its emf?

(b) This battery is then connected to a load resistance of 2.97 Ω. What is:

(i) the total resistance of the circuit

(ii) the current in the circuit

(iii) the terminal potential difference?

(c) A car battery is made up of lead–acid cells, and on starting up a car there may be a current of 200 A. If a car's headlamps are on while a car is being started, they are usually seen to dim appreciably. Explain why.

15 Figure 5.28 shows various arrangements of identical cells.

(a) From (i) – (iv), what can you deduce about the way internal resistances combine? (Do they combine like ordinary resistances?)

(b) What load resistance would you use to obtain maximum power transfer from the arrangement of cells in (v)?

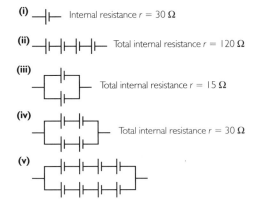

Figure 5.28 Combinations of cells for Question 15

Activity 13 Design challenge

This challenge is concerned with designing a solar power supply in order to achieve maximum power transfer to a given load. (It is not intended as a practical activity.)

A single solar cell, under certain conditions, has an emf of 0.45 V and a maximum power output of 0.1 W. An array of identical solar cells, under the same conditions, is required to supply 0.4 W to an external load of 0.506 Ω. If the cells are required each to supply the maximum possible power, how many cells are needed, and how must they be connected? (Hint: start by finding the internal resistance of a single cell, and assume that the internal resistances of these solar cells combine just like ordinary resistances.) Without doing any further calculations say how many cells would be needed to supply 0.9 W to the same load, and how they must be connected.

3 Solar cells

3.1 Solar cells in space

In the early days of space travel, satellites and spaceships used batteries or fuel cells but these had limited lifetimes and reliability. The two-man Gemini spacecraft in 1966 (Figure 5.29) was the first NASA ship powered by fuel cells. However, the accident that crippled Apollo 13 in 1970 was caused by a faulty connection in the service module's fuel cell system (Figure 5.30).

Figure 5.29 Gemini VI and VII rendezvous in orbit

Figure 5.30 The fuel cell system is hidden beneath the exposed section of the interior in this shot of Apollo 15 in lunar orbit in 1971 (photograph courtesy of NASA)

Space scientists and engineers then turned to solar cells to generate their electricity. The Soyuz spacecraft were the first to use solar cells and have been in operation since 1967. Although there were a number of early setbacks, the design has proved robust and reliable and Soyuz still ferry cosmonauts into orbit. Figure 5.31 shows a modern-day version.

Figure 5.31 A Soyuz spacecraft docking with the ISS in 2005 (photograph courtesy of NASA)

With early solar cell systems, there was still a chance of damage on lift off. Although the incident sunlight can provide a lot of energy for the spaceship, if the solar panels do not deploy properly then problems can follow. Figure 5.32 shows the space-station Skylab in 1973. Note how only one panel is deployed. Astronauts had to spend many hours trying to free the jammed panel in exhausting spacewalks. Note, as well, the improvised heat shield on top of the main module.

Figure 5.32 Solar panels on Skylab

Now solar cells are generated on thin metal sheets that can be rolled up for transport into orbit. NASA is in the process of delivering huge trusses of solar arrays to the International Space Station (ISS), which has to be completed before the retirement of the Space Shuttle in 2010. Each mission will deliver and install a 17.5 tonne truss segment to enhance the electrical system of the orbital station. The International Space Station (ISS) requires huge solar panels just to provide enough electricity for its three crew members. The computer-generated picture in Figure 5.33 shows an Orion craft outward bound for the Moon approaching the completed ISS. Note how similar the Orion is to the Apollo craft in Figure 5.30.

When unfurled, the 80 metre arrays provide power for the station (see Figure 5.34). Each of the 82 active array blankets that are grouped into 31 'bays' contains 16 400 silicon photovoltaic cells to convert sunlight into electricity. The truss also contains a rotary joint that rotates through 360 degrees, positioning the solar arrays to track the Sun.

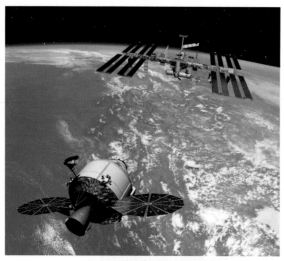

Figure 5.33 Approaching the completed ISS (photograph courtesy of NASA)

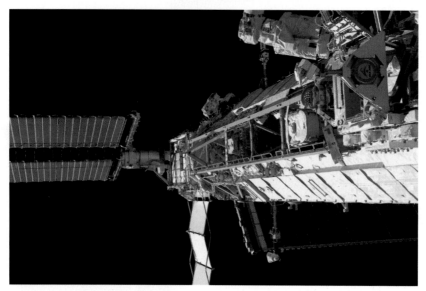

Figure 5.34 The International Space Station (ISS) deploying a solar cell truss in 2007 (photograph courtesy of NASA)

NASA and its contractor Lockheed-Martin developed a method of mounting the solar arrays on a blanket that can be folded like an accordion. The cells are made from purified crystal ingots of silicon that directly absorbs light to provide electricity for immediate use. These also charge up batteries to provide electricity in emergency.

The complete power system, consisting of American and Russian hardware, will generate 2000 kW h (kilowatt–hours) of total energy, about as much as 42 medium houses would typically use in a day.

Each unit is capable of generating nearly 32.8 kilowatts (kW) of direct current power. There are two units on each wing module, yielding a total power generation capacity of nearly 66 kW, enough power to meet the electrical needs of about 30 small houses consuming about 2 kW of power each. This does not come cheap – each array wing costs $370,000,000 (about £200 million).

A NASA official said 'Electrical power is the most critical resource for the station because it allows astronauts to live comfortably, safely operate the station, and perform complex scientific experiments. Since the only readily available source of energy for spacecraft is sunlight, technologies were developed to efficiently convert solar energy to electrical power.'

3.2 Explaining solar cells

In order to produce efficient designs for solar panels, scientists and engineers have worked on developing new materials. This requires an understanding of how solar panels use light from the Sun to release electrons within the material, which involves a process known as the **photoelectric effect**. When the photoelectric effect was first discovered in the 19th century, it required a new way of thinking about light to explain what was going on.

The photoelectric effect can be demonstrated by shining light onto a clean metal surface. Under certain conditions, electrons are given off. The electrons released in this way are called **photoelectrons**, and materials that readily release photoelectrons are said to be **photosensitive**.

Activity 14 Photoelectric effect

Observe the photoelectric effect produced by ultraviolet radiation with zinc or magnesium. One way to do this is to use a zinc plate and a gold leaf electroscope or an electrometer (devices that detect the presence of electric charge – see Figure 5.35). Give the electroscope or electrometer a negative charge. Shine ultraviolet radiation on to the (very clean) zinc and explain what happens to the charge indicated by the electrometer or electroscope. (Think what will happen if electrons can escape from the metal.)

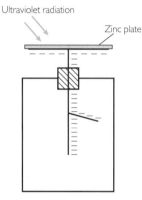

Figure 5.35 A negatively charged electroscope

Explaining the photoelectric effect

Experiments show that:

- for any given metal, with radiation below a certain **threshold frequency** no electrons are released even if the radiation is very intense
- provided the frequency is above the threshold, some electrons are released instantaneously, even if the radiation is very weak
- the more intense the radiation, the more electrons are released
- the kinetic energy of the individual photoelectrons depends only on the frequency of the radiation and not on its intensity.

These results can be explained using the photon model of light (see Figure 5.36). A wave model does not predict the observations correctly. In particular, the wave model would predict that weak radiation might eventually allow a large number of electrons to be released all with low energy, rather than the immediate release of a small number of electrons each with high kinetic energy (see Figure 5.37). The photoelectric effect gave one of the first indications, early in the 20th century, that the wave model of light is not always satisfactory. For explaining the photoelectric effect using a photon model, Einstein was awarded the Nobel Prize in 1905.

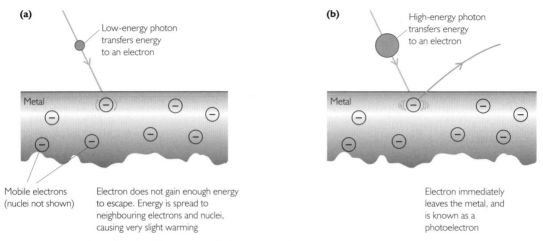

Figure 5.36 The photoelectric effect can easily be explained using a photon model ...

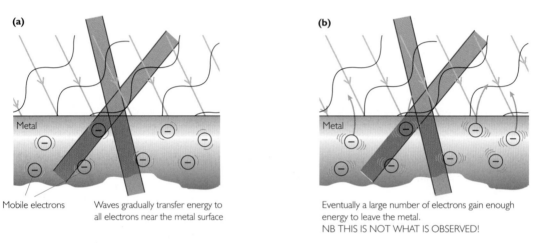

Figure 5.37 ... but not using a wave model

The release of photoelectrons is initiated by the impact of photons on the target. The energy, E_{ph}, of each photon is related to the wave frequency, f, by:

$$E_{ph} = hf \qquad (17)$$

An electron close to the surface of the target material absorbs the photon. In order to escape, the electron needs to do work against the electrical forces that bind it into the metal. The minimum amount of energy needed for this is called the **work function**, ϕ, of the material. In doing work the electron increases its potential energy by an amount ϕ. Any energy remaining is accounted for as the kinetic energy, E_k, of the escaped electron. Conservation of energy enables us to write the following expression:

$$E_{ph} = E_k + \phi \qquad (18)$$

Equation 18 describes the following features of the photoelectric effect:

- if $E_{ph} < \phi$, then an electron is unable to do sufficient work and so cannot escape
- if $E_{ph} = \phi$, then an electron may be released, but with no kinetic energy (i.e. it is free, but cannot move away)
- if $E_{ph} > \phi$, then an electron can escape with a maximum kinetic energy of $E_{ph} - \phi$.

Study note

See *The Sound of Music* for an introduction to photons.

We have labelled the photon energy E_{ph} here to distinguish from other energies also denoted by the letter E.

The smaller the work function, ϕ, of a material, the more photosensitive the material i.e. the lower the photon energy needed to release photoelectrons – or, put another way, the lower the threshold frequency. If the threshold frequency is f_0, then:

$$\phi = hf_0 \tag{19}$$

Equations 18 and 19 can be use to derive further equations describing the same situation for example:

$$hf = E_k + hf_0 \tag{20}$$

or

$$hf = \tfrac{1}{2}mv^2_{max} + \phi \tag{21}$$

where, as you saw in *Higher, Faster, Stronger,*

$$E_k = \tfrac{1}{2}mv^2 \tag{22}$$

is the kinetic energy of a particle mass m moving at speed v. The label 'max' is a reminder that this is the maximum kinetic energy that a photon can acquire; if it has to do more work than ϕ, then its kinetic energy will be less.

Measuring threshold frequency

A photocell needs to be made from a photosensitive material; if it is to respond to visible radiation, its threshold frequency needs to be lower than that of zinc or magnesium – as you saw in Activity 14, these metals only release photoelectrons when illuminated with ultraviolet radiation. In principle, the threshold frequency can be determined in a single measurement: shine light of a known frequency on to a material, measure the maximum kinetic energy of the photoelectrons, and then use Equation 20 to find f_0.

Figure 5.38 shows how, in principle, we can measure the electrons' kinetic energy using a photocell and an opposing potential difference that is adjustable using a circuit called a potential divider. If photoelectrons are able to move across the gap in the photocell, then there will be a continuous flow of charge around the circuit and the electrometer's picoammeter registers a current; if the applied potential difference, V, is zero, this is what will happen. But if the photoelectrons have to travel towards the negative terminal of the power supply, then they lose kinetic energy, and the higher the potential difference, the greater the initial kinetic energy the electron will need to have if it is to reach the other side. If no electrons cross the gap, then the ammeter will read zero.

As you have seen Part 2 of this the chapter, when charge moves through a potential difference, energy is transferred as described by the relationship:

$$\Delta W = q\Delta V \text{ or } \Delta E = q\Delta V \tag{Equation 3}$$

where ΔW or ΔE is the energy transferred, q the charge and ΔV the potential difference.

If an electron moved in the opposite direction across the gap, from negative to positive, then the power supply would be transferring energy to the electron; an electron starting from rest would gain kinetic energy:

$$E_k = e\Delta V \tag{23}$$

where e is the electron's charge. Travelling in the opposite direction, an electron would lose this amount of kinetic energy. So if the applied pd is adjusted so that it just stops the photoelectrons (i.e. so that the meter reading just becomes

> ### Study note
>
> Potential dividers are discussed in more detail in *Digging Up the Past.*

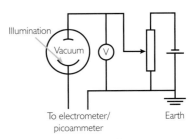

Figure 5.38 Apparatus for determining the threshold frequency of a photosensitive material

zero), then Equation 3 can be used to calculate the electron's initial kinetic energy. The potential difference that just stops the photoelectrons crossing the gap is called, unsurprisingly, the **stopping potential**.

Combining Equations 22 and 23, with V now representing the stopping potential, gives:

$$hf = eV + hf_0 \qquad\qquad (24)$$

Rather than making just one single measurement to determine f_0, it is much better to use several different values of f and record the corresponding stopping potential V. There is a linear relationship between f and V, and a graph of f plotted against V, or V against f, can then be used to determine f_0.

Maths reference

Linear relationships
See Maths note 5.2

Gradient of a linear graph
See Maths note 5.3

Units: the electronvolt

When we are dealing with individual electrons moving through a potential difference of a few volts, the energies involved are extremely small because the magnitude of the electron charge is only 1.60×10^{-19} C. For example, if an electron moves through a pd of 1 V, the energy transferred is only 1.60×10^{-19} J (see Equation 23). Rather than express such tiny energies in joules, we sometimes use an alternative unit, the **electronvolt** (eV), which is defined as the energy transferred when one electron moves through a pd of 1 V, so when an electron or a proton is accelerated through a potential difference of V, then the energy transferred is numerically the same as the pd measured in volts.

In other situations you might need to convert between joules and eV:

$$1 \text{ eV} = 1.60 \times 10^{-19} \text{ J}$$

Activity 15 Measuring threshold frequency

Use a white light source and set of coloured filters to find the threshold frequency, and hence the work function, of the photosensitive material in a photocell. Use a spreadsheet to plot and analyse a graph of your results.

3.3 Summing up Part 3

In this part of the chapter you have studied an important piece of physics that involves both electricity and light. You have seen that a photon model is needed to explain the photoelectric effect, and have learnt how the kinetic energy of electrons can be measured using an electrical technique. You have also seen how ideas about charge, potential difference and energy lead to the definition of a convenient unit for expressing very small energies: the electronvolt.

Questions

magnitude of electron charge, $e = 1.60 \times 10^{-19}$ C

electron mass, $m_e = 9.11 \times 10^{-31}$ kg

proton mass, $m_p = 1.67 \times 10^{-27}$ kg

Planck constant, $h = 6.63 \times 10^{-34}$ J s

speed of light, $c = 3.00 \times 10^8$ m s^{-1}

16 What is the energy transferred when (a) an electron and (b) a proton is accelerated through a pd of 5000 V? Give your answers in joules and in eV.

17 Calculate the speeds of the electron and proton in Question 16, assuming that the transfer of energy from power supply to the particles is 100% efficient.

18 Light with a wavelength 434 nm shines onto the surface of clean sodium, which has a work function of 2.28 eV. What is the maximum kinetic energy of the photoelectrons? Give your answer in eV and in joules.

19 Explain what would happen if

(a) the blue light in Question 18 shines on aluminium, whose work function is 4.08 eV

(b) ultraviolet light shines on sodium.

Activity 16 Photocell

Write a short account of the operation of a photocell, using the following terms: work function, threshold frequency, photosensitive, photon, photoelectron.

Further investigations

Using light passing through coloured filters, investigate the frequency response of some photosensitive devices (e.g. photographers' light meters, or light-dependent resistors). Try to determine the lowest-frequency (longest-wavelength) radiation to which they respond.

4 All under control

So far in this chapter you have dealt mainly with aspects of electrical power system. If you look back at what Jeremy Curtis said in Part 1 you will see that solar radiation is a bit of a mixed blessing for spacecraft: it supplies much-needed energy, but it also gives the engineer more problems to deal with. In this part of the chapter you will see how heating of electrical components has important implications for designing and operating a spacecraft.

4.1 Facing the right direction

In the summer of 1997, the Russian space station Mir (Figure 5.39) faced a series of problems that began when a docking manoeuvre with a supply module went wrong (rather than docking smoothly, the two vehicles crashed together, damaging part of the outer structure of the space station). While working on repairs, one of the crew accidentally pulled out a cable from the station's main computer. In the words of *New Scientist* magazine (27 July 1997): 'The error left the station almost totally without power and spinning out of control for more than 24 hours.'

Figure 5.39 The Mir space station

But why should disconnecting a *computer cable* leave the station without power? The clue comes in the phrase 'spinning out of control'. The computer controls motors that orient the space station and its solar panels. Unless they are correctly aligned with the Sun, the solar panels do not function effectively.

Power input and radiant energy flux

Figure 5.40 shows a solar cell and a beam of radiation. The 'strength' of the beam is usually described in terms of the rate at which it transfers energy across unit area square-on to the beam. This quantity is called the radiant energy flux, F, or the intensity, I, of the beam. (We will use F but you may find I in other books.) The SI units of F (or I) are watts per square metre (W m^{-2}). The rate at which energy is transferred to a surface therefore depends on the flux and the area, A, of the surface:

$$P_{in} = FA \tag{25}$$

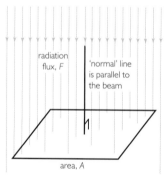

Figure 5.40 A solar cell square-on to a beam of radiation

Worked example

For example, if a beam has $F = 10$ W m^{-2} and shines square-on to a surface of area $A = 2$ m^2 (as in Figure 5.40), then (provided all the incident radiation is absorbed by the surface):

$$P_{in} = 10 \text{ W m}^{-2} \times 2 \text{ m}^2 = 20 \text{ W}$$

How large a solar array does a spacecraft need?

The size of solar array depends on the spacecraft's power demands. As you can see from Figure 5.41 the average power requirements of various spacecraft have risen appreciably since the 1960s. However, even when the power requirements are known, it is still necessary to have information about how efficiently the solar cells transfer energy. We can adapt Equations 15 and 16 to get:

$$\text{efficiency} = \frac{\text{useful energy output}}{\text{total energy input}} \qquad (15a)$$

and

$$\text{efficiency} = \frac{\text{power usefully transferred}}{\text{power input}} = \frac{P_{out}}{P_{in}} \qquad (16a)$$

Here, the input refers to the incoming solar radiation flux and the energy is that transferred by the cell.

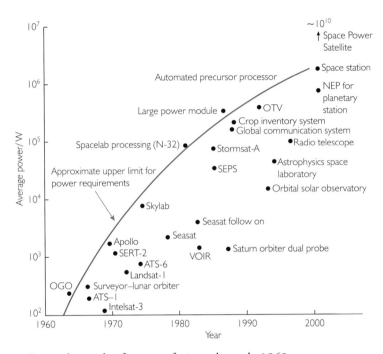

Figure 5.41 Power demands of spacecraft since the early 1960s

Worked example

The solar arrays on the Hubble Space Telescope have an area $A \approx 20$ m² and are about 10% efficient and the solar flux is $F \approx 1.4$ kW m⁻². To find the maximum possible output power:

$$P_{in} = FA$$
$$\approx 20 \text{ m}^2 \times 1.4 \text{ kW m}^{-2}$$
$$= 28 \text{ kW}.$$

P_{out} is 10% of P_{in} so:

$$P_{out} = 0.10 \times 28 \text{ kW}$$
$$= 2.8 \text{ kW}.$$

Activity 17 Sizing up solar arrays

Use data from Figure 5.41 to calculate the minimum area of solar array required by Intelsat-3, SERT-2 and Skylab. Assume that all their power is supplied from the solar arrays, and that the solar flux is $F \approx 1.4 \text{ kW m}^{-2}$.

Activity 18 Facing the right direction

Use a large solar cell and a lamp to explore how the power input to a solar cell would change if it were not square-on to the incoming radiation. Look at Figure 5.40 and think about the radiation that would by-pass the cell if it were tilted. Think what would happen if the cell were edge-on to the radiation.

Questions

20 Write a short paragraph, with at least one diagram, to explain why a breakdown in the computer guidance system of the Mir space station resulted in a loss of electrical power.

21 The International Space Station (Figures 5.33, 5.34) has solar arrays that must each produce a peak output power of 16 kW from an incident solar flux of 1.4 kW m^{-2}. If the efficiency of an array is 11%, what must be its area?

The importance of understanding radiation flux is not limited to solar panels on spacecraft. Question 22 uses the same ideas in another situation.

Question

22 In Germany the Daimler-Benz company has incorporated arrays of photovoltaic cells into one of its factories (Figure 5.42). The total area of the arrays is 5000 m² (about that of a soccer pitch). The company states that the peak power output from these arrays is 435 kW. If the peak solar flux at this location is 600 W m⁻², calculate the efficiency of these solar arrays.

4.2 Temperature changes in spacecraft

So far in this chapter we have been concerned with designing an electrical power supply system for a spacecraft. Temperature control is another important aspect of designing a spacecraft. All spacecraft, no matter what their task or position, have heating and cooling systems on board.

Figure 5.42 Solar arrays on a Daimler-Benz factory

Activity 19 Getting warmer

Spend a few minutes 'brainstorming' the following questions. Jot down a list of your ideas.

- Why might the temperature on board a spacecraft vary markedly if it was uncontrolled?
- What problems might be caused by large changes in temperature on board a spacecraft?

Think about space probes designed to study the Sun continuously, such as the Solar Heliospheric Observatory (SOHO), or to explore the outermost regions of the solar system, such as the *Voyager* and *Pioneer* probes. Think about manned space stations and moon landers, and about unmanned satellites of all types orbiting the Earth.

Many spacecraft spin about their own axes and so part, at least, keeps going into and out of the sunlight. Others orbit the Earth and so travel in sunlight on the day side and in the dark (cold) on the night side. While space can be as cold as 3 K (−270 °C), a spacecraft can quickly heat up when facing the Sun or through heating due to its own electrical equipment.

Uncontrolled heating and cooling can cause problems. Mechanical parts can suffer from expansion and contraction, as did the Hubble Space Telescope's first solar arrays which flexed as they moved in and out of the Earth's shadow, and if electrical components get too hot or too cold then they perform less well. For example, chemical batteries and solar cells tend to function better when they are cool. Table 5.2 shows typical temperature ranges within which various components can operate reasonably well.

Equipment	Temperature range/°C
Electrical equipment	−10 °C to +40
Chemical batteries	−5 °C to +15
Fuel (hydrazine)	+9 °C to +40
Microprocessors	−5 °C to +40
Mechanical parts	−45 °C to +65
Solar cells	−60 °C to +55
Solid state diodes	−60 °C to +95

Table 5.2 Typical operating temperatures for equipment used in spacecraft

Activity 20 Solar array temperature

Have a look at some satellite data and see how the temperature of the solar arrays changes with time. Comment on whether the temperatures lie within the ranges given in Table 5.2.

Sources of heating

As we have just seen, solar radiation is the main source of heating for satellites, but for the Space Shuttle and space probes that enter the atmosphere of the Earth or another planet (or moon), frictional heating is also important. The Space Shuttle, for example is covered with specially designed heat-resistant tiles.

Figure 5.43 is an article from a French magazine *Ciel et Espace* ('Sky and Space'). It describes the Cassini–Huygens mission, launched in 1997, that explored the planet Saturn. Like many space exploration projects, Cassini–Huygens is multinational. The Huygens probe, which is the European part of the project (with major contributions from the UK, France and Germany), was released from the main Cassini spacecraft into the atmosphere of Titan, one of Saturn's moons. This probe had a heat shield to protect its instruments from the high temperatures reached as it plunged through Titan's atmosphere.

P O I N T · F O C A L

Huygens dans la dernière ligne droite

LES préparatifs de Cassini-Huygens, ultime grande mission interplanétaire de ce siècle, sont entrés dans leur dernière ligne droite. Les nombreuses vicissitudes que les aléas budgétaires ont fait subir, du côté de la Nasa, à cette ambitieuse entreprise américano-européenne semblent désormais appartenir au passé. C'est du moins le vœu qu'exprimait ce printemps Roger-Maurice Bonnet, le directeur des programmes scientifiques à l'ESA. Quant à la petite sonde Huygens, la partie européenne du projet, elle est depuis la fin avril en cours d'assemblage à Ottobrun, chez Dornier, en Bavière. Dans un peu plus de six mois, Huygens devra être livrée à la Nasa, au centre spatial Kennedy, pour une dernière série de tests avant son arrimage sur l'imposante sonde Cassini et un lancement de l'ensemble par Titan 4, la plus puissante des fusées américaines. Date prévue pour le décollage : le 6 octobre 1997.

L'objectif de la mission, qui ne sera atteint qu'en juin 2004, est le monde glacé de Saturne et, plus particulièrement pour ce qui concerne la petite sonde européenne, l'exploration de Titan, le plus gros des satellites de la planète géante. C'est en principe le 27 novembre 2004, vingt et un jours après son largage par Cassini, que cet engin de 343 kg protégé par un bouclier de 2,7 m de diamètre pénétrera dans l'atmosphère de Titan. Un véritable saut dans l'inconnu, puisque les scientifiques ignorent totalement ce qu'ils trouveront à l'arrivée : des lacs ou des océans

de méthane, des terres secouées par des volcans ou grêlées de cratères d'impact. Au point, précise John Zarnecki, de l'université du Kent, en Grande-Bretagne, qu'il a fallu *"tenir compte de six ou sept modèles pour imaginer la surface et concevoir les instruments scientifiques"*. Au total, Huygens emportera six expériences destinées à l'étude de l'atmosphère, de la météo et de la surface de Titan. Parmi ces instruments, une mini-caméra à trois objectifs ultrasensibles qui devrait livrer des images de la surface durant les deux cents derniers mètres de descente. Une gageure.

Gageure également pour les industriels qui ont conçu cet engin de taille certes modeste mais qui devra résister à des conditions extrêmes. Pendant les deux heures et demie que durera le plongeon vers Titan, depuis une altitude de 1270 km et une vitesse de Mach 20 jusqu'à l'arrêt au contact de la surface, Huygens subira un freinage aérodynamique qui, en cinq minutes, fera chuter sa vitesse de 21 600 à 1 080 km/s. Échauffement estimé sur le bouclier de protection : 2 000 °C. Suivra une descente sous une batterie de trois parachutes pour atteindre la surface à 6 m/s. Temps de vie prévu au terme de cette descente : entre 3 et 30 minutes.

"De la protection thermique aux techniques de pilotage hypersonique et de rentrée atmosphérique en passant par les codes de calcul aérodynamique, Huygens est l'un des rares programmes qui ait mobilisé l'ensemble de nos compétences", souligne André Motet, directeur adjoint d'Aérospatiale, qui assure la maîtrise d'œuvre de la sonde. Le jeu en vaut la chandelle. Les scientifiques s'attendent à rencontrer dans l'atmosphère de Titan une chimie complexe, voire même de ces grosses molécules qui, sur Terre, ont servi de briques au vivant. Titan détient peut-être des clés qui permettront de percer un peu l'épais mystère qui entoure encore l'apparition de la vie.

Jean-Pierre Defait

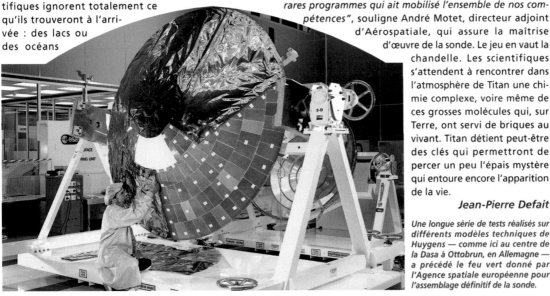

Une longue série de tests réalisés sur différents modèles techniques de Huygens — comme ici au centre de la Dasa à Ottobrun, en Allemagne — a précédé le feu vert donné par l'Agence spatiale européenne pour l'assemblage définitif de la sonde.

Figure 5.43 The Cassini–Huygens mission

Activity 21 The Cassini–Huygens mission

Have a look at Figure 5.43 and try to deduce some information about Cassini–Huygens. If you have ever learned any French (even if you did not think you were very good at it!) you will probably be able to guess at most of the meaning of the article. In particular, you should be able to find Cassini's mass, the size of its heat shield, the expected temperature as it is decelerated by Titan's atmosphere, and its initial and final speeds as it approaches Titan.

The article also outlines the scientific aims of the mission – try to work out from the article what aspects of Titan will be studied by the instruments carried by the Huygens probe.

Useful words: fusée = rocket, bouclier = shield, sonde = probe

Electrical components in space

As you can see from Table 5.2, batteries and solar cells are particularly sensitive to temperature. Questions 23–25 illustrate the problem in more detail and indicate the importance of careful choice of materials.

Questions

23 Figure 5.44 shows how one type of battery performed with temperature. 'Available capacity' is a measure of the energy that could be transferred from the battery, expressed as a percentage of the maximum transferable energy.

(a) What was the available capacity available at a temperature of

(i) 50 °C

(ii) −20 °C?

(b) By how much does the battery's capacity fall in going from 30 °C to 40 °C?

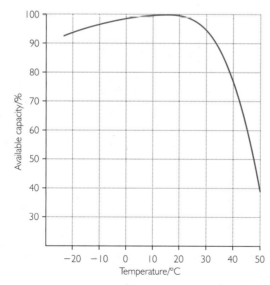

Figure 5.44 The performance of a VARTA RS nickel–cadmium battery

24 Figure 5.45 shows how the efficiency of solar cells made of different materials depends on temperature.

(a) Which material has the highest efficiency at each of the following temperatures?

(i)   0 °C

(ii) 200 °C

(iii) 400 °C

(b) Which material(s) would it not be sensible to use for solar cells if their temperatures were likely to rise above 30 °C?

(c) How does the efficiency of a cadmium sulphide (CdS) cell vary when the temperature changes from 0 to 400 °C?

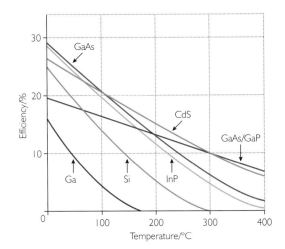

Figure 5.45 The effect of temperature on the efficiency of solar cells of various materials

25 Some proposed space stations of the near future are likely to need 1 MW of power (1 MW = 1 × 10^6 W). Suppose such a space station is to have solar arrays made from gallium arsenide (GaAs) and its operating temperature is to be kept below 50 °C. If the incident solar flux is 1.4 kW m^{-2}, what is the smallest area that its solar arrays must have?

Resistance and temperature

Not only does temperature affect the performance of solar arrays and batteries, but the resistance of many components also varies noticeably with temperature, as you will see in the following activity.

Activity 22 Changing resistance

You task is to obtain a set of readings for one component, showing how its resistance changes over the range 0 °C to 100 °C, and to write a short report of your findings to exchange with other students who have used a different component. See Figure 5.46

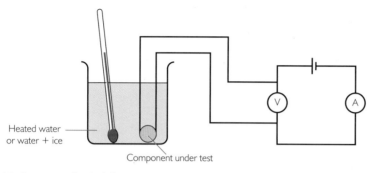

Figure 5.46 Apparatus for Activity 22

You will need to use your tables of results and calculations again later, in Activity 23.

Modelling current and resistance

The materials used nowadays to make electronic components have been developed to have particular electrical properties. In order to develop 'designer' materials, it is important to describe how particular materials behave and also to understand why. The change in resistance with temperature is one such aspect of material behaviour that can be explored with the help of scientific modelling.

Scientific modelling

The variation of resistance with temperature raises two questions: *Is there a simple relationship between resistance and temperature?* And *Why does resistance change with temperature?* The second question is of interest when we are trying to understand the behaviour of the natural world and perhaps develop new materials whose resistances change in particular ways. The first question is of particular interest to anyone wanting to know how an electrical system is going to behave. Both these questions involve the important scientific idea of **modelling**.

One of the main goals of science is to describe and understand the natural world – partly for the satisfaction of knowing for its own sake, and partly in order to make use of that knowledge through technology. Exploring the natural world in this way involves the use of **scientific models**. A scientific model is a way of thinking about and visualising objects or processes, often involving a mathematical description. A model in this sense does *not* normally mean a small- or large-scale replica.

Models that represent attempts to understand the world at a fundamental level often need to be adapted and refined depending on what we are using them for, and as our knowledge of the world develops. For example, if you picture atoms and molecules as small spheres as in Figure 5.47, you are using a model that can explain some large scale behaviour of materials (differences between solids, liquids and gases, for example). But it does not explain electric currents, nor does it explain how atoms combine chemically – a model of an atom consisting of a small nucleus surrounded by orbiting electrons is more helpful there. In turn, the nucleus-plus-electrons model has to be refined in order to explain radioactive decay. During this course, you will meet several examples of scientific models being developed and refined.

Figure 5.47 A simple model explains some of the properties of matter

Other models are empirical – that is, they are based on observation and experiment, rather than on any fundamental thinking about the underlying processes. Ohm's law ($V \propto I$) is an **empirical model** that describes mathematically the way that some materials behave. Sometimes empirical relationships give clues to something deeper. For example Isaac Newton observed that all falling objects have the same gravitational acceleration, which led him to an understanding of gravitational force without which we would not be able to launch satellites into orbit.

Even where they do not directly reveal fundamental insights into the laws of nature, empirical models are extremely useful to scientists and engineers. Ohm's law is just one example of such an empirical relationship. The relationship between resistance and temperature is another.

Model making and model fitting

In principle, making an empirical mathematical model is straightforward. It simply involves collecting some experimental measurements and looking for a mathematical way to describe them – ideally, a fairly simple relationship between two measured quantities. In practice, this can turn out to be less simple than it sounds.

More commonly, you are likely to be concerned with **model fitting**, where you start off with a model and see whether experimental measurements agree with it. You met this in Questions 6–8, where you had to decide whether a given material obeyed Ohm's law. Plotting a graph is often a good way to see whether measurements fit a given model – particularly if the expected graph is a straight line, and the experimental values are plotted complete with error bars.

Matching experimental measurements to a mathematical model can help you to find the values of unknown quantities. If a set of current and voltage measurements fit Ohm's law, then you can determine the conductor's resistance. In Section 2.3 you used a mathematical model to describe the behaviour of a power supply with internal resistance and hence were able to determine the internal resistance and the emf of the supply.

Is there a simple relationship between resistance and temperature?

Some conductors have a resistance that increases uniformly with temperature. This behaviour can be described by a mathematical model:

$$R_\theta = R_0 (1 + \alpha\theta) = R_0 + \alpha R_0 \theta \qquad (26)$$

where:

R_θ = resistance at temperature θ

R_0 = resistance at 0 °C

α is the **temperature coefficient of resistance**

The coefficient α can be described as 'the fractional increase in resistance compared with the value at 0 °C'. The units of α are °C^{-1}, so that the units of $\alpha\theta$ are °C^{-1} × °C, i.e. $\alpha\theta$ has no units – it is just a number.

The value of a can be positive or negative, depending on whether resistance increases or decreases with rising temperature. Resistors that are designed to have a large variation with temperature (i.e. **thermistors**) are often referred to as NTC (negative temperature coefficient) or PTC (positive temperature coefficient) thermistors, depending on the way their resistance changes.

Question

26 The temperature coefficient of resistance of copper is 4.28×10^{-3} °C^{-1}. A piece of copper connecting wire has a resistance of 0.50 Ω at 0 °C. What will be its resistance at 80 °C? In the light of your answer, say whether you think that the connecting wires in a circuit can be ignored when taking account of changes in electrical properties with temperature.

Table 5.3 lists some very precise values for the resistance of five different material samples, all designed to have a resistance of 1.00000 Ω at 0 °C (i.e. R_0 = 1.00000 Ω). By looking at the numbers in each column, you can see that the change in resistance with temperature is linear in each case.

Worked example

Figure 5.48 shows a graph of resistance R_θ plotted against temperature θ for copper. The graph is a straight line and cuts the vertical axis at $R_\theta = R_0$ = 1.00000 Ω. The gradient of the line is equal to αR. From Figure 5.48:

$\Delta\Omega = 0.40\ \Omega$

$= 0.0040\ \Omega\ °C^{-1}$

$= 4.0 \times 10^{-3}\ \Omega\ °C^{-1}$

$\alpha = \dfrac{\text{gradient}}{R_0}$

$= \dfrac{4.0 \times 10^{-3}\ \Omega\ °C^{-1}}{1.00000\ \Omega}$

$= 4.0 \times 10^{-3}\ °C^{-1}$

Study note

In this example the gradient is numerically the same as α, but that will not in general be the case.

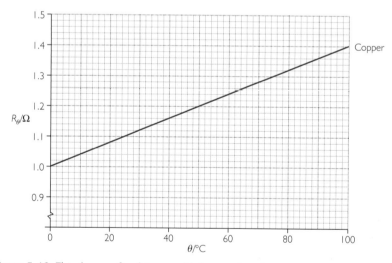

Figure 5.48 The change of resistance with temperature for copper

Maths reference

Linear relationships
See Maths note 5.2

Gradient of a linear graph
See Maths note 5.3

Question

27 (a) Which of the materials listed in Table 5.3 have a positive temperature coefficient of resistance, and which have a negative coefficient?

Temperature/°C	Resistance/Ω				
	Carbon	Copper	Constantan	Steel	Tungsten
0	1.00000	1.00000	1.00000	1.00000	1.00000
20	0.99000	1.00800	1.00002	1.06600	1.10400
40	0.98000	1.16000	1.00004	1.13200	1.20800
60	0.97000	1.24000	1.00006	1.19800	1.31200
80	0.96000	1.32000	1.00008	1.26400	1.41600
100	0.95000	1.40000	1.00010	1.33000	1.52000

Table 5.3 Temperature data for five resistors

(b) Suggest a reason why constantan is so called.

(c) Choose one of carbon, tungsten or steel. Plot a graph, similar to Figure 5.48, to show its behaviour. Calculate its temperature coefficient of resistance.

Activity 23 Model fitting

Your task is to determine whether your experimental measurement of resistance and temperature from Activity 22 match the mathematical model discussed above, and then to communicate your findings to students who have discussed other components.

Your graph of R_θ against θ that you plotted in Activity 22 will enable you to decide whether a particular set of measurements fits the model, and to find the values of R_0 and α.

4.3 Why does resistance change with temperature?

You will have seen that different components behave very differently on being heated. Why is this? We can explain the main features using a simple model for current and a so-called classical model of materials.

Modelling current

Within a conductor, there are some charged particles that are free to move. When a pd is applied, the particles that have negative charge (e.g. electrons) move towards the positive terminal and those with positive charge (e.g. positive ions in a solution) move towards the negative terminal. This movement forms an electric current as described by Equation 1a:

$$I = \frac{\Delta Q}{\Delta t}$$
(Equation 1a)

As the current is a measure of the rate at which charge passes a point, it will depend on the speed at which the particles move and on their individual charge. By thinking about what happens when the charged particles move, we can set up a model that relates these small-scale quantities to the current that we measure with an ammeter.

Figure 5.49 shows a section of conductor which contains freely moving particles each with charge q, moving along at speed v. There are n particles per unit volume – that is, their **number density** is n – and the conductor's cross-sectional area is A. We can use this model to derive an expression for I in terms of n, q, A and v.

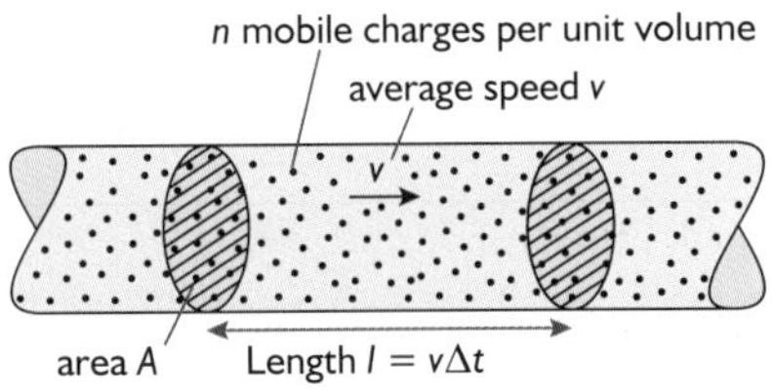

n mobile charges per unit volume
average speed *v*

area *A*   Length *l* = *v*Δ*t*

Figure 5.49 Charged particles in a conductor

In a time interval Δt, all the particles in a length l will pass a reference point X, where

$$l = v\Delta t \tag{27}$$

The number of particles, N, in this length of conductor is found by multiplying the volume, V, of the conductor by the particles' number density:

$$V = Al \tag{28}$$

so:

$$N = nAl \tag{29}$$

The charge ΔQ passing point X in time Δt is given by:

$$\Delta Q = Nq = nAlq \tag{30}$$

Substituting for l from Equation 27 gives:

$$\Delta Q = nAqv\Delta t \tag{31}$$

Dividing both sides by Δt and comparing with Equation 1a, we get

$$I = nAqv \tag{32}$$

Activity 24 Moving charges

How fast do you think the charged particles move in a current-carrying conductor? Write down your guess, then use the apparatus in Figure 5.50 to observe mobile coloured ions.

The results of Activity 24 may have surprised you. The speed v is known as the **drift speed** (or **drift velocity**) and in most conductors is quite small. The following questions explore the relationship between the various quantities in Equation 32.

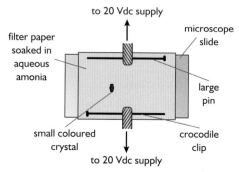

Figure 5.50 Apparatus for Activity 24

Questions

Electron charge $e = 1.60 \times 10^{-19}$ C

28 Referring to Equation 32, explain why increasing the pd across a conductor increases the current.

29 In a semiconductor, the number density of free-moving charges is much less than in a metal. If a sample of semiconductor is joined in series with a metal sample of the same cross-sectional area, and a pd is connected across the arrangement, what can you say about the drift speeds in the two samples?

30 (a) Suppose each atom in a metal occupies a cube of side 5×10^{-10} m and each contributes one electron that is free to move. What is the number density of the free electrons?

   (b) Suppose also that there is a current of 0.2 A in a wire made from the metal, with a square cross-section of side 0.5 mm. What must be the electrons' drift speed?

Modelling resistance

The model of current introduced above can be used when we try to explain changes in resistance with temperature. To reduce the resistance, we must do something that either increases the number-density of free-moving charged particles, or increases their drift speed for a given applied pd.

Our model for resistance starts with the ideas that materials are made up of atoms that are constantly vibrating. As the temperature rises, the atoms vibrate more vigorously. At the same time, a rise in temperature can result in the release of more electrons from atoms. It is the flow of these free electrons that forms an electric current.

As the electrons move through the material, they 'collide' with the vibrating atoms and are scattered, so their flow is disrupted. As the atoms vibrate more vigorously, so the frequency of 'collision' or interaction between the atoms and the electrons also increases. This reduces the rate of flow of the electrons and so the current falls. In other words, the resistance increases (Figure 5.51(a)). However, if more electrons are released from the atoms as the temperature rises, then there is a greater rate of flow of charge – the current increases. In other words, the resistance falls (Figure 5.51(b)).

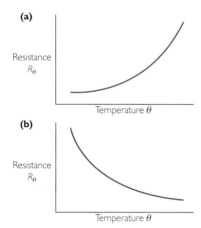

Figure 5.51 Schematic diagrams showing the change in resistance with temperature due to (a) thermal vibrations and (b) the release of electrons

The change in resistance with temperature, as illustrated in Question 27, depends on which of these two effects 'wins'. In a metal, the temperature has virtually no effect on the number of electrons that can move freely through the material, so the dominant effect is that of atomic vibrations. But in a semiconductor a small rise in temperature leads to the release of a large number of electrons, and this effect now outweighs that of atomic vibrations.

Electronic materials

Space technology relies heavily on 'designer' materials that have been developed to have particular electrical properties. Such materials are essential for the solar cells and the communication and control systems on board a spacecraft – just as they are for familiar Earth-based electronic devices. The development of electronic devices and materials is usually said to have started with the invention of the transistor in 1947. The following extracts from a newspaper article written fifty years later indicate the importance of this invention. But beware – as is all too often the case with such articles, this one contained a misleading explanation of the science.

Activity 25 Picking holes

Read the extract below from a feature in the *Guardian* newspaper carefully and look out for a misleading explanation of why the electric current in a semiconductor changes with temperature. Discuss what is wrong with the explanation given and suggest how it could be improved.

Huddled secretively in a corner of their laboratory, John Bardeen and Walter Brattain were building a primitive device (Figure 5.52) whose impact on the world not even these engineering geniuses appreciated … Cautiously they glued a fragment of gold foil to a wedge of plastic, pressed the wedge on to a sliver of germanium … When all was ready, Brattain switched on the battery, sending a trickle of power through the circuit … the faint signal leapt to 100 times its strength, a sudden powerful glow on the oscilloscope …

A week later the experiment was performed in front of their supervisor, William Shockley, and a handful of 'top brass' at their research centre … Shockley was quick to explain the significance of the work … No-one fully understood then how the crystalline structure of germanium provided different levels of resistance to electrical current that allowed small changes in input to cause large changes in output [but] for the first time you could amplify an electrical signal without the need for a glass vacuum tube … You were eliminating a fragile, costly and bulky limit to the scope and power of electronics.

The peculiar gadget that clicked into life in Bell Labs back in 1947 was, even then, a culmination of years of research into … a poorly understood group of materials, the semiconductors. The demonstration established solid-state physics (the study of the behaviour of electrons in solids) as a promising area of research, now one of the most important areas of science. The first step to producing a reliable transistor was to produce a more robust design … The second was to produce better quality semiconductor material. It was becoming clearer that the curious behaviour of semiconductors was due in part to naturally occurring impurities that disrupted the crystalline structure of the material, freeing some electrons to move around between atoms. The freed electrons left 'holes' through which current could flow, an effect that could be enhanced by temperature – or by artificially controlling the level of impurity.

Once these techniques were perfected, the transistor shrank in size and grew in significance. [It] became a cultural icon as the heart of the transistor radio, soon followed by TVs, cameras, hi-fi equipment and clocks. Mobile computing, the Internet and wireless communication will one day be joined by electronic business cards, watches that control our central heating, inexhaustible organ replacements and hand-held video-conferencing gear.

Figure 5.52 The first transistor

4.4 Summing up Part 4

In this part of the chapter you have learned why it is important to control the temperature on board a spacecraft, and seen how and why electrical resistance changes with temperature.

Activity 26 Summing up Part 4

Look through Part 4 and make sure you know the meaning of each of the terms printed in bold.

Write a short paragraph explaining why the resistance of copper increases with temperature whereas that of carbon decreases.

5 Remote sensing

As you saw in Part 1, spacecraft can be used to study the Earth's surface, to observe distant objects, and as part of modern telecommunications. Many of these applications involve measurements of speed and distance. For example, the Shuttle Radar Topography Mission (SRTM), launched in 2000, has mapped about 80% of the Earth's surface to provide detailed relief maps (Figure 5.53) with heights measured to a precision of 30 m, and sat nav and GPS systems involve making precise measurements of the position of the receiver relative to particular satellites.

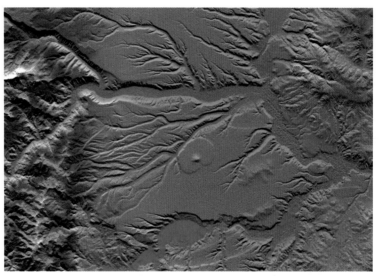

Figure 5.53 A colour-coded image showing height variations around the Corral de Piedra, Argentina, mapped by SRTM

5.1 Measuring distance

How can we measure how far away something is if we can't get to it? One way is by using a pulse–echo technique. A signal pulse is sent towards the object, and the time the reflection takes to return is measured. If we know the speed of the wave we can calculate the object's distance.

Activity 27 Ultrasound or 'laser' tape measure

Use a commercial ultrasound or 'laser' tape measure to measure the dimensions of the laboratory. In fact the laser is just to help you aim the device accurately. The signal is ultrasound. Why is this? (Hint: compare the time taken by a light beam and a sound beam to cross the room and return.)

Safety note

Do not allow the laser beam to shine into your own or anyone else's eyes.

Activity 28 Pulse-echo

Demonstrate distance measurement using a pulse–echo technique.

In space the distances are larger, and objects are separated by a vacuum, so electromagnetic waves are used rather than sound. One legacy of the Apollo moon landings is the retro-reflector left on the Moon by the astronauts, so that laser pulses can be sent from Earth and their reflection used to monitor the Earth–Moon distance accurately. See Figure 5.54.

Figure 5.54 Astronaut Buzz Aldrin placing the retro-reflector on the Moon

Questions

31 The mean Earth–Moon distance is approximately 378 000 km, and the speed of light is 3.00×10^8 m s^{-1}. How long does it take for a pulse of light to travel from Earth to the Moon and back?

32 This question will enable you to calculate the distance between Earth and Venus at Venus's position of 'greatest elongation', and hence the Earth–Sun distance. The orbit of Venus is shown in Figure 5.55. It is smaller than that of Earth, as Venus orbits the Sun its apparent distance from the Sun varies from our viewpoint.

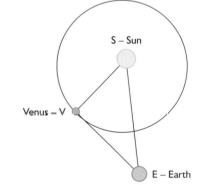

Figure 5.55 The orbit of Venus

(a) Explain why, at the position of greatest elongation, angle SVE = 90°.

(b) The time for a radar signal to return to Earth from Venus at its point of greatest elongation is 11.20 minutes. The speed of electromagnetic waves is 3.00×10^8 m s^{-1}. Calculate distance EV.

(c) Angle SEV has been measured to be 48°, approximately. How far away from Earth is the Sun?

(d) Suggest a reason why it is difficult to use pulse–echo reflection to measure the Earth–Sun distance directly.

(e) Venus is known as both the Morning star, and the Evening star. Why is this?

Activity 29 How far away is the Moon?

Explore the use of pulse–echo data to determine the distance of the Moon.

Pulse–echo techniques can also be used to measure speed. If an object's distance from us is measured twice, and the time between the intervals is known, then the average speed of the object is easy to calculate.

In some applications of pulse–echo techniques, the signal is partially reflected from several layers, providing information about an object's structure. In this case, the amount of detail that can be deduced is limited by the duration of the pulse, as Question 34 illustrates.

Questions

33 A rocket is approaching the Moon. The first reflected microwave pulse takes 1.00 seconds to return to the rocket. A pulse sent out 50 seconds later takes 0.94 seconds to return. What is the approach speed of the rocket?

34 Figure 5.56 shows how ultrasound pulse–echo measurements can be used in medicine to examine someone's heart. The ultrasound pulse is partially reflected at each interface, so a single emitted pulse gives rise to several pulses returning to the receiver. Table 5.4 lists typical thickness for each tissue in Figure 5.56 and the speed of sound in each.

Tissue	Typical thickness x/cm	Speed of sound v/m s^{-1}
soft tissue (inc. skin)	5.0	1500
fat	0.5	1450
muscle	1.0	1600
blood	5.0	1570

Table 5.4 Sound speed in body tissues

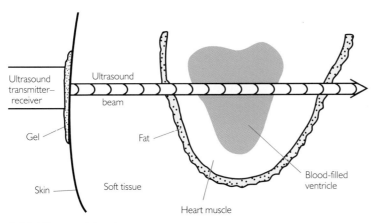

Figure 5.56 Ultrasound heart scan

(a) How long is the time delay between sending the pulse and receiving each echo, up to and including the one from the far side of the heart vessel? (You could set up a spreadsheet to perform these calculations.)

(b) Use your answers to (a) to draw a sketch showing the sequence of pulses detected by the receiver.

(c) Explain what would be detected if the duration of the pulse was

   (i)   about 1 µs

   (ii) about 10 µs.

5.2 Measuring speed

In the last section you learnt how to measure how far away a distant object might be and its average speed between two positions. But it is also possible to find out how fast, and in which direction, something is moving at any given instant. For example, it is possible to measure the speeds of distant stars moving towards or away from us, and to monitor the rotation of planets.

The speed of a remote object can be found using the **Doppler effect**. You have probably experienced the Doppler effect yourself. If an ambulance or police car has passed you with its siren sounding, you will have noticed that the pitch of the sound changed. It was higher as the siren approached you and lower when it was going away, and just as it passed you it was the same as it would have been if the siren had been stationary. The change in pitch is most noticeable if the source of the sound is moving at high speed directly towards or away from you.

Activity 30 The Doppler effect

The Doppler effect can be demonstrated by whirling a source of sound in a large horizontal circle (Figure 5.57). Listen for the change in frequency as the source of sound moves towards/away from you.

Use the *Audacity* software to record and display sounds that illustrate the Doppler effect.

Safety note

This should be done outside, and make sure that the listeners stand well away from the whirling object.

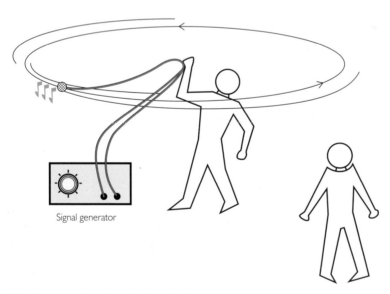

Signal generator

Figure 5.57 Demonstrating the Doppler effect

Activity 31 Doppler on the web

There are several websites that provide animations and information about the Doppler effect. Visit some and decide which you find most helpful in improving your understanding.

The Doppler effect

When a moving object emits or reflects waves, the frequency of the waves received by a stationary detector differs from the frequency at which they were produced. Figure 5.58 shows how this comes about. When the source is at rest, the wavelength and therefore the detected frequency are the same for the detectors placed each side of the source. With the source moving towards one detector and away from the other, the waves are squashed up on one side (decreased wavelength, increased frequency). This is because the source has moved to a new position by the time it produces each subsequent wave, but the waves themselves always travel at the same speed.

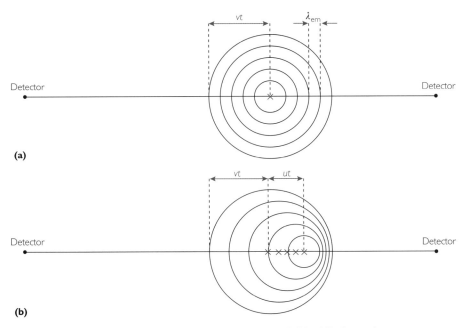

(a)

(b)

Figure 5.58 A source emitting waves (a) while at rest and (b) while in motion

The Doppler effect also occurs when a detector moves towards or away from a source of waves. Imagine one of the detectors in Figure 5.58 (a) moving towards the source; it would encounter more waves per second than if it were at rest, so it would measure a higher frequency (shorter apparent wavelength) than that of the source. Similarly if a detector is moving away from the source it measures a lower frequency (longer wavelength).

Provided the speed, v, of the source or detector is much less than the wave speed, c, the change in the received wavelength ($\Delta\lambda$) or frequency (Δf), known as the **Doppler shift**, is given by:

$$\frac{-\Delta f}{f_{em}} \approx \frac{\Delta\lambda}{\lambda} = \frac{v}{c} \qquad (33)$$

where:

$$\Delta f = f_{em} - f_{rec} \qquad (34)$$

and:

$$\Delta\lambda = \lambda_{em} - \lambda_{rec} \qquad (35)$$

Question 35 guides you step by step through deriving Equation 33.

> **Study note**
>
> By convention, a positive value of Δf corresponds to an increase in wavelength and hence to a decrease in frequency, which occurs when the source and receiver are moving apart, while a negative Δf corresponds to a decrease in wavelength and an increase in frequency.

Question

35 Suppose the source is emitting waves of wavelength λ_{em}, with speed c and frequency f_{em} and moving towards the receiver with speed v as shown in Figure 5.58.

(a) Write down the equation that links c, f_{em}, and λ_{em}.

(b) If source and observer were stationary, in one second f waves would be produced, spread over a distance c. If the source approaches the receiver at speed v, into what distance, x, will the f waves now be squashed?

(c) What will be the new wavelength of the waves, λ_{rec}?

(d) Show that the change in wavelength $\Delta\lambda = \lambda_{em} - \lambda_{rec} = \dfrac{v}{f}$.

(e) By substituting from $f = \dfrac{c}{\lambda}$, show that $\dfrac{\Delta\lambda}{\lambda} = \dfrac{v}{c}$.

(f) Calculate the new received frequency from $f_{rec} = \dfrac{c}{\lambda_{rec}}$, (use your answer to (c)).

(g) Show that $\dfrac{\Delta f}{f} \approx \dfrac{v}{c}$, if $v \ll c$.

The Doppler effect applies to *all* types of waves. With electromagnetic waves, we are generally not aware of the effect, because the wave speed (3.00×10^8 m s^{-1}) is so much greater than typical speeds of objects. However, even small changes in frequency can quite easily be detected using suitable equipment – police radar speed traps measure the Doppler shift in radio waves reflected from moving vehicles.

The Doppler effect is widely used in astronomy and space science. Activity 32 illustrates one example.

Activity 32 Binary stars

Binary stars are pairs of stars in orbit around their common centre of gravity. As each star approaches or recedes, the wavelengths of its spectral lines are Doppler shifted so the stars' speed can be measured. Look at explanations and animations of the Doppler shift can be seen on the US Naval Observatory Flagstaff Station website, and the University of Arizona website. Details of the URLs can be found at www.shaplinks.co.uk.

The Doppler effect also can be used to measure rotational speeds. For example, spectral lines from one 'edge' of the Sun (Figure 5.59) are Doppler shifted to a higher frequency and those from the opposite 'edge' are shifted to a lower frequency.

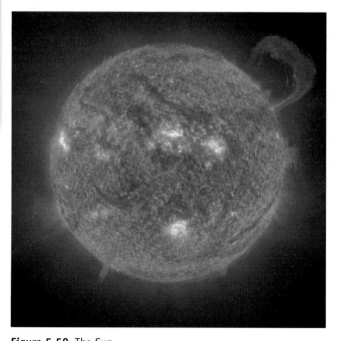

Figure 5.59 The Sun

Doppler disaster?

The Doppler effect is not always welcome; the Cassini–Huygens mission to Titan in 2005 (Figure 5.60) was nearly sabotaged by the Doppler effect. The Huygens probe that descended through Titan's atmosphere communicated with its parent spacecraft, Cassini, by sending radio signals encoded with a stream of binary on–off signals ('bits'). To decode the signals, Cassini's receivers were designed to pick up bits that arrived at a certain rate, give or take a certain margin.

Figure 5.60 The Cassini spacecraft releasing the Huygens probe

However, the speed of Huygens relative to Cassini would mean that the Doppler effect changed the rate at which the bits arrived, and this had not been accounted for when the receiver was built. During the long flight to Saturn, one of the communications engineers realised the problem and insisted that the receiver system was tested with frequency-shifted signals sent from Earth. His fears were justified.

Fortunately it was possible to reprogramme the probe release so as to minimise the Doppler effect during the descent (Figure 5.61). The signals were decoded successfully, providing a wealth of data about Titan's atmosphere, and the engineer was hailed as a hero who saved a 20-year, £400 million, mission from disaster.

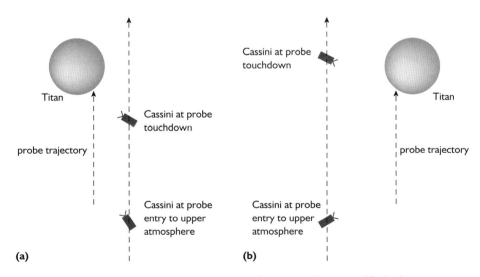

Figure 5.61 (a) The original plan for the release of Huygens (b) the modified release

Questions

36 Calculate the shift in frequency when a vehicle travelling at 20 m s^{-1} is producing a siren note of 3 kHz and approaching a listener. (The speed of sound in air is 330 m s^{-1}.)

37 Using the relative velocities given in Figure 5.61 for the descent of Huygens through Titan's atmosphere, calculate the maximum value of $\frac{\Delta f}{f}$

(a) in the descent as originally planned

(b) in the modified descent.

Double Doppler

When waves from a stationary transmitter/receiver are reflected from a moving target back towards their source, they are Doppler shifted twice. Suppose the target is approaching the source; it encounters waves at a higher frequency than those emitted, and the reflected signal has this higher frequency. But the reflected signal comes from something that is moving towards the transmitter/receiver, so the frequency of waves entering the receiver is greater than the frequency of waves leaving the target. The overall change in frequency is thus *double* that given by Equation 33:

$$\Delta f = \frac{2vf}{c} \qquad (36)$$

The rotation of planets can be studied by radar, using the double Dopper effect to analyse reflections from different parts of the planet's surface (Figure 5.62). The same effect is also used in radar speed traps.

Figure 5.62 Light reflected from Saturn's rings

Question

38 A radar speed trap emits a 10 GHz signal and records a Doppler shift of 2500 Hz for an approaching car. Was the car exceeding the speed limit of 70 mph (about 30 m s^{-1})? (Radar signals travel in air at 3.00 × 10^8 m s^{-1}:1 GHz (gigahertz) = 1 × 10^9 Hz.)

5.3 Summing up Part 5

In this part of the chapter you have seen how reflected radiation can be used to measure distance using a pulse–echo technique, and speed using the Doppler effect.

Pulse–echo reflection is used to measure many other distances. Activity 33 will help you to think about where the technique is applicable, and the advantages, and limitations of the technique.

Activity 33 On reflection

In small groups produce a brainstorm diagram about where the pulse-echo technique is useful, and for each application consider why it is suitable, what kind of wave is used, and what difficulties may occur. If there is time, extend your thinking by doing some research on the Internet, and produce some illustrations for your brainstorm poster.

Activity 34 Summing up Part 5

Look back through your work on this part of the chapter and make sure you know the meaning of all the terms printed in bold.

Then look back at Activity 36 in *The Sound of Music*. Add a new row to your table to include a summary of what you have learnt about the Doppler effect.

6 Mission accomplished

In this chapter you have studied some aspects of electric circuits and power supplies and some aspects of energy transfer processes. You have also studied the photoelectric effect which underlies the operation of solar cells, and learnt how pulse–echo and Doppler techniques enable speed and distance to be measured remotely. This concluding section is intended to help you look back over the whole chapter and consolidate your knowledge and understanding.

6.1 Conservation of …

In studying this chapter you have learned some fundamental pieces of physics that all relate to the important idea of conservation. In everyday language, conservation is often used to mean keeping something as it is, not damaging it, not wasting it, or not using it up. For example, you might talk of conserving the countryside, or conserving fuel.

In its scientific sense, conservation means that some measurable quantity remains unchanged. One example is the conservation of mass. For example, if you are dealing with a complex network of pipes (such as in a water or gas supply system) then conservation of mass is important to bear in mind when considering rates of flow through various parts of the system.

The conservation of charge is another example: no situation or process has ever been found in which the total amount of charge changes. You are probably used to the idea of positive and negative charge and that removing electrons (negatively charged) from an initially uncharged object leaves behind an equal amount of positive charge. The total amount of charge (found by adding up individual positive and negative charges with their correct signs) is always unchanged. We used this fundamental law of nature earlier in this chapter, when we said that the rate of flow of charge (i.e. the current) into a point in an electric circuit must be equal to the rate of flow of charge away from that point. You will probably also have used charge conservation in balancing nuclear and chemical equations.

The conservation of energy is another fundamental law of nature. Even through energy is not 'stuff', nor is it an easily measurable property of matter, it is still possible to define and measure amounts of energy transferred or stored, and no process has ever been found in which energy is either created or destroyed. The law of energy conservation underlies much of the work of this chapter. For example, in looking at energy in an electric circuit in Part 2 we used the fact that the energy supplied by a power source to each coulomb of charge must be equal to the energy delivered to the circuit by each coulomb.

Activity 35 Conservation rules – OK!

Many of the activities, diagrams and questions in this unit illustrate and use the conservation laws discussed above. By copying and completing Table 5.5, which lists several of these examples, you will make a chart showing conservation laws in use. Some of the rows of the table have been filled in to give you the idea – but you might decide to design your own chart in a different way, and perhaps to add sketch diagrams and extra notes.

Reference	Illustrates/uses conservation of ...	Notes
Section 2.1, Figure 5.12 How does a solar cell work?	Charge	Electrons drift across boundary leaving positive charge. Overall charge is still zero.
Activity 4	Energy	Emf (energy supplied to each coulomb) by cells connected in series is equal to sum of emfs of individual cells
Resistors in series and parallel	Charge	
Resistors in series and parallel	Energy	
Activity 7 Maximising the power	Energy	Total power supplied by cell = power in external circuit + power in internal resistance
Qs 16–18		
Energy and temperature change	Energy	
Activity 20	Mass	
Activity 20		
Q25		
Activity 15		

Table 5.5 Conservation laws in use

6.2 Modelling light

In Part 3 of this chapter you used a photon model to explain the photoelectric effect and in Part 5 you learnt about the Doppler effect, which was explained using waves. In some situations the behaviour of light (and other electromagnetic radiation) can best be explained using photon model, and sometimes a wave model works better. The nature of light has puzzled and intrigued scientists for many centuries, with sometimes one model being favoured and sometimes the other, depending on the experimental evidence available. Looking at the way ideas about light have changed over time illustrates an important aspect of how science works: theories and models are only as good as the experimental evidence that supports them, and may change over time as new evidence becomes available. Nowadays, we accept that we need two ways of modelling light, and that sometimes one is better and sometimes the other. But this view has only come about because of careful experiment and observation accompanied by deep thinking, with quite a bit of controversy along the way as you will see in Activity 36.

Activity 36 A brief history of light

Use the Internet and other resources to find out how ideas about light have changed over time, and something about the people involved in investigating and publicising them. Try to include the following:

- Ancient Greek and Roman ideas about light (e.g. Democritus, Lucretius, Augustine of Hippo)
- Arab scientists' ideas and work on the nature of light (e.g. ibn Al-Haytham)
- Isaac Newton's theory of corpuscles
- Thomas Young's experiments
- the photoelectric effect (Planck, Einstein).

Summarise your findings on a timeline that can be displayed as a poster.

> **Study note**
>
> You also used both the wave and photon model in *The Sound of Music*.

6.3 To boldly go …?

In this chapter, you have read about several examples of space missions, ranging from the small UoSATs via communications and weather satellites to the ambitious International Space Station and interplanetary probes. Since the mid 20th century, such missions have attracted multi-million pound investment from governments and commercial organisations, but they have also attracted controversy. The cost of space exploration is huge. NASA estimates that it will cost $100,000,000,000 to set up a permanent Moon base and the European Space Agency believes that it will need $27 billion to be a part of that presence. It costs roughly ten times more to send a human into space compared with a robot, so why bother to send people?

Currently the UK government contributes to a number of space projects through the European Space Agency but has no plans to undertake human spaceflight programmes. Recently, though, the government has indicated a willingness to look again at this. The Royal Astronomical Society published a report looking at the broader benefits of the UK's involvement in such projects.

Some people believe money used on space research is well spent, but others are concerned about ethical issues connected with space research.

Controversy has also arisen over the use of animals in space research. Both the Americans and the Soviets used animals (dogs, chimpanzees, turtles) in a number of their pre-human flights. Some, like Ham the chimp (Figure 5.63) and Belka the dog returned safely, but Leika was not so lucky: she burned up in orbit.

Figure 5.63 Ham the chimp after his journey in space

In Figure 5.64 Soviet scientists examine the first creatures from Earth to travel to the Moon. These turtles, along with some worms and fruit flies completed a circumlunar flight in Zond 5 (a modified Soyuz spacecraft) in October 1968. Because the

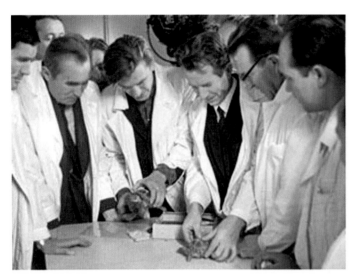

Figure 5.64 Soviet scientists examine some animals used in Moon voyages

spacecraft did not quite perform as expected on re-entry into the Earth's atmosphere it was decided not to risk a human crew for the next mission. In the United States NASA had no compunction against sending three astronauts on a circumlunar flight (Apollo 8 December 1968), even though the Saturn V rocket had not been tested with a crew before.

Another contentious issue is the military use of space. During the 1980s the United States government developed a number of space-based weapons under the umbrella of the Strategic Defence Initiative (or, as it was known in the press, 'Star Wars'). These included high-energy lasers, particle weapons, kinetic energy weapons ('Rods from the Gods') and interceptor rockets. While only a few of these reached any usable stage some of the research has been incorporated in modern weapons. The Patriot missile system used against SCUD missiles in the first Gulf war for example.

Recently the Chinese government used one satellite to destroy another and they have also used ground-based lasers to blind United States' Keyhole reconnaissance satellites as they pass over the Peoples' Republic. This latter action interferes with the Open Skies principle where satellites are able to over-fly other countries unmolested.

There has also been recent tension between Russia and NATO over the so-called 'Missile Shield' that the United States has been planning to install in eastern Europe (Figure 5.65). The Missile Shield violates the Anti Ballistic Missile Treaty (ABM) signed by the US and the Soviet Union in 1972.

Figure 5.65 Artist's concept of space-based interceptor system, the Strategic Defence Initiative (SDI)

What do you think of these issues? Use Activity 37 to explore your own views and those of other students.

Activity 37 Ethical issues and space

Explore one of the following questions relating to space research.

- Should we spend vast sums on space programmes? Or would the money be better spent on other things?
- Should we send humans to explore space when robots are far cheaper and are able to perform many of the same tasks that people can?
- Should animals be used in space research?
- Can weapons in space ever be justified?

Begin by finding out some factual background to your question and write a brief report. (How much does the UK government spend each year on space research and exploration? How have animals been used in space research?)

Then think about and write down your own views on the question. Try to relate these to one or more commonly used ethical frameworks.

Finally, share your views with other students in a discussion. Try to persuade them to your point of view, but also be prepared to listen and maybe to change your own mind.

Study note

Some widely used ethical frameworks were summarised in the chapter *Spare Part Surgery*.

As well as ethical issues, there are some perhaps unexpected environmental implications of exploring space: litter. In days gone by, seafarers threw their rubbish overboard without a second thought, believing the oceans to be so vast that a bit of litter would not matter, and industries deposited their waste into rivers and onto land. Similarly, space missions have resulted in debris being deposited in space. After all, space is vast and empty, isn't it? We now realise some of the environmental issues associated with pollution on Earth. Efforts are made to reduce it, and there are laws governing industrial processes and the disposal of waste. But what about space?

Activity 38 Space junk

Research and discuss the problem of 'space junk'.

Type 'space junk' into an Internet search engine and make a list of some of the items that have been 'dropped' in space.

Then list the possible hazards associated with these items.

Finally, list some of the solutions that have been proposed to deal with the problem. For each, say whether you think it would be effective, and whether it might be difficult or costly. Say which solutions you think should be put into practice.

6.4 Questions on the whole chapter

39 Electric vehicles were first introduced in the late 19th century. At the turn of that century, around 40% of all motor vehicles were powered by electricity, far more than by petrol. Recent years have seen a revival of interest in entirely battery-powered vehicles.

(a) The Elcat Cityvan 2000, a minivan developed in Finland, has a set of six lead–acid batteries of emf 12.0 V connected to provide 72.0 V. Draw a diagram to show how the batteries must be connected together.

(b) The vehicle is powered by a DC motor which, when under the greatest load, draws a current of 300 A from the batteries.

   (i) Given that each lead–acid battery has an internal resistance of 0.0065 Ω calculate the voltage (terminal potential difference) across all six batteries under this condition.

   (ii) Calculate the power transferred by the resistance of each battery while there is a current of 300 A in the circuit.

   (iii) What effect will this power transfer have on the batteries?

(c) The set-up described here does not maximise the power transferred by the batteries. Explain why it would in practice be unwise to design the van circuit to maximise the power transfer.

40 The total electrical power generated in the world today is about 10^9 kW. In the rich countries this works out at roughly 1 kW per person. During the next century the world's population is likely to grow to about 10^4 million. To provide everyone with the 'energy standard' of 1 kW, some 10^{10} kW of electricity will need to be produced.

One possible way to achieve this large increase in power might be to build Space Power Satellites (SPS) – a concept first conceived in the 1960s. This would involve building very large solar arrays in geostationary orbit and beaming the power developed by their solar cells to Earth by microwaves (see Figure 5.66).

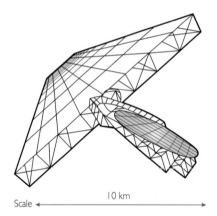

Scale ← 10 km →

Figure 5.66 Artist's impression of a Space Power Satellite

(a) Calculate the power that could be provided by an SPS array of area 5×10^7 m² (50 km²) if the combined efficiency of the array, transmitter and receiver is 8%, and the solar radiation flux is 1400 W m⁻².

(b) How many such arrays would be needed to provide 10^{10} kW?

41 Table 5.6 shows the energy transferred to a solar water heating panel of area 1 m² at latitude 52 °N for various times of year and 'angles of tilt' (Figure 5.67).

(a) What angle of tilt would maximise the transfer of energy

(i) in January

(ii) in July?

(b) Explain whether, for this site at latitude 52 °N, it would be best to have an angle of tilt of 40°, 50° or 60° to maximise the total energy transferred over a whole year. (It might be helpful to assume that all months are the same length.)

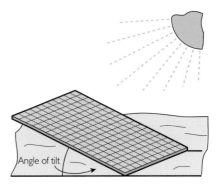

Figure 5.67 The 'angle of tilt' of a solar panel

Month	Angle of tilt									
	0°	10°	20°	30°	40°	50°	60°	70°	80°	90°
	Maximum energy transferred/MJ per day (for angle of tilt shown)									
Jan	4.0	5.8	7.2	8.6	9.7	10.4	11.2	11.5	11.2	10.8
Feb	6.8	9.0	10.8	12.6	13.7	14.8	15.1	15.1	14.8	14.0
Mar	13.3	15.5	17.6	19.1	20.2	20.5	20.5	19.8	18.4	16.6
Apr	20.5	22.3	23.8	24.8	24.8	24.1	22.7	20.6	18.4	15.1
May	26.3	27.7	28.4	28.8	27.4	25.2	23.0	19.8	16.6	13.0
Jun	28.4	28.8	29.2	29.2	27.4	25.2	22.3	19.1	15.1	11.2
Jul	28.1	28.4	28.8	29.2	27.4	25.6	23.0	20.2	16.2	12.2
Aug	23.0	24.8	25.6	25.9	26.3	24.8	22.7	20.5	17.3	13.7
Sep	16.2	18.7	20.5	21.6	22.3	22.7	21.6	20.5	18.7	16.2
Oct	9.0	11.5	13.7	15.1	16.6	17.3	17.6	17.3	16.6	15.5
Nov	5.0	6.8	8.6	10.1	11.2	12.2	12.6	13.0	12.6	12.2
Dec	3.2	4.7	6.1	7.2	8.3	9.0	9.7	10.1	10.1	9.7

Table 5.6 Data for Question 41

42 If a photoelectron is released from a layer of photosensitive material with no kinetic energy, then accelerated through 100 V, what will be its final kinetic energy? Give your answer in eV and in joules.

43 Suppose light produces a current of 6.5×10^{-11} A in a solar cell. How many electrons must be released every second?

44 In an incandescent light bulb, a thin piece of tungsten wire (the filament) is connected in series to the thick copper connecting wires. Two students are discussing the situation. Student A says 'the electrons must be moving at the same speed in all the wires'. Student B says 'the copper probably has a greater number-density of electrons because it's a good conductor'. Write a few sentences to explain whether you agree with what they are saying.

45 Amateur astronomers observing Saturn's rings found that at a wavelength of 616.0 nm there was a difference of 0.07 nm between the light reflected from Saturn and the light reflected from the outer edge of Saturn's ring.

(a) How fast was this edge of the ring moving, compared with the planet?

(b) How could measurements from different positions on the ring reveal whether Saturn's rings are solid or not?

6.5 Achievements

Now you have studied this chapter you should be able to achieve the outcomes listed in Table 5.7.

Table 5.7 Achievements for the chapter *Technology in Space*

Statement from examination specification	*Section(s) in this chapter*
50 describe electric current as the rate of flow of charged particles and use the expression $I = \dfrac{\Delta Q}{\Delta t}$	2.1
51 use the expression $V = \dfrac{W}{Q}$	2.1
52 recognise, investigate and use the relationships between current, voltage and resistance, for series and parallel circuits, and know that these relationships are a consequence of the conservation of charge and energy	2.1, 2.2, 6.1
53 investigate and use the expressions $P = VI$, $W = VIt$. Recognise and use related expressions e.g. $P = I^2R$ and $P = \dfrac{V^2}{R}$.	2.3
54 use the fact that resistance is defined by $R = \dfrac{V}{I}$ and that Ohm's law is a special case when $I \propto V$	2.2
55 demonstrate an understanding of how ICT may be used to obtain current–potential difference graphs, including non-ohmic materials and compare this with traditional techniques in terms of reliability and validity of data	2.2, 4.2
56 interpret current–potential difference graphs, including non-ohmic materials	2.2
59 define and use the concepts of emf and internal resistance and distinguish between emf and terminal potential difference	2.1, 2.2, 2.3
70 recognise and use the expression efficiency = [useful energy (or power) output]/[total energy (or power) input]	2.3, 4.1
60 investigate and recall that the resistance of metallic conductors increases with increasing temperature and that the resistance of negative temperature coefficient thermistors decreases with increasing temperature	4.2
61 use $I = nqvA$ to explain the large range of resistivities of different materials	4.3 (and see DIG)
62 explain, qualitatively, how changes of resistance with temperature may be modelled in terms of lattice vibrations and number of conduction electrons	4.3

Statement from examination specification	Section(s) in this chapter
63 explain how the behaviour of light can be described in terms of waves and photons	3.2, 5.2, 6.2
64 recall that the absorption of a photon can result in the emission of a photoelectron	3.2
65 understand and use the terms threshold frequency and work function and recognise and use the expression $hf = \phi + \frac{1}{2}mv^2_{max}$	3.2
66 use the non-SI unit, the electronvolt (eV) to express small energies	3.2
67 recognise and use the expression $E = hf$ to calculate the highest frequency of radiation that could be emitted in a transition across a known energy band gap or between known energy levels	3.2 (and see MUS)
69 define and use radiation flux as power per unit area	4.1
71 explain how wave and photon models have contributed to the understanding of the nature of light	6.2
72 explore how science is used by society to make decisions, e.g. the viability of solar cells as a replacement for other energy sources, the uses of remote sensing	6.3
29 identify the different regions of the electromagnetic spectrum and describe some of their applications	1.2 (and see MUS + DIG)
46 explore and explain how a pulse–echo technique can provide details of the position and/or speed of an object and describe applications that use this technique	5.1, 5.2
47 explain qualitatively how the movement of a source of sound or light relative to an observer/detector gives rise to a shift in frequency (Doppler effect) and explore applications that use this effect	5.2
48 explain how the amount of detail in a scan may be limited by the wavelength of the radiation or by the duration of pulses	5.1 (and see DIG)
49 discuss the social and ethical issues that need to be considered, e.g. when developing and trialling new medical techniques on patients or when funding a space mission	6.3

Answers

1   (a)   Fuel cells, with or without the addition of cryogenic engines. (The point representing 1 kW, 1 week, lies more or less on the line between these two regions of the graph.)

   (b)   Solar and nuclear dynamic systems. (Notice that 10^1 kW means 10 kW.)

2   (a) $\Delta Q = I\Delta t = 0.5$ C s^{-1} × 2.0 s = 1.0 C

   (b) $\Delta W = \mathcal{E}\Delta Q$

   (c) $\Delta W = 1.5$ J C^{-1} × 1.0 C = 1.5 J

3   In circuits (a), (c) and (e) the cells are in series so their voltages add: (a) 1.0 V, (c) 1.5 V (e) 2.0 V. Circuits (b), (d) and (f) all give 0.5 V as the cells are joined in parallel.

   In (g) each pair of cells in series gives 1.0 V. The two pairs are joined in parallel so the net output will also be 1.0 V.

4   A single row of 56 cells joined in series gives an output of 28 V. (Several rows of 56 cells could be connected in parallel to give the same output voltage.)

5   2 Ω. Read any pair of values for the graph e.g. when $I = 3$ A, $V = 6$ V.
   $R = \dfrac{V}{I} = \dfrac{6V}{3A} = 2\Omega.$

6   Graphs (b) and (c) show ohmic behaviour – they are straight lines through the origin.

7   The steeper the graph, the higher the resistance. (Graph (b) shows a higher resistance than graph (c).)

8   In (a) the resistance increases at large currents and voltages and in (d) the resistance decreases as the current is increased. You can check this by reading values from the graphs and calculating resistance. For example, in (a), when $I = 2$ A, $V = 1$ V and so $R = 0.5$ Ω; when $I = 3$ A, $V = 3$ V so $R = 1$ Ω (an increase). In (d), when $I = 1$ A, $V = 2$ V so $R = 2$ Ω, and when $I = 3$ A, $V = 3$ V and $R = 1$ Ω (a decrease).

9   Figure 5.68 shows a graph of the measurements. It is not possible to draw a straight line through the origin that also passes through all the error boxes, so the material does not obey Ohm's law. (Its resistance decreases as current and voltage increase.)

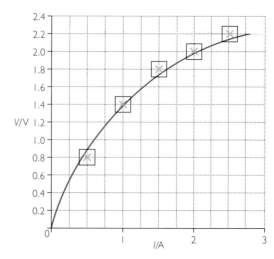

Figure 5.68 The answer to Question 9

10   (a) $\dfrac{1}{R} = \left(\dfrac{1}{1} + \dfrac{1}{2} + \dfrac{1}{4} + \dfrac{1}{8}\right)\Omega^{-1} = 1.875\ \Omega^{-1}$
   $R = \dfrac{1}{1.875}\ \Omega = 0.53\ \Omega$

   (b)   First replace the 3 Ω and 6 Ω combination:
   $\dfrac{1}{R} = \left(\dfrac{1}{3} + \dfrac{1}{6}\right)\Omega^{-1} = 0.5\ \Omega^{-1},\ R = 2\ \Omega$
   The complete arrangement is then equivalent to two 2 Ω resistors in series, giving a total resistance of 4 Ω.

   (c)   First replace the series of 4 Ω + 2 Ω by a single 6 Ω resistor. The complete arrangement is then equivalent to two 6 Ω resistors in parallel.
   $\dfrac{1}{R} = \left(\dfrac{1}{6} + \dfrac{1}{6}\right)\Omega^{-1},$ and so $R = 3\ \Omega.$

11   $P = IV,\ I = \dfrac{P}{V} = \dfrac{2000\ \text{W}}{230\ \text{V}} = 8.7$ A.

12   $P = \dfrac{V^2}{R},\ R = \dfrac{V^2}{P} = \dfrac{(230\ \text{V})^2}{60\ \text{W}} = 882\ \Omega.$

13   Current $I = \dfrac{\mathcal{E}}{R + r} = \dfrac{3.0\ \text{V}}{6\ \Omega} = 0.50$ A.

   $P_{\text{out}} = (0.50\ \text{A})^2 × 1\ \Omega = 0.25$ W.

   Total power is:

   $P_{\text{in}} = (0.50\ \text{A})^2 × 6\ \Omega = 1.50$ W.

   Efficiency $= \dfrac{0.25\ \text{W}}{1.50\ \text{W}} = 0.17 = 17\%.$
   (Alternatively, efficiency $= \dfrac{R}{R + r}$
   $= \dfrac{1\ \Omega}{6\ \Omega} = 0.17 = 17\%.$)

   This would not be a sensible arrangement for several reasons. The output power is less than it would be under conditions for maximum power (0.25 W compared with over 0.45 W when

$R = r$), but the total power transferred by the battery is greater than it would be if $R = r$ (1.5 W compared with 0.9 W), so the battery would run down more quickly. Most of the battery power (80% of it) is wasted due to internal heating in the battery.

14 (a) The cells are in series so the total emf and the total internal resistance are found by adding the separate emfs and internal resistances.

   (i) $6 \times 0.005\ \Omega = 0.03\ \Omega$.

   (ii) $6 \times 2.000\ V = 12.000\ V$.

   (b) (i) The internal resistance and the load form a series circuit, with resistance $R + r = (2.97 + 0.03)\ \Omega = 3.00\ \Omega$.

   (ii) $I = \dfrac{\mathscr{E}}{R + r} = \dfrac{12\ V}{3.00\ \Omega}$
         $= 4.0\ A$ (Equation 9).

   (iii) Terminal pd $V = IR = 4.0\ A \times 2.97\ \Omega$
         $= 11.9\ V$.

   (c) If the current in the battery is 200 A, then the 'lost volts' $Ir = 200\ A \times 0.03\ \Omega = 6\ V$. The terminal pd must therefore fall to 12 V − 6 V = 6 V. If the headlamps are designed to be connected to a pd of 12 V they will dim noticeably.

15 (a) Internal resistances combine just like ordinary resistances in series and parallel, as the following calculations for arrangements (ii) − (iv) show.

   (ii) Total resistance $r = 4 \times 30\ \Omega = 120\ \Omega$.

   (iii) $\dfrac{1}{r} = \left(\dfrac{1}{30} + \dfrac{1}{30}\right)\Omega^{-1} = \dfrac{1}{15}\ \Omega^{-1}$ so $r = 15\ \Omega$.

   (iv) Each pair of cells in series has resistance 60 Ω.
         $\dfrac{1}{r} = \left(\dfrac{1}{60} + \dfrac{1}{60}\right)\Omega^{-1} = \dfrac{1}{30}\ \Omega^{-1}$ so $r = 30\ \Omega$.

   (b) Power transfer will be maximum when the external load resistance is equal to the total internal resistance. Each set of four solar cells has a total internal resistance of 120 Ω. The two sets in parallel have a total resistance of 60 Ω so the external load must be 60 Ω.

16 (a) Energy transferred = 5000 eV
            $= 5000 \times 1.60 \times 10^{-19}\ J$
            $= 8.00 \times 10^{-16}\ J$.

   (b) The answer the same as (a), because the same amount of charge has been accelerated through the same pd.

17 Rearranging Equation 22, $v = \sqrt{\dfrac{2E_k}{m}}$. To be sure of getting speeds in m s^{-1}, you need to express the energies in joules, not eV.

Electron: $v = \sqrt{\dfrac{2E_k}{m_e}}$

$= \sqrt{\dfrac{2 \times 8.00 \times 10^{-16}\ J}{9.11 \times 10^{-31}\ kg}}$

$= 4.19 \times 10^{7}\ m\ s^{-1}$.

Proton: $v = \sqrt{\dfrac{2E_k}{m_p}}$

$= \sqrt{\dfrac{2 \times 8.00 \times 10^{-16}\ J}{1.67 \times 10^{-27}\ kg}}$

$= 9.79 \times 10^{5}\ m\ s^{-1}$.

18 Using Equation 17:

Photon energy

$E_{ph} = hf$

$= \dfrac{hc}{\lambda}$

$= \dfrac{6.63 \times 10^{-34}\ J\ s \times 3.00 \times 10^{8}\ m\ s^{-1}}{434 \times 10^{-9}\ m}$

$= 4.58 \times 10^{-19}\ J$.

$= \dfrac{4.58 \times 10^{-19}\ J}{1.60 \times 10^{-19}\ J\ eV^{-1}}$

$= 2.86\ eV$.

Then using Equation 18:

$E_k = E_{ph} - \phi$

$= 2.86\ eV - 2.28\ eV$

$= 0.54\ eV$.

$= 0.54 \times 1.60 \times 10^{-19}\ J$

$= 0.864 \times 10^{-19}\ J$

$= 8.64 \times 10^{-20}\ J$.

19 (a) The photon energy is less than the work function, so no photoelectrons will be released. (The absorbed radiation will just produce a slight heating of the metal.)

(b) Ultraviolet radiation has a shorter wavelength than visible light, hence a higher frequency and more energetic photons. The photoelectrons released from sodium will therefore more kinetic energy than those released by blue light.

20 Your answer should make the point that input power to solar panels varies with angle. In order to intercept as much radiation as possible, the panels must point directly at the Sun – a change of angle will result in a loss of power. You could go on to say that, provided the panels are within a few degrees of being square on, the loss of power will not be great. But once the misalignment is more than, say 20°, then the power drops by a large fraction. A diagram similar to Figure 5.40 would be suitable.

21 $P_{out} = 0.11 \times P_{in}$

so:

$P_{in} = \dfrac{P_{out}}{0.11}$

$= \dfrac{16 \text{ kW}}{0.11}$

$= 147 \text{ kW}$

$= 1.47 \times 10^5 \text{ W}.$

$P_{in} = FA$ $\qquad (\theta = 0°)$

so:

$A = \dfrac{P_{in}}{F}$

$= \dfrac{147 \text{ kW}}{1.4 \text{ kW m}^{-2}}$

$= 105 \text{ m}^2 \ (\approx 10^2 \text{ m}^2)$

or:

$A = \dfrac{P_{in}}{F}$

$= \dfrac{147 \times 10^5}{1400 \text{ W m}^{-2}}$

$= 105 \text{ m}^2 \ (\approx 10^2 \text{ m}^2)$

(Notice that you can either use W or kW but you must be consistent and not use a mixture.)

22 Input power is greatest when the arrays are square-on to the incident radiation.

$P_{in} = FA$

$= 600 \text{ W m}^{-2} \times 5000 \text{ m}$

$= 3.00 \times 10^6 \text{ W}$

$= 3.00 \times 10^3 \text{ kW}.$

Efficiency $= \dfrac{P_{out}}{P_{in}}$

$= \dfrac{435 \text{ kW}}{(3.00 \times 10^3 \text{ kW})}$

$= 0.145$

$= 14.5\%.$

23 (a) (i) about 38%

(ii) about 94%

(b) From about 95% to about 75%, i.e. by about 20% of the maximum available capacity.

24 (a) (i) gallium arsenide (GaAs)

(ii) cadmium sulphide (CdS)

(iii) gallium arsenide/gallium phosphide (GaAs/GaP)

(b) It would not be sensible to use gallium and silicon as their efficiencies have already fallen to zero and so no energy transfer would take place.

(c) At 0 °C cadmium sulphide cell's efficiency is near 29% but this falls to about 6% at 400 °C.

25 From Figure 5.45, a GaAs array would have an efficiency of about 25% at 50 °C. If P_{out} is to be 1×10^6 W, and P_{out} is 25% of P_{in}, then $P_{in} = 4 \times 10^6$ W.

If the entire array is square-on to the incident solar flux, then $\theta = 0°$ and $P_{in} = FA$ so $A = \dfrac{P_{in}}{F}$.

$\dfrac{P_{in}}{F} = \dfrac{4 \times 10^6 \text{ W}}{(1.4 \times 10^3 \text{ W m}^{-2})} = 2.8 \times 10^3 \text{ m}^2.$

26 $R_\theta = R_0(1 + \alpha\theta) = R_0 + \alpha R_0\theta$

$R_{80} = 50.0 \ \Omega \ (1 + 4.28 \times 10^{-3} \ °\text{C}^{-1} \times 80 \ °\text{C})$

$= 0.50 \ \Omega \times 1.324 = 0.67 \ \Omega.$

The percentage change in resistance is quite large:

$\dfrac{0.17 \ \Omega}{0.50 \ \Omega} \approx 34\%.$

However, the resistance of the connecting wires in a circuit is likely to be much less than that of other components, so can usually be ignored. In a circuit that contained a large amount of copper (in coils of electromagnets, for example) the effect of temperature would need to be considered.

27 (a) Copper, constantan, steel and tungsten all have positive temperature coefficients. Carbon is the odd one out with a negative coefficient.

(b) Its resistance is near coefficient over a fairly wide temperature range.

(c) Figure 5.69 shows a plot of all the data from Table 5.3. The temperature coefficients are: carbon $5.0 \times 10^{-4}\,°C^{-1}$, steel $3.3 \times 10^{-3}\,°C^{-1}$, tungsten $5.2 \times 10^{-3}\,°C^{-1}$.

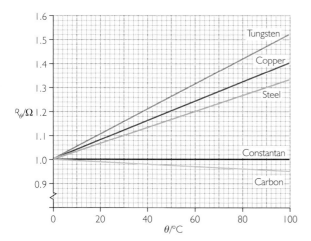

Figure 5.69 The answer to Question 27

28 Increasing the pd supplies more kinetic energy to each charged particle. The drift speed increases hence the current increases.

29 The samples are in series so there is the same current, I, in each. As each has the same area, A, and (we can assume) charges with the same charge, q, the drift speed, v, in the semiconductor must be much greater than in the metal to compensate for the much smaller number density, n.

30 (a) Volume occupied by each atom,
$V = (5 \times 10^{-10}\,m)^3 = 1.25 \times 10^{-28}\,m^3$.

Number density $n = \dfrac{\text{number of free charges}}{\text{volume occupied by that number}}$

$= \dfrac{1}{1.25} \times 10^{-28}\,m^3 = 8.00 \times 10^{27}\,m^{-3}$

(b) $A = (0.5 \times 10^{-3}\,m)^2 = 2.5 \times 10^{-7}\,m^2$

From Equation 32,

$v = \dfrac{I}{nAq}$

$= \dfrac{0.2\,A}{8.00 \times 10^{27}\,m^{-3} \times 2.5 \times 10^{-7}\,m^2 \times 1.60 \times 10^{-19}\,C}$

$= 6.25 \times 10^{-4}\,m\,s^{-1}$.

31 Distance $x = 378\,000\,km = 3.78 \times 10^8\,m$

Distance there and back, $2x = 2 \times 3.78 \times 10^8\,m$

time $t = \dfrac{2x}{v}$

$= \dfrac{2 \times 3.78 \times 10^8\,m}{3.00 \times 10^8\,m\,s^{-1}}$

$= 2.52\,s$

32 (a) Angle SVE = 90° because it is the angle between a radius and a tangent.

(b) 11.2 minutes = $11.2 \times 60\,s = 672\,s$

so distance EV $= \dfrac{3.00 \times 10^8\,m\,s^{-1} \times 672\,s}{2}$

$= 1.008 \times 10^{11}\,m$.

(c) $\dfrac{EV}{ES} = \cos 48° = 0.669$

so ES $= \dfrac{1.008 \times 10^{11}\,m}{0.669}$

$= 1.5 \times 10^{11}\,m$.

(d) The Sun emits so much radiation that it would completely swamp the reflected signal. Also, reflections would occur from several levels in the gaseous atmosphere in the Sun so the reflected pulse would be smeared out.

(e) Venus and Mercury can only be seen close to the Sun, because they are nearer to the Sun than we are – we never see them in the night sky opposite the Sun. Mercury is very faint. Venus appears as the Sun sets, or rises just before the Sun – of course it is there in the daytime sky but not visible.

33 From first position of rocket, distance to Moon

$x_1 = \dfrac{1\,s \times 3.00 \times 10^8\,m\,s^{-1}}{2} = 1.50 \times 10^8\,m$.

From second position, distance

$x_2 = \dfrac{0.94\,s \times 3 \times 10^8\,m\,s^{-1}}{2} = 1.41 \times 10^8\,m$.

Distance travelled, $x = x_1 - x_2 = 0.09 \times 10^8\,m$

Speed $v = \dfrac{0.09 \times 10^8\,m}{50s} = 1.8 \times 10^5\,m\,s^{-1}$

34 (a) The sound travel time, t, in each layer is $t = \dfrac{2x}{v}$. Table 5.8 lists the travel times, and the time delay before receiving the each from the far side of each layer. For example, the echo from the far side of the fat layer is received after the pulse has travelled 'there and back' through the fat *and* the soft tissue.

Tissue	Thickness x/cm	Speed v/m s^{-1}	Travel time/ 10^{-5} s	Time delay of echo/ 10^{-5} s
soft tissue	5.0	1500	6.7	6.7
fat	0.5	1450	0.7	7.4
muscle	1.0	1600	1.3	8.7
blood	5.0	1570	6.4	15.1

Table 5.8 The answers to Question 34(a)

(b) See Figure 5.70.

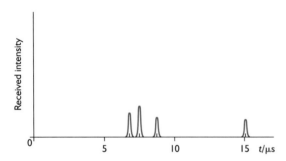

Figure 5.70 The answer to Question 34(b)

(c) (i) The reflected pulses would be distinct from one another, approximately as indicated in Figure 5.70.

(ii) Some of the reflected pulses would overlap, as the interval between them is less than the duration of the pulse. The scan would not be able to provide such detailed information as in (i).

35 (a) $c = f_{em}\lambda_{em}$ or $f_{em} = \dfrac{c}{\lambda_{em}}$ or $\lambda_{em} = \dfrac{c}{f_{em}}$

(b) $x = c - v$

(c) $\lambda_{rec} = \dfrac{(c-v)}{f_{em}}$

(d) $\Delta\lambda = \lambda_{em} - \lambda_{rec} = \dfrac{c}{f_{em}} - \dfrac{(c-v)}{f_{em}} = \dfrac{v}{f_{em}}$

(e) $\Delta\lambda = \dfrac{v}{f_{em}} = v \times \dfrac{\lambda_{em}}{c}$, so $\dfrac{\Delta\lambda}{\lambda_{em}} = \dfrac{v}{c}$

(f) $f_{rec} = \dfrac{c}{\lambda_{rec}} = \dfrac{cf_{em}}{(c-v)}$

(g) $\Delta f = f_{em} - f_{rec}$

$= f_{em} - \dfrac{cf_{em}}{(c-v)}$

$= f_{em} \times \dfrac{(c-v-c)}{(c-v)}$

$= \dfrac{-vf_{em}}{(c-v)}$.

If $v \ll c$, then $c - v \approx c$, so $\Delta f \approx \dfrac{-vf_{em}}{c}$

and $\dfrac{\Delta f}{f} \approx \dfrac{v}{c}$.

36 From Equation 33, $\dfrac{\Delta f}{f_{em}} \approx \dfrac{v}{c}$, so

$\Delta f \approx \dfrac{3000 \text{ Hz} \times 20 \text{ m s}^{-1}}{330 \text{ m s}^{-1}} = 182 \text{ Hz}.$

37 (a) Using Equation 33, $\dfrac{\Delta f}{f}$

$= \dfrac{-5.8 \times 10^3 \text{ m s}^{-1}}{3.0 = 10^8 \text{ m s}^{-1}}$

$= -1.93 \times 10^{-5}.$

(b) $\dfrac{\Delta f}{f} = \dfrac{-2.7 \times 10^3 \text{ m s}^{-1}}{3.00 = 10^8 \text{ m s}^{-1}}$

$= 0.90 \times 10^{-5}.$

38 Using Equation 36:

$\dfrac{\Delta f}{f_{em}} \approx \dfrac{2u}{v}$

and so speed of car,

$u = \dfrac{v\Delta f}{2f_{em}}$

$= \dfrac{3.0 \times 10^8 \text{ m s}^{-1} \times 2500 \text{ Hz}}{2 \times 10^{10} \text{ Hz}}$

$= 37.5 \text{ m s}^{-1}$

so it *is* exceeding the speed limit.

CHAPTER 6

Digging Up the Past

Why a chapter called *Digging Up the Past*?

Imagine the scene: it is a frosty Sunday morning in the middle of the English countryside. From a collection of parked cars and bikes emerge a group of people in warm clothing, their breaths dissipating as clouds through the blazing sunlight as they proceed to pace purposefully around a field of grass. This is an archaeological team at work. And who is that person in the middle? Well, believe it or not, it is a physicist! Modern physics now finds a place at the heart of historical research. This chapter will reveal how archaeologists use physics to locate, date and investigate.

The story starts with some non-destructive investigations that use electrical measurements to locate likely areas with buried features. You will discover how the secrets of England's greatest battle, the Norman Invasion, are being revealed using physics in Sussex. Resistance, resistivity and electric potential can all be understood in the context of probing beneath the soil.

Figure 6.1 shows three artefacts from archaeological finds. Consider the recent history of each. How did anyone know it was under the ground and how did they know where to look? How can we be sure about its age? What about the structure and material composition? How could we answer these questions? These are the very questions that require an understanding of physics. History and physics meet in these three photographs.

Figure 6.1a Archaeological artefacts: ceremonial iron dagger

Figure 6.1b Archaeological artefacts: stone wall

The story concludes with techniques used to investigate and analyse artefacts that have emerged from an archaeological site. The passage of X-rays through objects uncovers secrets of material and hidden structure, and electron microscopy reveals information about the composition and origin of specimens.

Figure 6.1c Archaeological artefacts: terracotta urn

Overview of physics principles and techniques

In this chapter you will build on earlier work to learn more about electric circuits, and about the physics of electromagnetic radiation and diffraction. In doing so, you will be developing your skills in devising and carrying out experimental work, and in using mathematical techniques to display and analyse data. There are also opportunities to use spreadsheets and the internet.

In this chapter you will extend your knowledge of:

- properties of waves from *The Sound of Music*
- properties of materials, resistance and DC circuits from *Spare Part Surgery* and *Technology in Space*
- energy and using graphs from *Higher, Faster, Stronger*.

In other chapters you will do more work on:

- resistance and DC circuits in *Transport on Track*
- properties of materials in *The Medium is the Message*
- waves in *Build or Bust?*
- using graphs in *The Medium is the Message* and *Probing the Heart of the Matter*.

But back to the beginning. The interesting objects are under the soil somewhere, but where? Locating the correct site is the challenge ahead of us. You are now a geophysicist with an interest in archaeology – read on.

1 The secrets of resistance

The trained forensic scientist will look for clues at the scene of the crime, and these clues will be readily available on the surface. A hair on a table or a blood stain on a carpet might well be minute fragments of evidence but they are visible and they are accessible. The archaeologist has a much harder task. Time erases surface traces. The clues to the past often lie buried under layers of soil. The only indications that something might lie below the surface are aerial photographs showing perhaps a hint of a ditch or a wall and of course historic records from the day.

When such historical research does hint that there might be something interesting below your feet, it is not a signal to grab your spades and dig. Careless digging may disturb valuable hidden artefacts and, anyway, all the land in the UK is owned by somebody. Your local Archaeological Society is your first port of call if you are interested in delving deeper.

Figure 6.2 Carrying out a resistive survey

1.1 Resistive surveying

Digging is an invasive activity. It changes the site permanently, rather like the destructive testing of a car just to test if the air bag works in the event of a crash. There are other preliminary techniques that archaeologists will use before picking up their spades and trowels. One of the most common is a resistive survey (Figures 6.2 and 6.3).

Figure 6.3 A physicist advises an archaeologist where to dig

The water content of soil helps it to conduct an electric current. Rocks are poorer conductors (more resistive). By sinking two metal probes into the soil and measuring the resistance between them, archaeologists can make a tentative guess at where buried rocks and stone, and so walls, are likely to be. Later in this chapter you will see just how this technique helped reveal the truth about the exact landing site for William the Conqueror's invading army of 1066, perhaps the most important event in the history of England.

Activity 1 Archaeology on the Internet

Use the Internet to find out about archaeological survey techniques and for background information on archaeology in general.

Activity 2 Probing resistance

Use a resistance meter (an ohm meter), or an ammeter and voltmeter, to explore the resistance of various conducting objects. Include some non-uniform conductors such as that shown in Figure 6.4. Discuss your results in a group and make a summary describing how resistance depends on the length and cross-section of a conductor.

The search for the site of the Norman invasion

Local Hastings historian Nick Austin has used soil resistance measurements to uncover what he believes to be the site of the first Norman encampment and landing site prior to the famous battle of Hastings. His survey was carried out between 1993 and 1994, and covered an area 560 m by 320 m that his research had led him to believe was the site of the main Norman fort. Aerial photographs of the area indicated a possible fort outline but gave no clue to subsoil details.

He used a resistance meter to measure soil resistance at 1 m spaces, and used a computer to display the results. Dark patches indicate areas of high resistance, i.e. the likely presence of walls, and light patches indicate low resistance, i.e. the likely presence of damp ditches. Figure 6.5 shows some of his results.

Nick Austin's interpretation of the results identifies a series of ditches and sections of the fort perimeter. The survey corroborates the aerial photographs and Nick's own dowsing results. However, the major finding was the square structure at the centre of the area. This is clearly a man-made structure and is most likely to be the keep of the original Norman fort.

Activity 3 uses a simple model to help you appreciate how resistance measurements can give information about hidden structures. In place of soil you will use conducting Teledeltos paper.

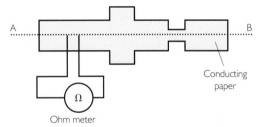
Figure 6.4 Probing the resistance of a non-uniform conductor

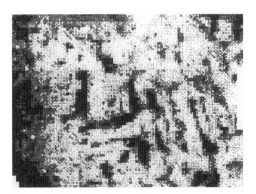
Figure 6.5 Results of a resistance survey

Activity 3 Modelling a resistive survey

In each of two practical challenges you will be presented with the task of saying something about what lies below the surface by probing the resistance between pairs of points. The conducting paper has been shaped to represent places of lower and higher resistance. Your task is to locate the hidden structure simply by taking resistance measurements.

In real fieldwork a system of four probes is used. The outer two supply an emf to drive current through the soil and the inner two measure the potential difference across a standard distance. The measured pd depends on the resistance of the subsoil.

Resistance surveys can be seen in action on the *Time Team* DVDs (Channel 4). They help to confirm the speculation about subsoil structures that aerial surveys and historic document research indicate might be present. Only when three, or more, indicators point towards a definite structure will the diggers start their work.

However, even if there is something hidden mysteriously below the topsoil, it is not always easy to predict its shape. As you saw in Activity 2, different shapes will give different resistances. Just how does shape affect resistance?

1.2 Resistivity

Activity 4 uses conducting putty as a material with a shape which can be easily changed. It will help you to focus your attention on the factors influencing the resistance of a conductor.

Activity 4 Size and shape

Investigate how the resistance of a piece of conducting putty with a fixed area of cross-section depends on its length. Combine your results with those of other students who have used a different area. Use a spreadsheet to plot graphs showing how the resistance varies with length and with area of cross-section.

Resistivity defined

Gathering results from Activity 4 you can show that for a piece of material of length l and cross-sectional area A, resistance is directly proportional to the length of the conductor:

$$R \propto l \qquad (1)$$

and is inversely proportional to the cross-sectional area:

$$R \propto \frac{1}{A} \qquad (2)$$

Combining these two:

$$R \propto \frac{l}{A} \qquad (3)$$

Provided we assume that temperature is constant, the only other variable is the type of material itself. For a given size and shape of sample, some materials have higher

Maths reference

Inverse proportionality
See Maths note 5.4

resistance than others. The number describing this property of the material is called its **resistivity**, and is given the symbol ρ (the greek letter rho). Equation 3 then becomes:

$$R = \frac{\rho l}{A} \qquad (4)$$

If you rearrange Equation 4, it becomes:

$$\rho = \frac{AR}{l} \qquad (4a)$$

Resistivity therefore has units of resistance $\times$ area $\div$ length, in other words resistance $\times$ length, so its SI units are Ω m. The numerical value of a material's resistivity is the same as the resistance of a 1 m long piece with a cross sectional area of 1 m^2, i.e. a 1 metre cube.

Note that the resistivity of a material is a property of that type of material. It is a quantity like density in that it doesn't depend on the particular shape of material. It is pointless looking up the resistance of copper in a data book. You might as well look up the mass of copper. These answers will not be listed because the answer depends on the size and shape of the sample. However, it is quite likely that you will be able to look up a value for the resistivity of copper as well as its density. No other information is required, just the name of the material.

The other side of the resistivity coin is the **conductivity**, symbolised σ (the Greek letter sigma). A good conductor will have a high conductivity and a low resistivity. Resistivity and conductivity are simply the reciprocals of one another:

$$\sigma = \frac{1}{\rho} \qquad (5)$$

The SI units of conductivity are $\frac{1}{\Omega m}$, or $\Omega^{-1} m^{-1}$.

Resistors in circuits

You have studied DC circuits and will know something about resistors connected together either in series or in parallel. We can relate an understanding of resistivity to such combinations.

Maths reference

Index notation and units
See Maths note 2.2

Reciprocals
See Maths note 3.3

Figure 6.6 shows how four blocks of conducting material (resistive putty perhaps), each with resistance R, might be combined. In Figure 6.5(b) the four blocks are in series and can be considered as one long block. By increasing the length, we know the resistance increases, and this is in keeping with the resistor combination formula:

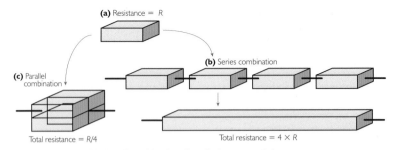

Figure 6.6 Combining four blocks of resistive material

$$R = R_1 + R_2 + R_3 \ldots \qquad (6)$$

Figure 6.6(c) shows how four blocks are equivalent to a single block with four times the cross-sectional area of the original. The four blocks are placed in parallel. We know from experience that the total resistance is reduced. We would expect the resistance to be divided by four, and this is in keeping with the parallel resistance formula:

$$\frac{1}{R} = \frac{1}{R_1} + \frac{1}{R_2} + \frac{1}{R_3} \ldots \qquad (7)$$

> **Study note**
>
> The chapter *Technology in Space* included expressions for combinations of resistors.

Activity 5 Measuring resistivity

You task is to determine an accurate value for the resistivity of a metal made into a wire, using a micrometer to measure its diameter and using electrical measurements when it is connected in a circuit.

Questions

1 A student found the resistance of a 20 cm length of cable to be 8 Ω. What would you expect the resistance of a metre of the cable to be? What assumption have you made?

2 An archaeologist found that two probes placed in soil 0.5 m apart showed a resistance of 300 Ω. What would you expect the resistance to be if the separation of the probes was increased to 5 m? The probe spacing in archaeology is typically 0.5 m rather than 5 m. Suggest a reason for this.

3 (a) The resistance of a 10 cm length of resistive putty was found to be 48 Ω. The putty was then rolled to twice its length. Suggest a value for the resistance between its ends and explain your answer.

(b) The original cross sectioned area of the putty in part (a) was 1 cm². Use this and the data from part (a) to calculate a value for the resistivity of the putty.

4 Derive an expression for the resistance R of a material sample in terms of its length, l, area of cross section A and conductivity σ.

5 The resistance readings in Table 6.1 were taken from an archaeologist's notebook. She suspected that a salt water ditch and a stone wall crossed the path of her linear resistance readings but lie buried beneath the topsoil. Suggest a place for the location of each feature.

Distance from gate to probes / m	Ohm meter reading of resistance between probes / kΩ
1	0.35
2	0.33
3	0.35
4	0.37
5	0.22
6	0.34
7	0.35
8	0.36
9	0.77
10	0.80
11	0.34
12	0.30

Table 6.1 Resistive survey data for Question 5

Comparing resistivities

Table 6.2 lists the resistivities of some materials, and Figure 6.7 is a partially completed bar chart displaying the data.

Material	Resistivity, ρ / Ωm
Copper	1.7×10^{-8}
Glass	1.0×10^{12}
Carbon	1.4×10^{-5}
Perspex	1.0×10^{16}
Lead	2.1×10^{-7}

Table 6.2 Resistivities of some materials

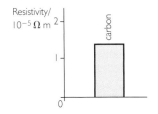

Figure 6.7 A bar chart to display the data from Table 6.2

Question

6 The bar in Figure 6.7, for carbon, is 1.4 cm long. On the same scale, how long would be the bars representing glass and copper? Comment on your answers.

As your answers to Question 6 have shown, there are problems using a linear scale to represent data that range over several orders of magnitude. To compress the range of the numbers, such data are often plotted using a **logarithmic scale** (often shortened to log scale); the logarithm of the value is plotted rather than the value itself.

> **Maths reference**
>
> Combining powers
> see Maths note 1.3

 Activity 6 Plotting on a log scale

Copy Table 6.2, and add another column headed 'log₁₀ of resistivity'. Use your calculator to find the values of $\log_{10}$ resistivity. Now display these values on a bar chart. Select your own scale factor so that the chart fits into a half page of your notes.

> **Maths reference**
>
> Logs and powers of 10
> see Maths note 8.1
> Logs on a calculator
> see Maths note 8.2

Explaining resistivity

A good electrical conductor such as the metal wire you used in Activity 5 has a low resistivity. Electrical insulators have much higher resistivities. **Semiconductors**, as their name implies, are a group of materials in the middle of the resistivity spectrum. Table 6.3 lists some typical resistivities of each of these types of material.

Year	Material	Resistivity / Ω m
good conductors	copper	1.7×10^{-8}
	lead	2.1×10^{-7}
semiconductors	silicon	2.3×10^{3}
	germanium	4.7×10^{-1}
insulators	glass	1.0×10^{12}
	Perspex	1.0×10^{16}

Table 6.3 Conductors, semiconductors and insulators

In the chapter *Technology in Space* you used an expression for the current in a conductor of cross-sectional area A, with n mobile charges per unit volume, each with charge q, drifting with an average speed v:

$$I = nAvq \tag{8}$$

This is a slightly odd expression as it contains a mixture of things that depend only on the material (n and q), along with one that depends on the dimensions of the sample (A) and the average drift speed.

What does drift speed depend on? It must depend on the potential difference, V, applied between the ends of the sample, and it turns out that:

$$v = \frac{kV}{l} \tag{9}$$

where l is the length of the conductor and k is a constant (sometimes known as the 'mobility') that depends on the material. Substituting this expression for v, Equation 8 then becomes

$$I = \frac{nAkVq}{l} \tag{10}$$

We can compare this with the resistance equation:

$$I = \frac{V}{R} \tag{11}$$

to deduce that:

$$R = \frac{l}{nAkq} \tag{12}$$

Then using Equations 4 and 5 we can write new expressions for resistivity and conductivity:

$$\rho = \frac{1}{nkq} \tag{13}$$

$$\sigma = nkq \tag{14}$$

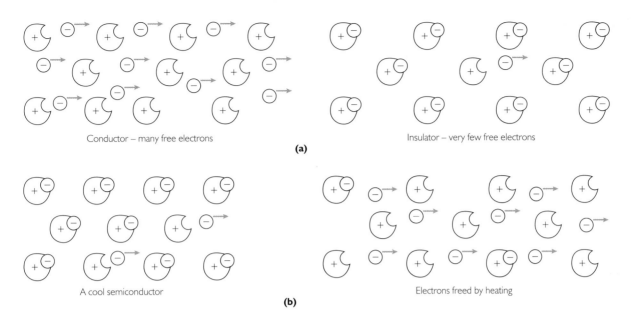

Figure 6.8 Sketches illustrating electrons (a) in a conductor and in an insulator, and (b) in a cool and a warm semiconductor

A simplified model enables us to explain these vast differences in resistivity. The model involves electrons being free from atoms and available to move through the lattice. Figure 6.8(a) illustrates the difference between conductors and insulators.

To explain why some electrons are free to move and others are bound in atoms demands a deep appreciation of the energetics of atoms and is beyond this course. However, it is not difficult to appreciate that, given enough extra energy, a bound electron can become a free electron. The most common example of this effect occurs in semiconductor materials, where a small increase in temperature can have the effect of releasing a large proportion of electrons, vastly increasing their number density n, and so reducing the resistivity of a material dramatically. Semiconductors are very temperature-sensitive (Figure 6.8b) and so find a use in thermistors and temperature-sensing circuits. LDRs (light dependent resistors) free their electrons through energy supplied by light. The result is the same: better conduction.

Question

7 Look back at Section 4.2 of *Technology in Space*, then refer to Equations 9, 13 and 14 above and explain how heating a metal conductor is able to *increase* its resistance.

1.3 Potential difference and potential dividing

Resistance surveying essentially involves measuring the potential difference between parts of an electric circuit, as described below. This section looks more closely at the meaning of a voltage and involves estimating and calculating potential differences in simple circuits.

Multi-probe techniques

To introduce you to the idea of a resistive survey, the earlier work in this section was restricted to the simple case of two resistance probes sunk into the soil. A more advanced technique, pioneered by Frank Wenner in 1916, uses a voltmeter to probe positions along a line of soil in which there are two other probes which are connected to a power supply. Figure 6.9 shows a simplified version of the arrangement.

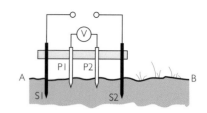

Figure 6.9 Multi-probe resistance surveying

Probes S1 and S2 are the electrodes connected to the supply. (In the field, an alternating supply is used to avoid problems of gaseous build up at the electrodes due to electrolysis.) P1 and P2 are a second pair of probes at a fixed separation and which are connected to a voltmeter. By using the voltmeter probes along the line AB, a map of potential differences can be built up. As the pd between fixed points is related to the resistance between the points, the voltmeter readings indicate patches of high and low resistance.

The Wenner arrangement was used in 1971 to probe the site of a prehistoric ditch on the Hog's Back in Surrey. The results are shown schematically in Figure 6.10 together with a sketch of the ditch.

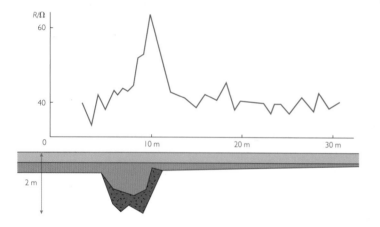

Figure 6.10 A survey of the Hog's Back revealed a prehistoric ditch

> **Study note**
>
> The word 'tomography' means 'drawing a slice' – the same word is used in some types of medical imaging e.g. computerised tomography (CT) scanning.

Resistive tomography

The availability of fast computer sampling has made possible a new technique in resistive surveying known as 'resistive tomography' (Figure 6.11). This uses a linear array of probes, 14 in this example, which are placed in location and the potential differences between all possible probe combinations is sampled using an interfaced field computer. The computer will then generate a resistive map of the ground in a vertical place immediately below the row of probes. As the array of probes is advanced across a site, the computer can eventually generate a three-dimensional resistive image of the entire area to a depth of a few metres. This can be done 'live' on site.

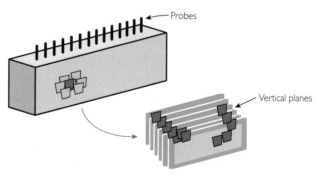

Figure 6.11 Resistive tomography

Dividing the potential

Before going any further, you would be advised to recall that voltmeters are connected 'across' components and that a voltage represents the energy transferred as electric charges pass through a component.

Activity 7 Energy in circuits

Consider the two circuits shown in Figure 6.12. Discuss with a friend whether each circuit diagram makes sense. Consider in particular the conservation of energy.

Your discussion in Activity 7 ought to have identified that in the second circuit there appears to be more energy 'transformed' by the charges as they pass through the resistors than the cell could supply (remember that $1 V = 1 J C^{-1}$). If the cell supplies each coulomb of charge with 12 J of energy, then no more than 12 J can be transferred by each coulomb. A simple statement which summarises this conservation of energy is to say that, in such a circuit, the sum of pd values across the resistors in a series circuit must match the pd from the cell. Activity 8 explores this idea further.

Activity 8 Splitting the potential difference

Use circuits such as those in Figure 6.13 to explore the relationships between resistances and potential differences. Look in particular at the ratios of the resistances and of the potential differences.

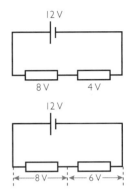

Figure 6.12 Two circuit diagrams for Activity 7

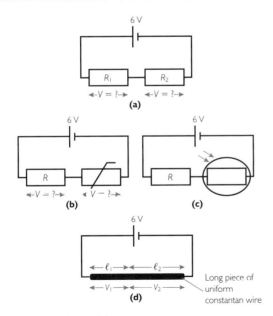

Figure 6.13 Circuit diagrams for Activity 8

The type of circuit you have used in Activity 8 is known as a **potential divider**: the terminal potential difference of a supply is divided between series resistors in the ratio of their resistances or, in the case of a uniform wire, the ratio of the distances from the contact point to the ends.

What difference?

Have you ever wondered why voltage is called potential *difference*? The difference between what and what? Consider the simple arrangement of two resistors in Figure 6.14. Without setting up the circuit you ought to be able to identify the value of the pd across each resistor.

The answer is 4 V and 8 V. The values 4 V and 8 V are potential differences. The word 'difference' implies a subtraction of two values. What two values could you use to produce 8 V? Obviously 8 V and 0 V come to mind. But also 9 V and 1 V, or 15 V and 7 V. They will each produce a difference of 8 V.

These pairs of numbers suggested are known as **potentials**. Their difference is naturally enough called a **potential difference**. It is conventional to label the negative terminal of a battery or cell as having a potential of zero. If the battery provides a potential difference of 12 V then the positive terminal will be labelled +12 V, hence a potential difference of 12 V.

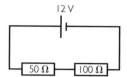

Figure 6.14 Two resistors in a circuit

Activity 9 Potential

Consider the two circuits shown in Figure 6.15. The circuit in Figure 6.15(a) has a number of places labelled with potentials. The potential differences are also shown. Look over this circuit so that you can follow the reason for the potential values.

Copy this circuit in Figure 6.15(b) and label the places marked X with suitable potentials. Take the cell negative terminal as zero. One of the potential differences is included as a clue.

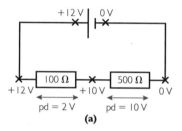

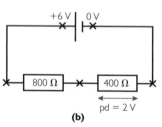

Figure 6.15 Circuit diagrams for Activity 9

In resistive surveying, the surveyor is mapping potential and potential differences in a three dimensional object (the ground), so the situation is somewhat harder to analyse theoretically than that in Activity 9. Activity 10 is an extension activity in which you can explore variations of potential in two dimensions.

Activity 10 Potential in two dimensions

Use an arrangement such as that shown in Figure 6.16 to explore the variation in potential in a two dimensional object. Plot a map of potential on a piece of graph paper. Look for points that have the same potential; you can link these with a line called an **equipotential** line.

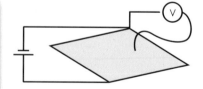

Figure 6.16 Diagram for Activity 10

1.4 To dig or not to dig?

Archaeology involves much more than simply surveying a site and deciding to dig. In the early days of archaeology, many archaeologists considered that their work was so important that the feelings of local people about their ancestors could be ignored. This lack of sensitivity has parallels in other areas where science is used. Today our society expects scientists to recognise when they are dealing with sensitive issues and to take into account the feelings of all involved.

Archaeological ethics refers to a number of moral issues raised through the study of the material past. A common ethical issue in modern archaeology has been the treatment of human remains found during excavations, as occurred with Kennewick Man, a 9000-year-old skeleton found in America (Figure 6.17).

In 1996, a skeleton was found on the bank of the Columbia River in the state of Washington. The local Umatilla Indians and four other tribes claimed the bones, which came to be called Kennewick Man, as their ancestral remains. Scientists also staked a claim to carry out analysis of the bones to questions about how and when early man first came to the Americas. A legal battle ensued over the next eight years concerning the rights to the bones, which concluded with the scientists eventually being able to study the bones.

The legal battle developed through a code of ethics that cover property laws, health and safety as well as moral considerations. Previously sites of great significance to native peoples had been excavated and any burials and artefacts taken to be stored in museums or sold. Now, however, archaeologists are taking a more respectful approach. This issue is not just limited to ancient remains; more recent burial sites such as First and Second World War graves and cemeteries have been disturbed by property developments which have seen the remains of people with living relatives being exhumed and taken away. Ships that were sunk and have been declared war graves are another sensitive area.

The trade in artefacts has also raised ethical questions regarding the ownership. The market for imported artefacts has encouraged damage to archaeological sites and often led to appeals for the recall. Pieces of archaeological material have been removed from their place of origin and caused damage to historic sites such as Pompeii. There has also been controversy over the return of the so-called Elgin Marbles (Figure 6.18), which were collected from Greece and are now in the British Museum.

Another issue is the question of whether unthreatened archaeological remains should be excavated and potentially destroyed or preserved intact for future generations to investigate. There has begun to be a strong ethical argument for only excavating sites threatened with destruction.

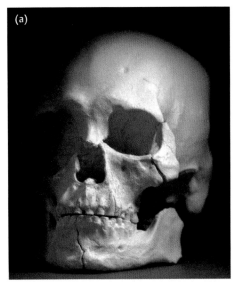

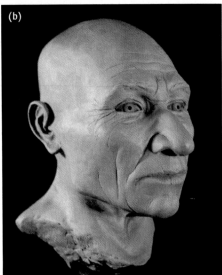

Figure 6.17 (a) The Kennewick Man skull (b) a reconstruction based on the skull

Activity 11 Kennewick Man

Use the Internet to research the case of Kennewick Man. Using the ethical frameworks introduced in *Spare Part Surgery*, summarise ethical arguments for and against the archaeological investigation of Kennewick Man.

Figure 6.18 The Elgin Marbles

1.5 Summing up Part 1

In this section you have reviewed and extended your knowledge of electric circuits and learned how the idea of resistivity is used in archaeological surveys. You have also considered some ethical issues relating to archaeological investigation. Activity 12 is designed to help you to consolidate what you have learned and also develop your communication skills.

Activity 12 Summing up Part 1: surveying the field

Imagine that you are a member of an archaeological team – perhaps a friend has dragged you along to a summer vacation 'holiday' site for company. The historical evidence suggests that a nearby field belonging to a local farmer, Mr Muncastle, is a likely site for the location of a Roman villa. The team you are working with want to carry out a resistive survey of the field. They have persuaded you to write to the farmer asking permission to gain access to his field for a survey, lasting about a day. (He will have to usher his bull into another field and is not too sure what damage you might do).

Write a letter to Mr Muncastle explaining the non-destructive nature of the test in a language that he will appreciate and which explains what will happen during the day.

Prepare for a meeting with Mr Muncastle to do a deal with him in the event that the survey is positive and the team wants to spend four days excavating a corner of his field. What do you anticipate he will say? How will your team respond?

Further investigations

If you have an opportunity for further practical work, you might consider the following suggestion.

Use Teledeltos paper to explore the equipotentials and electric field around conductors with a variety of shapes. In particular, look at the potential gradient around sharp points and think how this might relate to the operation of a lightning conductor, and to guidelines on safe behaviour during thunderstorms.

Questions

8 For the circuits shown in Figure 6.19

(a) calculate the value of the pd when the voltmeter is attached as shown by the broken lines

(b) state the potential at each of the places marked X.

9 Heavy vehicles can distort older steel and iron bridges. To monitor these distortions, scientists use an instrument called a strain gauge. A strain gauge is a thin strip of metal (Figure 6.20). When it stretches it gets longer and thinner and so its resistance changes. This change can be monitored using a potential divider circuit.

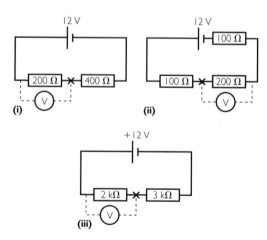

Figure 6.19 Circuit diagrams for Question 8

(a) If a strain gauge is stretched, in what way will its resistance change? Will it increase or decrease?

(b) If the length of a strain gauge increases by 10%, by what factor will the resistance change? Explain your answer.

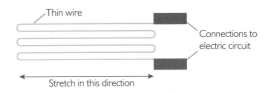

Figure 6.20 Schematic diagram of a strain gauge

The gauge shown in Figure 6.20 has a total uncoiled length of 28 cm. The thin strip is 0.6 mm wide and 0.002 mm thick. The resistivity of the metal used is 2.6×10^{-6} Ω m.

(c) Calculate the resistance of the strain gauge.

In the potential divider circuit, the gauge is placed in series with a fixed resistor, R. They are connected to a 6 V supply and R must be chosen so that the pd across the strain gauge is 1.2 V.

(d) Calculate the value of the fixed resistor, R.

2 The inside story

2.1 X-Rays

Once artefacts have been recovered then the real investigation begins. **X-rays** are invaluable in probing the finds both to discover what lies hidden within and to determine from what materials they are made and how best to conserve them.

Activity 13 The electromagnetic spectrum

Table 6.4 lists typical wavelengths of the radiation that makes up the electromagnetic spectrum. Use these values to produce a labelled 'number line' showing the various regions of the spectrum. First try plotting the wavelengths on a linear scale and note any problems this creates. Then copy Table 6.4 and add another column headed '$\log_{10}$ of wavelength'. Complete this column (you should be able to do most of this without using a calculator) and then plot the values of $\log_{10}$ of wavelength on a number line. Choose a scale so that the entire spectrum fits across the page in your notes.

Type of radiation	Approximate wavelength range / m
γ-rays	10^{-14} (or less) – 10^{-11}
X-rays	10^{-11} – 10^{-8}
ultraviolet	10^{-8} – 4×10^{-7}
visible	4×10^{-7} – 7×10^{-7}
infrared	7×10^{-7} – 10^{-3}
microwaves	10^{-3} – 10^{-1}
radio	10^{-1} – 10^{3} or more

Table 6.4 The electromagnetic spectrum

Maths reference

Logs and powers of 10
see Maths note 8.1

Logs on a calculator
see Maths note 8.2

In Activity 13 you used a wave model for X-rays. In *The Sound of Music* and *Technology in Space* you used a photon model for electromagnetic radiation. An X-ray photon has higher energy than a photon of visible light, which explains why X-rays are so penetrating and cause ionisation.

Activity 14 X-rays in archaeology

When you have completed your work on this part of the chapter, prepare a short information leaflet, aimed at the general public, giving a brief outline of how X-rays can be used in archaeology. Imagine that people might pick up this leaflet, so it must be visually attractive and make them want to open it to see what's inside.

Such leaflets are often a single sheet of A4 folded into three (or A3 folded into four). Use diagrams and graphics to get your message across as well as text. As you work through this part of the chapter, be thinking about what you will include in your

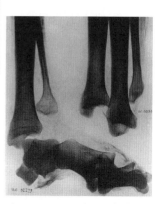

Figure 6.21 A mummified leg

2.2 What's inside?

Soon after Wilhelm Röntgen discovered X-rays in 1895, Sir Flinders Petrie obtained the first X-ray of a mummified leg (Figure 6.21); Mr Elliot Smith and Mr Howard Carter told of their experience transferring the rigid Pharaoh Tuthmosis VI by taxicab to a private X-ray unit in Cairo Hospital. Surprisingly almost no further use was made of X-rays by archaeologists until the 1960s. Now conservators consider X-rays an essential tool in probing archaeological finds to assess their condition and determine the best techniques of conservation.

The greatest benefit of X-radiography of archaeological artefacts is that it is a completely non-destructive technique. X-rays applied to intact mummies can identify the contents (Figures 6.22 and 6.23) through conventional transmission (shadow) radiographs, while a radiograph of bones can provide information on skeletal maturity, the effects of illness and nutrition, and injuries that may have been the cause of death. Generally such X-rays are undertaken in hospitals at night to avoid interfering with normal working procedures and alarming the patients.

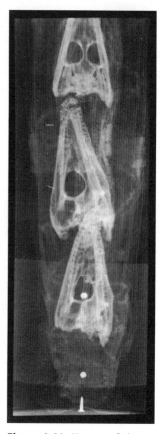

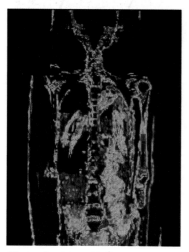
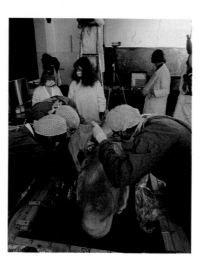

Figure 6.22 X-radiography of an entire mummy can show the contents without the need to destroy or severely damage it

Figure 6.23 Mummy of three crocodile skulls

Another technique, 'fluoroscopy', involves the transmission of X-rays onto a fluorescent screen and provides an instant dynamic view of the object. Since the apparatus allows the object to be moved during examination and viewed in real time, a three-dimensional effect is introduced as the object moves on the screen.

'Pete Marsh', the Iron Age human body found in Lindow Moss in Cheshire, was X-rayed using computer aided tomography (CAT) scans whereby images of 'slices' of the body can be studied. The image is achieved by blurring out the unwanted shadows above and below the place of interest. CAT scans have been invaluable in understanding pathological changes and mummification techniques.

Sometimes a group of archaeological objects, such as the contents of a grave, are lifted en bloc, then the components can be X-rayed in situ. This may reveal traces of former organic components related to the original metal objects for example.

X-rays will obviously show up the shape of metal objects but they can also show the extent of any deterioration as the decorative metals that were used corrode differently from the base metal of the main object and may now appear 'suspended' in the surface corrosion products. The remains of a scabbard may also be preserved in the corrosion and will only be visible in the radiograph (Figure 6.24).

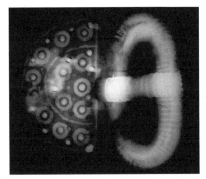

Figure 6.24 A radiograph of an Anglo-Saxon buckle showing bands of silver 'floating' 2 mm above the surface

The radiographs also reveal a wealth of information about the technological skills that went into the production of metal objects. Bronze Age castings of sword and daggers can be studied radiographically, and the distribution and orientation of the gas voids in the castings give an insight into the development of the casting process.

The conversion of iron into steel by carburisation is a means of improving the quality of the metal products. Anglo-Saxon smiths developed a range of techniques to make the most of the various properties of each metal, and the techniques they used to combine metals by 'pattern welding' can be seen radiographically on swords and other weapons (Figure 6.25).

The corrosion of iron was well understood, and in medieval times tin or copper plating was often used for its protection. Such non-ferrous platings are seen on radiographs as bright lines surrounding the object.

The construction techniques revealed through X-rays of artefacts can help in determining whether the artefact is genuinely of the period and whether there has been any attempt at repair or restoration in the past. Radiography was used in the authenticity studies of a pair of Egyptian bronze cats. They were shown to be made from copper filings and resin rather than cast bronze as would be expected of authentic articles from the period.

Radiography has thus become an essential tool for the study of a range of archaeological materials, and few conservators would risk conserving iron objects in particular and, increasingly, objects of other materials without the benefit of an X-ray to assess their condition.

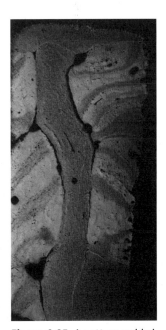

Figure 6.25 A pattern welded dagger

Activity 15 The Coppergate Helmet

Read the following article, and then answer Questions 10 to 14 that follow. Skim read the article first, look at the questions and then read more carefully to find the answers. Use a dictionary to look up any unfamiliar words.

In 1976 a major archaeological dig of a Viking settlement in the centre of York produced a wealth of finds that were so exciting that it was decided to build the Jorvik Viking Centre on the site. Now thousands of people visit the site every year and enjoy the re-creation of life in Viking times and of the excavations which were undertaken. However another amazing find came to light quite by accident – the Coppergate Helmet (Figure 6.26).

Andy Shaw was using a mechanical excavator to level the site ready for the construction of a large commercial centre when his supervisor, Chris Wade, noticed that the machine had struck something solid. Immediately archaeologists from the York Archaeological Trust were called in. It was apparent that the object was a helmet from the Anglo-Saxon period and of immense importance. It took archaeologists over 5 hours to excavate the area and record meticulously the precise location of the helmet and the surrounding features.

The helmet was lifted without removing the clay filling (Figure 6.27) in an attempt to prevent damage to any objects inside the helmet. This was finally accomplished at 9 o'clock in the evening of the day it was discovered.

To prevent corrosion of the copper alloy and iron, the helmet and contents were transferred to a closed container to be stored in an atmosphere of nitrogen giving the conservators time to plan the conservation strategy and investigation of the helmet.

In order to see what was inside the helmet it was taken to York District Hospital and radiographed (X-rayed) with the help of hospital staff. They made a series of horizontal transmission shadow radiographs and the subsequent images were digitally enhanced. The horizontal radiographs revealed a substantial amount of chain mail and one of the cheek-pieces inside the helmet.

The helmet was too fragile to be turned upright so to obtain vertical radiographs it was subjected to circular motion tomography using the sophisticated CAT (computer-aided tomography) scanner. Tomography is a means of generating images of 'slices' of the body by blurring out the unwanted shadows above and below the area of interest. This produced a series of vertical radiographs of the whole helmet at 10 mm internals which gave information on the precise location of any objects inside.

Nothing of this nature had been attempted before as the machine was designed to cope with the human body and the helmet, being metal, had a density outside the scope of the scanner. However the images were successful (Figure 6.28) and armed with the X-ray sections the archaeologists excavated the interior of the helmet, knowing in advance exactly where the chain mail and cheek-piece lay.

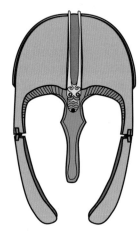

Figure 6.26 The Coppergate Helmet

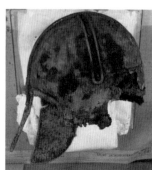

Figure 6.27 The helmet shortly after excavation

Later, high-performance industrial radiography of the helmet and microfocus radiography of the chain mail rings enabled conservators to resolve the complex technology involved in the helmet's original construction. The manufacture of the chain mail was an incredible feat of skills on the part of the Anglo-Saxon metalworkers who must have taken many hundreds of hours to make it. It was surely one of the most costly components of the helmet.

Evidence from the shape of the helmet, the inscriptions, the form of the animal ornaments and the methods of construction (compared with similar finds such as the Sutton Hoo Helmet from a royal grave in Suffolk and the Benty Grange Helmet from Derbyshire) led to the conclusion that it was made between ad 750 and 775 for a member of a noble family, possibly named Oshere. It was probably in use for a number of years as it has been well polished and suffered some damage. The fact that it was not found in a burial site but had apparently been placed in the pit deliberately with the chain mail and cheek-piece placed inside indicates that it may have been hidden with the intention of recovering it later. Perhaps the helmet was used and lost during the struggle for York between the Saxons and the Vikings in AD 866.

Radiocarbon dating of the timbers in the pit showed that their age was 1250 ± 70 years, suggesting that the wood was cut in about 732 ad ± 70 years.

The spot where the Coppergate Helmet was found is marked by a plaque in the entrance of the Jorvik Centre in Coppergate, York. The helmet itself can be seen at the nearby Castle Museum.

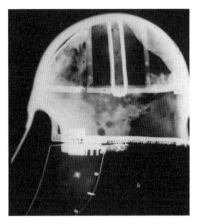
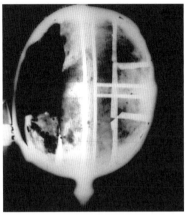

Figure 6.28 Some of the original radiographs

Questions

10 Explain why the term 'shadow radiograph' is appropriate for a conventional X-ray image.

11 Explain the difference between conventional X-ray images and X-ray tomography.

12 Explain why there was concern that the hospital CAT scanner might not give satisfactory images of the helmet and its contents.

13 Explain, using diagrams if necessary, how taking both horizontal and vertical radiographs enabled the archaeologists to pinpoint the precise location of the items inside the helmet.

14 In addition to determining the nature and location of the contents of the helmet, what other information did archaeologists obtain through using X-rays?

2.3 X-ray diffraction

As well as using X-rays to probe the content of artefacts, another technique, X-ray diffraction, is a powerful tool in determining the precise composition of the materials that make up the object. This can lead archaeologists to a better understanding of the sources of the materials used, the methods used in manufacturing objects and how best to prevent further corrosion.

The archaeologist's view of X-ray diffraction

As well as using X-rays to 'see' what an artefact consists of, it is possible to use X-rays to analyse precisely what materials were used to create it. X-ray diffraction is a technique that use **monochromatic** (i.e. single-wavelength) X-rays to identify which crystalline minerals are present, and it will reveal not only whether an element is present as an oxide or other compound but also the proportions of each mineral present and the phases of the minerals (i.e. the forms of crystal lattice, see Figure 6.29). It can be applied to metal, ceramic or stone artefacts and the pigments used to decorate them.

The use of X-ray diffraction to identify the minerals present in an object has a number of strengths:

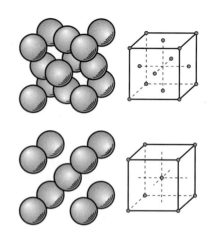

Figure 6.29 Different phases of a mineral

- First, conservators are interested in the corrosion products present. This allows them to know more about the burial history of the artefact – its exposure to water or oxides present in the soil surrounding it, for example. From this information conservators can determine the best conservation techniques to prevent further deterioration of the object.

- Archaeologists can determine the raw materials used to create the object. This may give an indication of the sources of these raw materials, which is helpful in determining ancient trade routes and other aspects of ancient economy and society. More importantly the precise composition of alloys, often determined by investigation of the slags and residues left in metal workshops, can indicate the techniques used to manufacture them, which then indicates the archaeological period and the economy of the site – was metal working undertaken only in the construction of the site or did the site include a metal workshop or even a mint where silver was extracted as coins were recycled, for example?

- X-ray diffraction analysis will provide information on the thermal treatments to which metal artefacts were subjected during their fabrication. For example, cold working a metal results in fragmentation and distortion of the crystal grains, whereas annealing results in the growth of large, undistorted crystal grains that give very distinctive X-ray diffraction patterns.

- Study of the pigments used to decorate objects can provide an insight into the techniques of manufacture of the pigment and its application to the surface of the object. Different oxides were used to produce a particular colour in different places across the ancient world at different times. A study of similarities and differences in techniques for producing pigments can indicate whether groups were developing in isolation or influenced each other. The same technique can identify the true colours used in, for example, 17th century works of art (Figure 6.30) to aid accurate restoration.

- X-ray diffraction can be used to estimate the firing temperatures used when making pottery because both the minerals present and their phases depend on the temperature to which the sample has been heated.

- Finally, the fluorine content of bones decreases with age, and thus X-ray diffraction techniques can be used to date bones by measuring their fluorine content – although the levels are dependent on the area in which the animal lived and therefore have to be calibrated accordingly.

Figure 6.30 A 17th century work of art

How does X-ray diffraction work?

The simplest version of X-ray diffraction takes a thin slice of material and allows the X-rays to pass through it, or else passes the X-rays through a powdered sample of the material. You can simulate this using a fine powder sandwiched between two microscope slides. Another simulation uses a diffraction grating (a regular array of very narrow lines ruled on transparent plastic or glass) in front of the laser – this produces a very different pattern.

Activity 16 Simulating X-ray diffraction

Observe the patterns produced on a screen when a laser shines through some fine powder and through one or more diffraction gratings (Figure 6.31). Try using two gratings together, crossed at various angles to one another. With other students, 'brainstorm' your ideas about how these different patterns are produced.

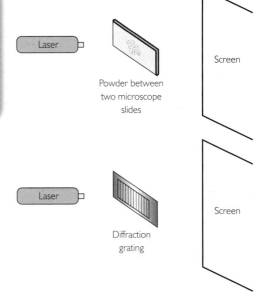

To explain the two effects you have just seen you will need to understand three phenomena: **diffraction**, **superposition** and **interference**.

Diffraction (the spreading of waves) and interference (when coherent waves undergo superposition and reinforce or cancel each other) combine to give the effects seen when a laser is shone through a single crystal or a powdered polycrystalline sample giving intense points of light at certain well-defined points as seen in Activity 16. If we know more about the precise mechanisms involved, we can use this phenomenon to find out more about the sample.

Figure 6.31 Diagrams for Activity 16

Activity 17 Doing diffraction – investigating interference

Explore the effects of diffraction and interference of different waves through a series of three experiments: a ripple tank, a laser and microwaves.

Activity 17 illustrates diffraction – the spreading of waves into unexpected areas. For significant diffraction to happen the gap or obstacle must be comparable in size to the wavelength of the waves used. If coherent waves travel through different distances and then meet up again, then superposition produces an interference pattern. If waves pass through a regular array of gaps or obstacles, such as a diffraction grating, then diffraction gives rise to many sets of overlapping waves that produce a **diffraction pattern** when they superpose. You can demonstrate that by shining a laser through fine gauze (Figure 6.32). The finer the gauze, the greater the spacing of the dots in the pattern. The spacing of the dots also depends on the wavelength of the light: the longer the wavelength, the greater the spacing – though this cannot be demonstrated with a laser since it is monochromatic.

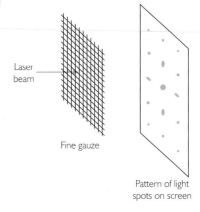

Figure 6.32 Producing an optical diffraction pattern

Von Laue's experiment

The use of X-rays to analyse materials is based on an experiment performed early in the 20th century by the German physicist Max von Laue.

After X-rays were discovered in 1896, their nature was the subject of much speculation and experiment. They were found not to be charged particles because they were not deflected by electric or magnetic fields, unlike electrons (whose deflection was discovered and explained by J. J. Thomson in 1897). In 1912 Max von Laue succeeded in showing that X-rays were electromagnetic waves by obtaining an interference effect. All attempts had so far failed as the gap size used was far greater than the wavelength of the X-rays, then a crystal was used to diffract the X-rays and cause interference patterns. Figure 6.33 shows the basic set up of his experiments.

The central bright spot surrounded by fainter pattern of surrounding spots (Figure 6.34) confirmed that X-rays were waves, and measurements showed that their wavelength was of the order 10^{-10} m. It is precisely this effect that you saw with the laser shone through crossed gratings in Activity 16.

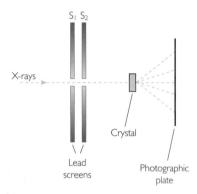

Figure 6.33 Von Laue's experimental set up

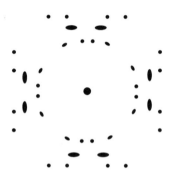

Figure 6.34 An X-ray diffraction photograph

Activity 18 Diffraction by crystals

Use a ripple tank to see the diffraction effects produced by a two-dimensional 'crystal' and use microwave apparatus to see the same effect in three dimensions.

Powder photos

Von Laue's experiments worked because the atomic spacing in crystals is similar in size to the wavelength of the X-rays. The precise pattern generated can be used to make measurements of the interatomic spacing in the crystal and this is used to analyse the composition of materials. Broadly speaking, the smaller the interatomic spacing, and the shorter the wavelength, the more widely spaced the bright parts of the diffraction pattern.

Most X-ray diffraction analysis is carried out on a powdered sample, which is turned into a rod-shaped specimen either by mixing it with an adhesive or by sealing it in a glass capillary tube. The rod is placed in a cylindrical X-ray camera and illuminated with a beam of monochromatic X-rays (Figure 6.35). As the X-rays are reflected off successive layers of atoms, they undergo interference, and the film in the camera records a number of bright lines (areas of constructive interference) (Figure 6.36). The distances between these lines give information on the crystal lattice spacing of the crystals present, which then is used to identify the actual composition of the object by comparing the patterns of lines with the lines formed by known substances. (This powder technique cannot be used to determine the crystal form of metals as the powdering distorts the crystal lattice; instead a thin slice of the artefact has to be used.)

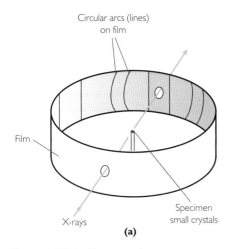

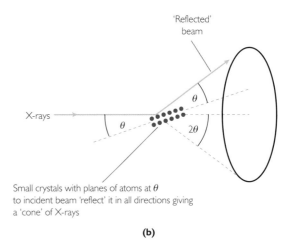

Figure 6.35 An X-ray powder camera

Activity 19 Where has it been?

The Coppergate Helmet was subjected to X-ray diffraction analysis in order to determine what it was made from and the likely conditions in which it was buried. Study the relevant X-ray diffraction spectra and write a short report on the deductions that can be made.

Activity 20 Tales of the unexpected

Write an essay entitled 'Diffraction – tales of the unexpected'. This should not only explain diffraction (briefly) but also discuss why diffraction can be associated with the phrase 'the unexpected'.

At the end of your essay note the points on which you would like feedback from your teacher.

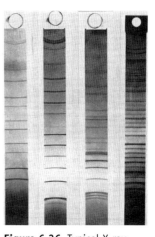

Figure 6.36 Typical X-ray powder photos

2.4 Summing up Part 2

In this section you have seen how the composition of artefacts can be analysed using X-rays. You have also had several opportunities to develop skills of communication.

Activity 21 Summing up Part 2

Spend a few minutes checking through your notes and make sure you understand all the terms printed in bold. Then use what you have learned to complete your leaflet for Activity 14.

Questions

15 Explain why a sound from the far side of a doorway can be heard in all parts of a room but light passing through the same doorway casts a sharp shadow.

16 Steel components corrode (rust) in various ways when exposed to air, rain and heating. The compounds FeO (iron oxide), Fe_2O_2 (iron(III) oxide), Fe_3O_4 (tri-iron tetroxide) and $Fe_2O_3 \cdot H_2O$ (hydrated iron(III) oxide) can be formed, depending on the type of steel and its situation. Knowledge of which oxide(s) are present can help the manufacturer and the user of the steel.

X-ray diffraction is frequently used to analyse the corrosion products. The corroded surface layer of the steel component is scraped off, ground into a powder, placed in the X-ray camera and exposed to X-rays.

A drawing of an X-ray diffraction photograph for one particular sample is shown in Figure 6.37, together with those of pure iron and some known oxides. The original photographs were all negatives and so the dark lines indicate a high exposure to X-radiation.

Using the images in Figure 6.37, state which oxide was present in the corroded surface layer (in addition to iron itself).

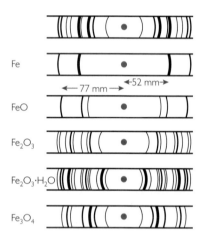

Figure 6.37 X-ray diffraction images for Question 16

3 Taking a closer look

When an artefact is uncovered at an archaeological dig, one of the first investigations may be to use an optical microscope to look at it in more detail. This can be done on the site, as optical microscopes are very portable. But optical microscopes have their limitations so artefacts are sometimes taken to a laboratory for more detailed examination using an electron microscope. This part of the chapter is about how optical and electron microscopes are used in archaeology.

3.1 Seeing the detail

A typical human eye can only **resolve** (distinguish between) two point-like objects if they are separated by at least 0.1 mm when viewed from 25 cm away. The **resolution** of the eye is limited by diffraction. As light waves pass through the pupil of the eye, they are diffracted, so the image is slightly blurred. The wavelength of visible light is very much smaller than the width of the pupil so the amount of diffraction is very small. But if the two images are very close together on the retina, they merge together and it is impossible to distinguish between them. Figure 6.38 shows this effect with light passing through a much smaller aperture than the pupil of your eye, and Activity 22 illustrates the limits to the resolution of your own eyes.

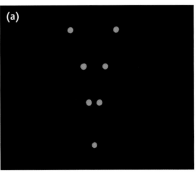 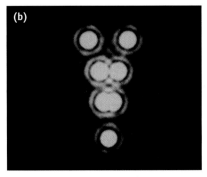 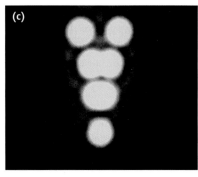

Figure 6.38 (a) A set of light sources (b) the sources viewed through an aperture (the two closest sources are barely resolved) (c) the same sources viewed through a smaller aperture (the two closest sources are not resolved)

 Activity 22 Seeing the detail

Use a fine dark-coloured pen to draw two parallel lines a few mm apart on a piece of white card. Place the card several metres away and walk slowly towards it. Record the distance at which you are first able to see the two lines separately. From these measurements, calculate the smallest separation that you would be able to resolve from a distance of 25 cm.

(a)

An optical microscope with a magnification of 500 can be used to observe detail down to a scale of 0.2 μm (200 nm). A feature 0.2 μm wide will have an image 0.1 mm wide, which you can resolve when you observe it from a distance of about 25 cm using the microscope.

 Activity 23 Optical microscope

Use an optical microscope to look at some artefacts.

(b)

Figure 6.39 shows optical microscope pictures of some archaeological artefacts. If archaeologists could see smaller objects clearly, such as details of pollen grains, this would give them more information to help work out, for example, the age of the site and whether it has been disturbed. A grain of pollen is about 40 μm in size (though they vary between about 10 μm and 100 μm).

A feature that is 0.1 μm wide and magnified 1000 times would give the same size image as an object 0.2 μm wide and magnified 500 times, but there is a problem. The wavelength of visible light is between about 750 and 400 nm, so features of about 750 nm and smaller will themselves produce noticeable diffraction of the light waves. This will blur the image produced by the microscope. As you saw in Activity 17, an object that is much smaller than the wavelength will not cast a shadow, so this limits the size of objects that can be seen with an optical microscope; objects smaller than 200 nm (0.2 μm) simply do not produce an image.

(c)

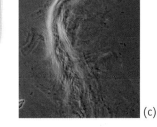

(d)

Figure 6.39 Optical microscope pictures of archaeological specimens

Also, there is a limit to the detail that can be seen within a larger object. The smaller the feature, the greater the blurring, and magnifying the image makes no difference to its clarity. For detail on a scale of a few hundred nm, the blurring is considerable, and below the 200 nm limit it is no longer possible to distinguish features at all.

A design brief for a microscope similar to an optical microscope but with a greater magnification would say that it needed to use radiation with a shorter wavelength than visible light. The obvious choice would ultraviolet or X-rays, but there are difficulties in designing lenses for these regions of the electromagnetic spectrum. Fortunately, another solution, the electron microscope, is available that uses different waves and can be used to study pollen grains (Figure 6.40) and other minute objects.

Figure 6.40 Electron micrographs of pollen grains

Activity 24 Pollen

Find out about a pollen reference collection – what is it and how is it useful? Compare the electron microscope images with the optical microscope images. What are the advantages and disadvantages of each type of microscope?

3.2 Electron waves

Beams of electrons can be diffracted. You probably find this quite surprising since diffraction is something that waves do, whereas you have probably always thought of electrons as particles.

The wave-like side of the electron's character was discovered in the 1920s, and the discovery came about in two ways. First came a theoretical prediction by the French scientist Prince Louis de Broglie (1892–1987). Louis de Broglie knew that light can behave as either waves or 'particles' (photons), and in 1924, in his doctoral thesis, he put forward the idea that electrons (and other particles) might have a similar dual nature and sometimes behave as waves. He predicted that the wavelength, λ, of a particle of mass m moving at speed v would be given by:

$$\lambda = \frac{h}{mv} \tag{15}$$

where *h* is the Planck constant. This wavelength is sometimes known as the **de Broglie wavelength** of a particle. Experimental support for this revolutionary idea was obtained within the next three years by American scientists Clinton Davisson and Lester Germer, and independently by British scientist George Thomson, who produced diffraction patterns using electron beams directed at metals and crystals. All these scientists received Nobel prizes for their work. George Thomson was the son of J. J. (Joseph) Thomson who, a generation earlier, had received a Nobel prize for demonstrating conclusively that electrons were particles.

Electron diffraction can be demonstrated using the apparatus shown in Figure 6.41. In the electron gun (Figure 6.42), electrons are 'boiled off' a heated filament by a process called **thermionic emission**. They are then attracted towards a positive anode, accelerated by a potential difference of a few hundred volts. The glass tube containing the gun and target has a near-vacuum inside, so there are few air molecules to impede the motion of the electrons. Most of the electrons pass straight through the anode and fly straight on to hit the graphite target. The electron beam passes through the thin layer of graphite and hits the fluorescent layer inside the rounded end of the tube, transferring energy to it and making it glow. Graphite consists of layers of carbon atoms arranged in regular hexagonal patterns (Figure 6.43), and so produces a diffraction pattern when waves pass through whose wavelength is comparable to the atomic spacing.

> **Study note**
>
> You met the Planck constant in *The Sound of Music* when dealing with atomic energy levels.

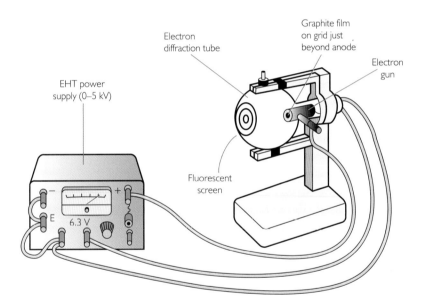

Figure 6.41 An electron diffraction tube

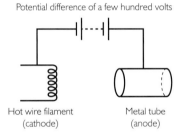

Figure 6.42 An electron gun

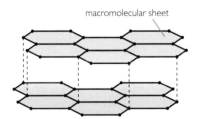

Figure 6.43 The arrangement of carbon atoms in graphite

Activity 25 Electron diffraction

Use the apparatus shown in Figure 6.41 to observe electron diffraction. Predict and observe what happens to the electrons' speed, and hence to their wavelength and to the spacing of the diffraction pattern, as the accelerating voltage is increased.

Hold a magnet close to the tube and observe what happens to the beam.

As you have seen in *Technology in Space* and in Part 1 of this chapter, when charge moves through a potential difference, energy is transferred as described by the relationship

$$\Delta E = q\Delta V \tag{16}$$

where ΔE is the energy transferred, q is the charge and ΔV the potential difference.

If an electron moved in the opposite direction across the gap, from negative to positive, then the power supply would be transferring energy to the electron; an electron starting from rest would gain kinetic energy:

$$E_k = e\Delta V \tag{17}$$

where e is the electron's charge.

Question

electron charge $e = 1.60 \times 10^{-19}$ C

electron mass $m_e = 9.11 \times 10^{-31}$ kg

Planck constant $h = 6.63 \times 10^{-34}$ J s

17 Electrons are accelerated in an electron gun by a potential difference, V, of 100 V.

(a) What is each electron's maximum kinetic energy? Give your answer in units of joules and in eV.

(b) Calculate the speed of the electrons.

(c) What is the electrons' de Broglie wavelength?

The very small wavelength of electrons with energies of a few hundred eV is comparable with X-rays, which explains why the electrons are diffracted when they are passed through crystals. X-ray diffraction patterns rely on X-rays being scattered by relatively heavy atoms, whereas electrons are diffracted by light elements such as carbon, nitrogen and hydrogen, so electron diffraction patterns are used to analyse materials made of lighter elements (Figure 6.44). And there is another very big, and useful, difference between X-rays and beams of electrons: electrons are deflected by magnetic and electric fields whereas X-rays are unaffected. This is made use of in the design of electron microscopes.

> **Study note**
>
> You met the energy unit eV in *Technology in Space* when dealing with the photoelectric effect.

> **Study note**
>
> You will learn more about the behaviour of charged particles in electric and magnetic fields if you study *Transport on Track*, *The Medium is the Message* and *Probing the Heart of Matter*.

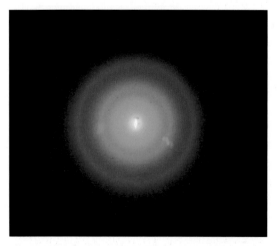 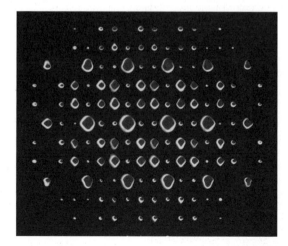

Figure 6.44 Electron diffraction patterns of (a) graphite and (b) titanium-nickel alloy

3.3 Electron microscopes

Nowadays there are several types of microscope that make use of electron beams. Two widely used designs are the Transmission Electron Microscope (TEM) and Scanning Electron Microscope (SEM).

TEM

The first type of electron microscope to be developed, and still widely used, is the transmission electron microscope (TEM). A TEM can be used to look at samples that are very thin (typically less than 1 μm thick). The electrons pass through the sample and form an image.

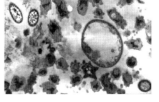

Figure 6.45 TEM image of microbial consortia within a degraded archaeological hair shaft

Some samples have to be specially prepared to withstand being placed in a vacuum and bombarded by high-energy electrons. Biological samples are particularly difficult to prepare as they contain water, (which does not diffract electrons) and would be damaged by the conditions in the microscope. Methods include embedding them in plastic, freezing with liquid nitrogen, and staining with heavy metal stains to provide image contrast. Very thin slices can then be shaved off to look at with the TEM. Figure 6.45 shows a TEM image.

How does passing a beam of electrons through a thin sample of material result in an image? The technique depends on the fact that not all of the electrons pass through the sample unaffected. There are three possibilities:

- electrons pass through undeflected
- electrons are deflected (scattered) without losing energy
- electrons lose energy, and most of these are also scattered.

If all of these electrons continued on through the microscope and were focused on a fluorescent screen there would be no contrast in the image formed. An aperture placed in the beam allows only the electrons that are either undeflected or scattered through small angles to pass through. Areas of the sample that cause a lot of scattering will appear dark.

In many ways the TEM is like an optical projection microscope. The parts are compared in Figure 6.46. Instead of a lamp, the electron microscope has an electron gun, as described in Section 3.1. This requires a high-voltage supply. The smaller the detail to be observed, the smaller the wavelength required, and so the higher the energy of the electrons used in the beam. The whole microscope is enclosed in a vacuum, which requires vacuum pumps to maintain the very low pressure.

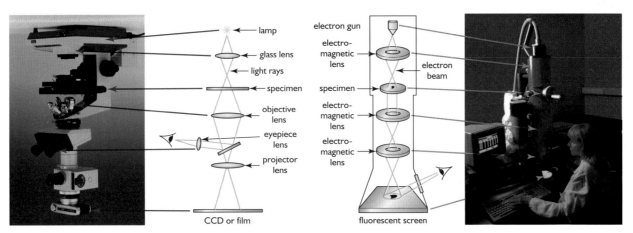

Figure 6.46 Schematic diagrams of (a) Optical microscope (upside down) (b) TEM

Activity 26 TEM

Use a computer simulator to try out a TEM. Find out more about its design and operation, about sample preparation, and look at some images.

In the optical microscope light is focused onto the sample and onto a ground glass screen using glass lenses. The electron microscope uses magnetic 'lenses'. Electromagnets can be used to focus the electron beam because electrons are deflected as they pass through a magnetic field – as you saw in Activity 25. To focus the electron beam the strength of the magnetic field is changed by altering the current in the electromagnets. Electrons can't be seen, so the electron microscope has a fluorescent screen to reveal the image. When the electrons strike the screen it glows. If a CCD camera is used to produce pictures and store them on computer, then 'false colour' pictures can be produced in which different intensities are given different colours to enhance the contrast of the image.

SEM

Not all materials can be made into the very thin samples needed to get a TEM image. The scanning electron microscope (SEM) uses the beam of electrons in a different way (Figure 6.47). The electrons that are back-scattered from the sample are collected and used to build up the image (these are sometimes called secondary electrons). The scanning coils are supplied with a varying electric current to produce a varying magnetic field. This sweeps the electron beam across the sample repeatedly to build up an image, in the same way that an old TV or PC monitor sweeps an electron beam across the screen to build up the picture.

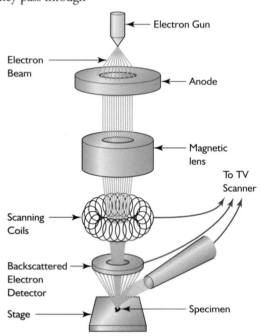

Figure 6.47 Schematic diagram of an SEM

Activity 27 SEM

Investigate the SEM. Find out how it is used and look at some images. Use a simulator.

3.4 Summing up Part 3

In this part of the chapter you have seen how artefacts can be examined in more detail using microscopes to magnify a specimen. You have learned that:

- there is a minimum size feature that can be seen with a microscope, because of diffraction effects
- beams of electrons can behave as waves and you have seen electron diffraction patterns that are proof of this
- electrons have a wavelength that is related to their speed
- electron microscopes can be used to see much smaller features than an optical microscope.

You have looked at the design and operation of transmission electron microscopes (TEMs) and at scanning electron microscopes (SEMs).

Activity 28 Summing up Part 3

Spend a few minutes checking through your notes, making sure you know all the meanings of the key terms printed in bold. If you are developing a summary of key terms, make sure that it is up to date. In particular look at some of the terms that you met in earlier chapter and see whether you need to refine or extend your definitions.

Activity 29 Microscopes

Write a paragraph explaining the advantages of using an electron microscope rather than an optical microscope. Use the following terms in your explanation:

- de Broglie wavelength
- resolution (or resolve)
- diffraction
- image

Question

electron charge $e = 1.60 \times 10^{-19}$ C

electron mass $m_e = 9.11 \times 10^{-31}$ kg

Planck constant $h = 6.63 \times 10^{-34}$ J s

18 A particular electron microscope is designed to study objects down to about 50 nm in size. For the images to provide useful information, the electron wavelength should be no more than twice the size of the objects.

(a) Derive an expression relating the accelerating voltage, V, to the electron wavelength λ and the charge and mass of the electron.

(b) Calculate the accelerating voltage that should be used.

(c) Explain what will happen to the electron wavelength and the resolution of the microscope if the accelerating voltage is doubled.

4 Archaeologists at work

Many years ago archaeology might have been considered to be closely related to history, and to require a lot of time cataloguing artefacts. The help that science could offer a 19th-century archaeologist was limited. Today science, and especially physics, provides many useful tools for locating and identifying objects. In addition to this, one of the key facts we want to know about an artefact is how old it is, and here physics can help in many ways.

In this section you will learn about how the community of archaeological scientists interacts with society and how it regulates itself to ensure that the evidence it presents can be trusted.

4.1 The dating game

In Parts 1 to 3 of this chapter you saw how archaeologists can locate ancient remains and how they can determine the composition of artefacts. But *when* they were made is also of vital concern. By dating artefacts, the archaeologist can begin to build up the time sequence of events which can then be used to provide an accurate analysis of the past.

Before the application of scientific methods and techniques, the dating of archaeological sites and artefacts was pretty much a hit and miss affair. For example, if one item was found at a deeper level in the dig than another then it was assumed to be older; the Stone Age was supposed to be earlier than the Bronze Age and this in turn preceded the Iron Age. There are hidden assumptions here. Has the site been disturbed, mixing up the artefacts? Were no stone articles produced in the Bronze Age?

Over the past half century or so, archaeological dating has become more sophisticated. Measurements involving a deep understanding of nuclear physics, magnetism in materials and chemical analysis (amongst others) are now regularly used by archaeologists. As a result, a much greater understanding of the relationships between past events has emerged, though even these modern methods have their limitations. Fluorine testing, for example, is used for relative dating. It is useful for testing whether bones were buried at the same site at the same time, but cannot be used to compare bones from different sites, or to give a definite age. It relies on the fact that fluorides are found in most groundwater and are absorbed by bones. The amount of fluoride absorbed depends on the water, the soil and the time the bone has been buried, so a modern bone can be distinguished from an ancient bone.

Activity 30 allows you to explore the basic principles behind two very important dating methods: dendrochronology and radiocarbon dating. See Figure 6.48.

Figure 6.48 (a) A timber frame house dated by dendrochronology (b) a skeleton dated by radiocarbon dating

Activity 30 Dating techniques

Prepare a poster on a sheet of A3 paper that describes dendrochronology or radiocarbon dating.

Activity 31 Why so many?

What is the point of developing several methods of dating? Isn't just one sufficient? In a group, spend a few minutes 'brainstorming' possible responses.

4.2 Is it genuine?

Hoaxes have been quite common in archaeology as people used them to seek fame and financial gain. However, with better testing and authentication techniques, hoaxes tend to be less common now than in the past. Previous archaeological hoaxes have often involved burying an artefact and then 'discovering' it. In Japan in 2002, an archaeologist, Shinichi Fujimura was photographed digging the holes to bury the artefacts. One of the most famous hoaxes involved a skull found in Sussex in 1911.

(a)

(b)

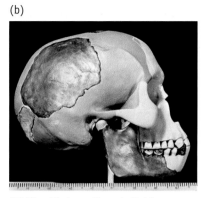

(c)

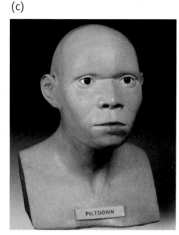

Figure 6.49 (a) working on the Piltdown site (b) The Piltdown Man skull (c) a reconstruction based on the skull

Piltdown Man

Charles Dawson, an amateur archaeologist, was involved in the discovery of a human skull and jaw alongside some animal bones. The find was of great importance at the time, as it seemed to confirm Darwin's theory of evolution. The jaw was ape-like, fitting predictions of how our ancestors had looked and was believed to be 500 000 years old (Figure 6.49). Arthur Woodward, a respected palaeontologist (fossil expert), named the skull Dawson's Dawn Man or *Eoanthropus dawsoni*, and it subsequently became known as the 'Piltdown Man' after the location of the gravel pits where it was found.

Despite some doubts being raised (for example, someone said it was an orang-utan skull), for the next forty years the skull was widely believed to be authentic. It supported the developing theory that linked apes to humans: an ape-like jaw, but a skull containing a human sized brain. Further finds near the site seemed to confirm the authenticity of the find and it was accepted by the Natural History Museum. However by 1953, evidence from other parts of the world seemed not to fit the find. Fossils found elsewhere in Europe and Africa were dismissed as they didn't match Piltdown Man, and had a much smaller skull instead. One find in South Africa, 'Taung's child' was later recognised as a member of the human family tree after initially being dismissed.

With the advent of new dating methods and more rigorous analytical techniques, Piltdown Man was re-assessed by the Natural History Museum. From these tests, it became clear that the skull was made up of two parts – a chimpanzee jaw and a human skull. To make them look old, the bones had been stained brown. In fact, all the bones from Piltdown had undergone similar staining. More detailed analysis also revealed that the teeth of the skull had been filed down to remove the dental patterns of the female ape. Dating techniques contributed further important evidence, and Piltdown Man was exposed as a hoax.

Charles Dawson was strongly implicated in the hoax, and some of his other finds were almost certainly fake. However, other people have also been suspected. These include Martin Hinton, keeper of the zoology collection at the Natural History Museum, an assistant for Woodward, Charles Chatwin and even the writer Arthur Conan Doyle. Additional evidence of a hoax came from a strange shaped tool, nicknamed the 'cricket bat', which was found three years later. This was thought by some scientists to be a joke. With Dawson's death in 1915, no more finds were made, but the true perpetrators have never been identified conclusively.

Activity 32 Piltdown Man

Use the Internet to find out more about the Piltdown Man hoax. Find out which dating techniques were used and what they revealed, then summarise the scientific evidence that led to Piltdown Man being declared a hoax.

4.3 Peer review

Scientists communicate their work by publishing reports (usually known as papers) in specialist journals, which can be either paper-based or electronic. As you saw in *Spare Part Surgery*, one way the scientific community tries to prevent hoaxes is by using the **peer review** process. 'Peer' means 'equal', and a key feature of the process is that scientists, in this case archaeologists, review each other's research work before it is published.

The reviewers are expected to comment on the work in the light of their own experience in the same area. They are asked to say whether:

- the science is correct
- the observations or experimental results appear to be authentic
- the conclusions are justified by the results
- the work is original
- other people's relevant work is acknowledged
- the findings are communicated clearly.

If a journal editor receives unfavourable reports from the peer reviewers, the paper is usually sent back to the authors for revision or, in extreme cases, may be rejected. Magazines and journals which are peer reviewed declare this, so that their readers know how much weight to give the information they are reading (Figure 6.50). Scientists use the published information to devise their own investigations, and if their results are not consistent with what's reported, then questions are asked. This is what eventually led to the Piltdown hoax being uncovered – the find did not fit in with all the other finds, and discussions eventually led to more rigorous tests, which proved that it was not genuine.

Some journals publish **review papers**. These do not necessarily report original research, but they are written by an expert in the field to give a full and up-to-date account of recent work in a particular area of research. Such papers are peer reviewed to check that the science is correctly reported and clearly communicated to the intended readers.

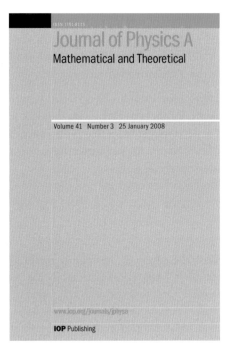

Figure 6.50 Title page of a peer-reviewed journal

Activity 33 Peer review

Using the checklist given above, review a write-up of a piece of experimental practical work that has been carried out by another student following this course.

Activity 34 Reviewing an article

Read a review article and comment on its suitability for publication.

The peer review process is also used when allocating research funds. In the UK there are eight Research Councils that are responsible for allocating government money for research. Each year the Councils invest around £1.3 billion in research in UK universities, around £500 million in their own Research Institutes, and around £300 million for UK researchers to access international facilities. This money is used to fund the very best world-class research as judged by independent, expert peer review. The research funding covers the medical and biological sciences, astronomy, physics, chemistry and engineering, social sciences, economics, and the arts and humanities. An archaeologist who wanted funding for a project would put forward a detailed written proposal to the relevant council, which would then be peer reviewed, along with other proposals, to decide which projects should be funded.

Activity 35 Funding for research

Look at the guidance for referees given by the Engineering and Physical Sciences Research Council (EPSRC) and discuss the advantages of this peer review process. Are there any disadvantages?

5 Reconstruction

5.1 Putting the pieces together

An archaeologist collects and uses diverse pieces of evidence that provide information about a site – these might include survey results, physical objects (from walls to minute fragments of bone or pottery), and written historical evidence. These then need to be pieced together to construct a coherent and self-consistent story.

Rather in a similar way, this chapter has contributed several pieces to your study of physics, some of which need to be brought together with things you have learned, or will be learning, in other chapters. In Activity 36, you will bring together some pieces from separate parts of this chapter, and in Activity 37 you will collect together information from this and some other chapters.

Activity 36 Using electromagnetic radiation

In this chapter and in *Technology in Space* and *The Sound of Music*, you have seen several uses for various types of electromagnetic waves. Using a copy of Table 6.5, make a summary listing at least two uses of radiation in each part of the spectrum given in the table. Look back through your work on all these three units, and use books or the Internet to find some additional information.

Type of radiation	Approximate wavelength range / m	Examples of uses
γ-rays	10^{-14} (or less) $-$ 10^{-11}	
X-rays	$10^{-11} - 10^{-8}$	
ultraviolet	$10^{-8} - 4 \times 10^{-7}$	
visible	$4 \times 10^{-7} - 7 \times 10^{-7}$	
infrared	$7 \times 10^{-7} - 10^{-3}$	
microwaves	$10^{-3} - 10^{-1}$	
radio	$10^{-1} - 10^3$ or more	

Table 6.5 Using electromagnetic radiation

Activity 37 Putting the pieces together

You learned something about each of the following in at least two parts of this chapter. Look back through your work and make brief notes under each heading, listing the examples that were used in different parts of the chapter and noting any ways in which the ideas were developed or refined. You could also add cross-references to other chapters where you have studied these same areas.

- DC circuits: resistance, resistivity, potential difference, potential divider, charge and energy
- Waves: electromagnetic spectrum, diffraction, superposition
- Electrons: particle and wave models
- Ethical issues in science
- Logarithmic scales

5.2 Questions on the whole chapter

19 Scientists planned to use a length of steel wire, 12.35 m long, diameter 1.0 mm, stretched between two rocks either side of a geological fault line to detect small changes in movement of the Earth. As the fault moved, the wire would extend and its resistance change.

(a) The resistivity of the wire chosen was 1.5×10^{-8} Ω m. Calculate the resistance of the length of wire used.

(b) The scientists were interested in changes in length greater than 5%. For such a change:

  (i) what would be the new length of the stretched wire?

  (ii) what would its new resistance be?

(c) In practice the 'noise' inherent in the experiment prevented meaningful results. Suggest what might be the cause of 'noise' in this case.

20 Some electronic devices need connections to –4.5 V, 0 V and +4.5 V. To achieve this, they are operated from so-called *split rail supplies*. Figure 6.51 shows how a 9 V battery can be used for such a supply.

(a) (i) Match the labels –4.5 V, 0V and +4.5 V to the points A, B and C.

(ii) What name is given to this type of arrangement of resistors and battery?

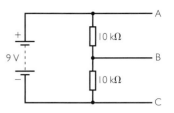

Figure 6.51 A 9 V battery is used to provide a split rail supply

(b) To check that the 25 kV overhead supply has been disconnected before starting work, railway engineers use a device consisting of a probe of resistance R_p and a voltage detector of resistance R_d. One end of the device is hooked on to the power line and the other end attached to Earth (usually the rail). See Figure 6.52. If the 25 kV supply is left on, then the detector registers a voltage of 25 V.

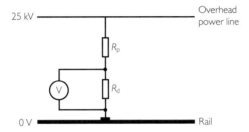

Figure 6.52 A device for checking overhead power lines

(i) If the resistance of the whole arrangement is 39 MΩ, what are the resistances of the detector, R_d, and the probe, R_p?

(ii) If there was poor contact with the rail or overhead cable (e.g. due to rust), explain the effect this would have on the voltage across the detector.

21 This poem was printed in *Punch* magazine in January 1896. Write a paragraph explaining what it is about.

> *O Röntgen, then the news is true,*
> *And not a trick of idle rumour,*
> *That bids us each beware of you,*
> *And of your grim and graveyard humour.*
>
> *We do not want, like Dr Swift,*
> *To take our flesh off and to pose in,*
> *Our bones, or show each little rift,*
> *And joint for you to poke your nose in.*
>
> *We only crave to contemplate,*
> *Each other's usual full-dress photo,*
> *Your worse than 'altogether' state,*
> *Of portraiture we bar in toto!*

22 Some artefacts from an archaeological site need to be investigated in more detail.

The features the archaeologists want to study are:

A pollen grains with features 0.1 μm in size

B weevil holes in grains of flour, which are 0.5 mm in size

C microfossils with features 1 μm in size

D ships' timbers with features 150 nm in size

E damage to bones with features 0.002 mm in size

(a) Which features can be resolved with an optical microscope?

(b) Which features can only be resolved with an electron microscope?

(c) Explain why the electron microscope can resolve smaller features than the optical microscope.

23 In an electron microscope, the electrons are accelerated through a potential difference of 10 kV.

(a) What is the kinetic energy of the electrons in eV and in joules?

(b) Calculate the speed of the electrons.

(c) What is the electrons' de Broglie wavelength?

$e = 1.60 \times 10^{-19}\,C$

$m = 9.11 \times 10^{-31}\,kg$

$h = 6.63 \times 10^{-34}\,J\,s$

24 A group of people are speculating about the finds on an archaeological site, and suggest hiding an artefact – a bone from another site for archaeologists to find. One of them speculates whether this may have been done in the past – and whether you can believe anything you read about archaeological finds.

(a) Write a paragraph for the local paper explaining some of the different tests that could be done on a bone found at an archaeological site and at the site where it is found to check that evidence is consistent.

(b) Write a paragraph to explain how journals use the peer review process to check on the material they are publishing – and how a reader could tell whether a journal uses the process or not.

5.3 Achievements

Now you have studied this chapter you should be able to achieve the outcomes listed in Table 6.6.

Table 6.6 Achievements for the chapter *Digging Up the Past*

	Statement from examination specification	Section(s) in this chapter
57	investigate and use the relationship $R = \dfrac{\rho l}{A}$	1.2
58	investigate and explain how the potential along a uniform current-carrying wire varies with the distance along it and how this variation can be made use of in a potential divider	1.3
61	use $I = nqvA$ to explain the large range of resistivities of different materials	1.2 (and see SPC)
29	identify the different regions of the electromagnetic spectrum and describe some of their applications	2.1, 2.2, 3.1 (and see SPC + MUS)
41	investigate and recall that waves can be diffracted and that substantial diffraction occurs when the size of the gap or obstacle is similar to the wavelength of the wave	2.3
42	explain how diffraction experiments provide evidence for the wave nature of electrons	3.2
43	discuss how scientific ideas may change over time, for example, our ideas on the particle/wave nature of electrons	3.2
48	explain how the amount of detail in a scan may be limited by *the wavelength of the radiation* or by the duration of pulses	3.1, 3.3 (and see SPC)

Answers

1   A metre length would be five times longer, so its resistance is $5 \times 8\ \Omega = 40\ \Omega$. This assumes that the cable cross-section is uniform throughout and so resistance is proportional to length.

2   The separation of 5 m is ten times greater, so you would expect a resistance about ten times greater i.e. about 3000 Ω. A 5 m separation would not detect small-scale structures such as a 1 m wide door opening, or a 0.5 m wall.

3   (a) If the length is doubled, then the area of cross-section is halved (to keep the same volume). On its own, doubling the length doubles the resistance, so does halving the area; both of these together multiply the resistance by four.

$4 \times 48\ \Omega = 192\ \Omega.$

(b) $l = 10$ cm $= 0.1$ m and

$A = 1\ \text{cm}^2$

$= (1 \times 10^{-2}\ \text{m})^2 = 1 \times 10^{-4}\ \text{m}^2.$

Using Equation 4:

$$\rho = \frac{RA}{l} = \frac{48\ \Omega \times 1.0 \times 10^{-4}\ \text{m}^2}{0.1\ \text{m}}$$

$= 4.8 \times 10^{-2}\ \Omega$ m.

4   Combining Equations 4 and 5:

$$R = \frac{l}{\sigma A}$$

5   The salty ditch was about 5 m from the gate (the resistance is lowest there) and the wall about 9 to 10 m from the gate where the resistance is highest.

6   On this scale, 1 cm represents $10^{-5}\ \Omega$ m.

$1.0 \times 10^{12} = 1.0 \times 10^{17} \times 10^{-5}$ so the bar for glass would be 1.0×10^{17} cm long (1.0×10^{15} m).

$1.7 \times 10^{-8} = 1.7 \times 10^{-3} \times 10^{-5}$, so the bar for copper would be 1.7×10^{-3} cm long (1.7×10^{-5} m).

The shortest bar would be too small to see, while the longest would not fit on the Earth (in fact, it would extend beyond the limits of the Solar System).

7   Heating a metal does not release a significant number of extra electrons, but it increases the vibrations of the atomic lattice through which the free electrons move. This scatters the electrons, reducing their average drift speed. In terms of Equations 9, 13 and 14 it reduces the constant k, reducing the material's conductivity and increasing its resistivity.

8   (a) (i)   4 V

(ii)   6 V

(ii)   4.8 V

(b) (i)   +8 V

(ii)   +9 V

(iii)   +7.2 V

9   (a) The wire will get longer and thinner, and on both counts the resistance will increase.

(b) If the length increases by 10%, then the area must decrease by 10% to keep the volume constant.

If $R_{\text{original}} = \dfrac{\rho l}{A}$, then

$$R_{\text{stretched}} = \frac{\rho(1.1l)}{(A \div 1.1)}$$

$= R_{\text{original}} \times (1.1)^2$

$= 1.21 R_{\text{original}}$

i.e. the resistance increases by 21%.

(c) Using Equation 4:

$R = \dfrac{\rho l}{A}$, then

$$= \frac{2.6 \times 10^{-6}\ \Omega\ \text{m} \times 28 \times 10^{-2}\ \text{m}}{0.6 \times 10^{-3}\ \text{m} \times 0.002 \times 10^{-3}\ \text{m}}$$

$= 606\ \Omega$

(d) See Figure 6.53. The pd across the fixed resistor is four times that across the gauge, so its resistance must be $4 \times 606\ \Omega = 2424\ \Omega$.

Figure 6.53 Diagram for the answer to Question 9 (d)

10   The term 'shadow radiograph' is appropriate because the photographic film has a shadow cast upon it of the more dense parts of the object being X-rayed. The X-rays pass through the less dense materials and are absorbed by the more dense ones such as metals or bones. Thus the film is blackened except where the X-rays were prevented from getting to it. These shadow areas remain white on the film.

11 X-ray tomography is a technique whereby an image of a 'slice' of the body is acquired by blurring out the unwanted shadows below and above the plane of interest. A conventional X-ray image is a single 'shadow' produced by the transmission of X-rays through the thickness of the body and thus may result in certain elements being 'hidden' by those in front.

12 There was concern that the CAT (Computer Aided Tomography) scanner might not give satisfactory images because the machine was calibrated for objects of the density of the human body. The helmet, being made of metal had a density, beyond the scope of the scanner. (The effect was to give images with less definition than might be required.)

13 The use of horizontal and vertical radiographs allowed the position of items within the helmet to be pinpointed as it provided two crossed slices by which the position of the item was defined – much as two co-ordinates define the position of a point on a graph (see Figure 6.54).

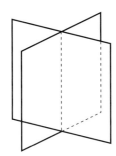

Figure 6.54 Crossed 'slices' pinpoint location

14 The high performance industrial radiography of the helmet gave conservators information on the technology involved in the helmet's construction while microfocus radiography of the chain mail provided information on its construction.

15 The typical width of a doorway is 1 m which is comparable with the wavelength of sound thus the sound wave can be diffracted and spread out into the room. The wavelength of light is much smaller so the beam of light is not noticeably diffracted.

16 Fe_3O_4

17 (a) Maximum kinetic energy

$E_k = 100$ eV

$= 100 \times 1.60 \times 10^{-19}$ J e V^{-1}

$= 1.60 \times 10^{-17}$ J.

(b) Assuming the electrons reach their maximum possible kinetic energy, we can write:

$\frac{1}{2}mv^2 = eV$

$v^2 = \frac{2eV}{m}$

$= \frac{2 \times 1.6 \times 10^{-19} \text{ C} \times 100 \text{ J C}^{-1}}{9.11 \times 10^{-31} \text{ kg}}$

$= 3.52 \times 10^{13}$ m^2 s^{-2}

Units of v^2 are J kg^{-1}, which is equivalent to m^2 s^{-2}.

$v = 5.93 \times 10^6$ m s^{-1}

> **Maths reference**
>
> See Maths Note 2.3 Derived Units

(c) Using Equation 15,

$\lambda = \frac{h}{mv}$

$= \frac{6.6 \times 10^{-34} \text{ J s}}{9.1 \times 10^{-31} \text{ kg} \times 5.93 \times 10^6 \text{ m s}^{-1}}$

$= 1.2 \times 10^{-10}$ m.

Again notice the units of λ: J s kg^{-1} m^{-1} s, which is equivalent to m.

18 (a) Wavelength $\lambda = 100$ nm. Rearranging Equation 15 to find the electron speed v:

$v = \frac{h}{m\lambda}$

Accelerating voltage V is related to the electron kinetic energy:

$eV = E_k = \frac{1}{2}mv^2$

so $V = \frac{E_k}{e} = \frac{\frac{1}{2}m\left(\frac{h}{m\lambda}\right)^2}{e} = \frac{h^2}{2me\lambda^2}$

(b) $V = \frac{(6.63 \times 10^{-34} \text{ J s})^2}{2 \times 9.11 \times 10^{-31} \text{ kg} \times 1.60 \times 10^{-19} \text{ C} \times (100 \times 10^{-9} \text{ m})^2}$

$= 6.03 \times 10^4$ V

(c) Doubling the accelerating voltage doubles the electrons' kinetic energy. This multiplies their speed by $\sqrt{2}$ (= 1.414). This reduces their wavelength to $\frac{100 \text{ nm}}{\sqrt{2}}$ and the smallest objects that can be observed now have a size of $\frac{50 \text{ nm}}{\sqrt{2}} = 35$ nm.

Maths Notes

0 | Signs and symbols

0.1 Equations and comparisons

In physics, we are often interested in whether two quantities are exactly equal, or almost equal, or whether one is greater than the other. Table 1 lists the signs used for expressing such relationships.

Symbol	Meaning	Notes
=	is equal to	
≡	is exactly the same as	used to emphasis the point that two expressions are two ways of writing exactly the same thing (as opposed to two different things being the same size)
≠	is not equal to	
≈	is approximately equal to	
~	is the same order of magnitude as	
<	is less than	the smaller quantity is written at the narrow end of the symbol
>	is greater than	
≤	is less than or equal to	
≥	is greater than or equal to	
≪	is much less than	
≫	is much greater than	

Table 1 Signs for equations and comparisons

> **Maths reference**
>
> Order of magnitude
> See Maths note 7.4

0.2 The delta symbol

The symbol Δ (the capital Greek letter delta) is used to mean 'a small amount of' or 'a change in'. Notice that Δ does *not* represent a number, so resist the temptation to cancel Δ, e.g. if it appears on the top and bottom of an expression.

For example, the symbol Δt represents a time interval and is often used when describing rates of flow or rates of change. For example, if an amount of charge ΔQ flows past a point in a time interval Δt, then the current I can be written

$$I = \frac{\Delta Q}{\Delta t}$$

The delta symbol is also used to denote an experimental uncertainty. For example, if a distance x is measured as 23 mm but could be out by 1 mm in either direction, then the uncertainty in the measurement is $\Delta x = 1$ mm. The measurement is written as $x \pm \Delta x$, i.e. 23 mm ± 1 mm.

0.3 Summation

The symbol Σ (the capital Greek letter sigma) means 'the sum of'. Note that Σ does *not* represent a number so resist the temptation to cancel it, e.g. if it appears on both sides of an equation.

The symbol Σ is usually used with other symbols respresenting the items to be added together. For example, the mass of a sample of gas is Σm, where m represents the mass of an individual gas molecule and the sum includes all the molecules in the sample.

When dealing with vectors, the sum is a vector sum i.e. it takes account of direction as well as magnitude. ΣF, meaning the sum of forces in a given situation, can be represented by a vetor diagram.

1 Index notation

An **index** (plural **indices**) or **power** is the superscript number which, when a positive whole number, means squared, cubed, etc. For example

$5^2 = 5 \times 5 = 25$

$7^3 = 7 \times 7 \times 7 = 343$

$0.6^4 = 0.6 \times 0.6 \times 0.6 \times 0.6 = 0.1296$

1.1 Index notation and powers of 10

Table 2 shows 'powers of 10'. The number in any row is found by dividing the number in the row above by 10.

$100\,000 =$	$10 \times 10 \times 10 \times 10 \times 10 =$	10^5
$10\,000 =$	$10 \times 10 \times 10 \times 10 =$	10^4
$1\,000 =$	$10 \times 10 \times 10 =$	10^3
$100 =$	$10 \times 10 =$	10^2
$10 =$	$10 =$	10^1
$1 =$	$1 =$	10^0
$0.1 =$	$\dfrac{1}{10} =$	10^{-1}
$0.01 =$	$\dfrac{1}{10 \times 10} = \dfrac{1}{10^2} =$	10^{-2}
$0.001 =$	$\dfrac{1}{10 \times 10 \times 10} = \dfrac{1}{10^3} =$	10^{-3}

Table 2 Positive and negative powers of 10

Extending the pattern gives a meaning to zero and negative indices. If you replace all the 10s in Table 2 by any other number that you choose, you should be able to convince yourself that

$\qquad x^0 = 1$ for *any* value of *x*.

> **Maths reference**
>
> Units and physical quantitites
> See Maths note 2.1

1.2 Standard form

To represent very large and very small numbers, we generally use **standard form**, also called **scientific notation**.

A number written in standard form consists of a number with a single digit (not zero) before the decimal point, multiplied by a power of 10.

Large numbers

$5\,620\,000$ (five million six hundred and twenty thousand) becomes 5.62×10^6

$407\,300$ (four hundred and seven thousand, three hundred) becomes 4.073×10^5.

Small numbers

$0.5680 = 5.680 \times 0.1 = 5.68 \times 10^{-1}$

$0.000\,702\,3 = 7.023 \times 0.0001 = 7.023 \times 10^{-4}$

1.3 Combining powers

Powers of the same number

When multiplying two numbers expressed as 'powers' of the same number, the powers add:

$$10^2 \times 10^3 = (10 \times 10) \times (10 \times 10 \times 10) = 10^5$$

i.e. $\qquad 10^2 \times 10^3 = 10^{(2 + 3)}$

$$6^2 \times 6^2 = (6 \times 6) \times (6 \times 6) = 6^4$$

When dividing, the powers subtract

$$10^6 \div 10^2 = (10 \times 10 \times 10 \times 10 \times 10 \times 10) \div (10 \times 10) = 10^4$$

i.e. $\qquad 10^6 \div 10^2 = 10^{(6 - 2)}$

The rules still work when negative powers are involved:

$$10^5 \times 10^{-2} - 10^5 \times \left(\frac{1}{10^2}\right) = 10^5 \div 10^2 = 10^3$$

i.e. $\qquad 10^5 \times 10^{-2} = 10^{(5 - 2)}$

$$x^4 \times x^{-3} = x^{(4 - 3)} = x$$

$$4^3 \div 4^{-2} = 4^3 \div \left(\frac{1}{4^2}\right) = 4^3 \times 4^2 = 4^5$$

i.e. $\qquad 4^3 \div 4^{-2} = 4^{(3 - -2)} = 4^{(3 + 2)}$

> **Maths reference**
>
> Reciprocals
> See Maths note 3.3

Powers of different numbers

When dealing with a mixture of numbers of different type, collect together all numbers of the same type and combine their powers by adding or subtracting:

$$2 \times 10^4 \times 3 \times 10^5 = (2 \times 3) \times (10^4 \times 10^5) = 6 \times 10^9$$

$$1.38 \times 10^{-23} \times 2.3 \times 10^3 = 1.38 \times 2.3 \times 10^{(-23 + 3)}$$
$$= 3.174 \times 10^{-20}$$

$$3y^2 \times 7y^5 = 21y^7$$
$$5z^2 \times 3z^{-2} = 15z^0 = 15$$

1.4 Manipulating powers on a calculator

Powers of 10

Think of the EXP or EE key as 'times 10 to the power of'.

To enter 7.54×10^9: enter 7.54, press EXP and enter 9. (Notice that you do *not* type in 10 – if you do, you will multiply your number by 10, making it 10 times too big.)

Your calculator might use its own shorthand to display this number as 7.54 09, or 7.54^9, or 7.54 EE 9 (or similar). But you should always *write* it as 7.54×10^9.

Negative powers of 10

To enter a negative index, use the ± or +/− key (*not* the 'minus' key, because that will subtract the next number from the one you have just entered).

To enter 1.38×10^{-23}: enter 1.38, press EXP, enter 23 and press ±.

Squares, etc.

To square a number, use the x^2 key. For example, to work out 1.3^2, enter 1.3 and press x^2 to get 1.69.

Pressing x^2 again squares the answer, i.e. calculates your original number to the power of 4. Pressing x^2 three times altogether gives you your original number to the power of 8, and so on – each time you press x^2, you double the power.

Other powers

Use the y^x key to raise one number to the power of a second number. y is the first number you enter, and x the second.

To calculate 2.5^3: enter 2.5, press y^x, enter 3, press =.

Other negative powers

As with powers of 10, use the ± or +/− key to enter negative numbers.

To calculate 2.5^{-3}: enter 2.5, press y^x, enter 3, press ±, press =.

1.5 Powers that are not whole numbers

The square root of a number x can be written as $x^{\frac{1}{2}}$ or $x^{1/2}$:

$$x^{\frac{1}{2}} \times x^{\frac{1}{2}} = x^{\left(\frac{1}{2}+\frac{1}{2}\right)} = x^1 = x$$

so $x^{\frac{1}{2}} = \sqrt{x}$.

Similarly, $x^{\frac{1}{3}} = \sqrt[3]{x}$ (the cube root of x); $x^{\frac{1}{4}} = \sqrt[4]{x}$ and so on.

Other fractional powers can also be interpreted in terms of roots, for example:

$$x^{\frac{3}{2}} = \sqrt{(x^3)} \text{ (the square root of } x\text{-cubed)}$$
$$= (\sqrt{x})^3 \text{ (the cube of the square root of } x\text{)}$$

and

$$x^{-\frac{1}{2}} = \frac{1}{x^{\frac{1}{2}}} = \frac{1}{\sqrt{x}}$$

Fractional powers can also be written using decimal numbers, for example:

$$x^{\frac{1}{2}} = x^{0.5}$$
$$x^{\frac{3}{2}} = x^{1.5}$$

Powers that are neither simple fractions nor whole numbers are less easy to interpret, but they still exist and can be calculated (e.g. using the y^x key of a calculator. For example:

$$10^{0.333} = 2.153$$
$$10^{0.6021} = 4.000$$
$$5.6^{\pi} = 224.1$$
$$9.34^{-0.83} = 0.1565$$

(All these answers are given to four significant figures.)

> **Maths reference**
>
> Significant figures
> See Maths note 7.2

2 Units

The SI system of units (Système Internationale d'Unités) has been established by international agreement. In your study of physics you will use mainly SI units. The basic SI units are listed in Table 3. Notice that, when a unit is named after a person, the unit symbol has a capital but the *name* of the unit does not.

Quantity	SI unit	Notes
mass	kilogram, kg	
time	second, s	
length	metre, m	
electric current	ampere, A	used to define the unit of charge, the coulomb
temperature	kelvin, K	
luminous intensity	candela, cd	not used in this course, but included here for completeness
amount of substance	mole, mol	

Table 3 The basic SI units

2.1 Units and physical quantities; graphs and tables

A physical quantity consists of a number and a unit. Without the unit, the quantity is incomplete. When a symbol represents a physical quantity, it represents the *complete* quantity – units and all. For example, suppose v represents speed, and a particular speed is found to be 5 m s⁻¹. You should write

$$v = 5 \text{ m s}^{-1}$$

(*not* just $v = 5$ and *not* v (m s⁻¹) = 5).

Units can be manipulated just like numbers and other symbols. When labelling axes of graphs, and when listing physical quantities in tables, it is conventional to divide each quantity by its unit to get a pure number.

For example, you can divide both sides of the expression for v above by m s⁻¹ and write

$$v/(\text{m s}^{-1}) = 5$$

If you are plotting values of v on a graph, or listing them in a table, you should label the graph axis, or the table column, as v/m s⁻¹.

Large and small numbers

Suppose you were dealing with speeds that were all several million metres per second:

$$v = 2 \times 10^6 \text{ m s}^{-1}, \qquad v = 7 \times 10^6 \text{ m s}^{-1}, \text{ etc.}$$

To make the numbers more manageable, you could use the same rule as above to write $v/(10^6 \text{ m s}^{-1}) = 2$, etc., and label your graph and table as shown in Figure 1.

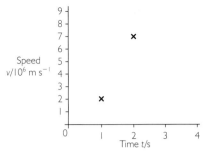

Figure 1 Labelling graphs and tables

2.2 Manipulating units; index notation and units

In calculations, the units should be manipulated as well as the numbers. This can help you keep track of what you are doing as well as being correct – so it is a good habit to get into.

Indices can be used with units and with algebraic symbols. For example,

$$4^{-1} = \frac{1}{4} = 0.25, \qquad x^{-2} = \frac{1}{x^2}$$

Units such as coulombs per second, or joules per coulomb, can be written either as C/s and J/C or using index notation: C s^{-1} and J C^{-1}. Similarly, metres per second, in calculations of unit of speed, can be written as m/s or m s^{-1}. For example,

$$70 \text{ m} \div 20 \text{ s} = 3.5 \text{ m s}^{-1}$$

Using the index notation helps prevent table headings and graph labels having too many oblique strokes.

When multiplying numbers, units or symbols, collect together all those of the same type and add their indices. For example:

$$2 \text{ C s}^{-1} \times 4 \text{ s} = 8 \text{ C}$$
$$10 \text{ m s}^{-1} \div 5 \text{ s} = 2 \text{ m s}^{-2}$$

Maths reference

Index notation and powers of 10
See Maths note 1.1

Units and physical quantities; graphs and tables
See Maths note 2.1

2.3 Derived units

Table 4 shows how SI units are combined to give units of various quantities. Some common combinations are given 'shorthand' names.

Quantity	Unit name	Symbol	Equivalent
speed			m s^{-1}
acceleration			m s^{-2}
force	newton	N	$1 \text{ N} = 1 \text{ kg m s}^{-2}$
gravitational field strength			$1 \text{ N kg}^{-1} = 1 \text{ m s}^{-2}$
energy, work	joule	J	$1 \text{ J} = 1 \text{ N m} = 1 \text{ kg m}^2 \text{ s}^{-2}$
power	watt	W	$1 \text{ W} = 1 \text{ J s}^{-1} (= 1 \text{ kg m}^2 \text{ s}^{-3})$
frequency	hertz	Hz	$1 \text{ Hz} = 1 \text{ s}^{-1}$
electric charge	coulomb	C	$1 \text{ C} = 1 \text{ A s}$ $1 \text{ A} = 1 \text{ C s}^{-1}$
potential difference, emf	volt	V	$1 \text{ V} = 1 \text{ J C}^{-1} (= 1 \text{ kg m}^2 \text{ C}^{-1} \text{ s}^{-2})$
electrical resistance	ohm	Ω	$1 \text{ Ω} = 1 \text{ V A}^{-1} (= 1 \text{ kg m}^2 \text{ C}^{-2} \text{ s}^{-1})$

Table 4 Some common derived SI units

Study note

In writing units, the coulomb is often treated as if it were the basic unit rather that the ampere.

2.4 SI prefixes

When dealing with quantities that are large or small, we often use prefixes as an alternative to standard form. For example, a distance of 1.3×10^4 m could be written as 13 km, and a distance of 0.0037 m could be written as 3.7 mm. The official SI prefixes go up and down in steps of 10^3. Table 5 lists the SI prefixes that you are likely to encounter in your study of physics.

Prefix	Symbol	Equivalent in powers of 10
tera	T	10^{12}
giga	G	10^{9}
mega	M	10^{6}
kilo	k	10^{3}
centi	c	10^{-2}
milli	m	10^{-3}
micro	μ	10^{-6}
nano	n	10^{-9}
pico	p	10^{-12}
femto	f	10^{-15}

Table 5 SI prefixes

Study note

The centimetre is not officially an SI unit [because 'centi' (10^{-2}) does not fit the pattern] but it is widely used.

When dealing with conversions involving prefixes, it is wise to write down each step using appropriate powers of 10, *and include the units at each stage*. For example, suppose light of a certain colour has a wavelength of 468 nm and you want to use standard form to write the wavelength in metres:

$$468 \text{ nm} = 468 \times 10^{-9} \text{ m}$$
$$= 4.68 \times 10^2 \times 10^{-9} \text{ m}$$
$$= 4.68 \times 10^{-7} \text{ m}$$

Suppose the tension in a rope is 1.35×10^5 N and you want to express it in kN:

$$1 \text{ kN} = 10^3 \text{ N, so } 1 \text{ N} = \frac{1}{10^3} \text{ kN} = 10^{-3} \text{ kN}$$

$$1.35 \times 10^5 \text{ N} = 1.35 \times 10^5 \times 10^{-3} \text{ kN}$$
$$= 1.35 \times 10^2 \text{ kN}$$
$$= 135 \text{ kN}$$

Suppose an electric current is 4.56×10^{-4} A and you want to express it in μA:

$$1 \text{ }\mu\text{A} = 10^{-6} \text{ A, so } 1 \text{ A} = \frac{1}{10^{-6}} \mu\text{A} = 10^6 \text{ }\mu\text{A}$$

$$4.56 \times 10^{-4} \text{ A} = 4.56 \times 10^{-4} \times 10^6 \text{ }\mu\text{A}$$
$$= 4.56 \times 10^2 \text{ }\mu\text{A}$$
$$= 456 \text{ }\mu\text{A}$$

2.5 Dimensions

The **dimensions** of a quantity show how it is related to the basic quantities listed in Table 3. Symbols M, L and T are used to represent the dimensions of mass, length and time.

For example, volume is calculated from length × breadth × height so has dimension of length^3 or L^3; speed is found from distance ÷ time so has dimensions of L/T or LT^{-1}. The dimensions of force are those of mass × acceleration: MLT^{-2}.

Square brackets are used to denote the dimensions of a quantity. For example

$$[\text{velocity}] = LT^{-1}$$
$$[\text{force}] = [\text{mass}] \times [\text{acceleration}] = MLT^{-2}$$

Dimensions are more fundamental than units. You might, for example, choose to express a speed in miles per hour rather than SI units of m s^{-1}, but the dimensions are still LT^{-1}, i.e. length (miles) ÷ time (hours).

Any equation must be dimensionally consistent, that is, the dimensions of the left-hand side must be the same as those of the right-hand side. This can help you check whether a particular equation is correct, and can also enable you to derive relationships between quantities.

3 Arithmetic and algebra

3.1 Fractions, decimals and percentages

A fraction is really a division sum, e.g.

$$\frac{4}{5} = 4 \div 5; \qquad \frac{7}{3} = 7 \div 3.$$

You can express a fraction as a decimal number by doing the division on a calculator.

When fractions are multiplied together, you can often simplify the arithmetic by using the fact that the multiplication and division can be carried out in any order, e.g.

$$\frac{7}{5} \times \frac{3}{14} = \frac{7 \times 3}{5 \times 14}$$

and cancelling any common factors, e.g.

$$\frac{7 \times 3}{5 \times 14} = \frac{3}{5 \times 2} = \frac{3}{10} = 0.3.$$

You can think of the **percentage** sign, %, as being made up of a 1, 0, 0 to remind you that it is a fraction of 100 parts. To calculate a percentage from a number expressed as a fraction or a decimal, you multiply by 100:

$$\frac{1}{2} = 0.5 \text{ and } 100 \times 0.5 = 50 \text{ so } \frac{1}{2} = 50\% \text{ (or } 50/100)$$

$$\frac{1}{4} = 0.25 \text{ and } 100 \times 0.25 = 25 \text{ so } \frac{1}{4} = 25\% \text{ (or } 25/100)$$

$$\frac{7}{8} = 0.875 \text{ and } 100 \times 0.875 = 87.5 \text{ so } \frac{7}{8} = 87.5\%$$

For example, if a solar array produces an output power of 600 W from an input power of 4 kW (4000 W), its efficiency is

$$\frac{600\text{W}}{4000\text{W}} = 0.15 = 15\%.$$

To find a percentage of a quantity, you *multiply* the quantity by the percentage expressed as an ordinary fraction or decimal number. For example, to find 15% of 60 multiply 60 by 15/100 (or by 0.15)

$$\frac{15}{100} \times 60 = \frac{90}{10} = 9$$

or

$$0.15 \times 60 = 9.$$

3.2 Brackets and common factors

To evaluate an expression such as

$$6(2 + 3 - 4 + 5), \quad \frac{12 + 8}{4} \quad \text{or} \quad I(R_1 + R_2 + R_3)$$

you usually first deal with the additions and subtractions inside the bracket and then multiply or divide the result by the number or symbol outside. Alternatively you can carry out several separate multiplications or divisions on each number or symbol inside the bracket in turn, then do the additions or subtractions. For example

either $6(2 + 3 - 4 + 5) = 6 \times 6 = 36$

or $6(2 + 3 - 4 + 5) = 12 + 18 - 24 + 30 = 36$

either $\dfrac{12 + 8}{4} = \dfrac{20}{4} = 5$

or $\dfrac{12 + 8}{4} = \dfrac{12}{4} + \dfrac{8}{4} = 3 + 2 = 5$

A calculation that involves several multiplications or divisions using the same number and then adding or subtracting the results can be simplified if it is rewritten using brackets with the **common factor** outside. For example

$$25 + 30 + 35 = 5(5 + 6 + 7)$$

$$3x + 3y + 3z = 3(x + y + z)$$

$$IR_1 + IR_2 + IR_3 = I(R_1 + R_2 + R_3)$$

$$\frac{7}{2} + \frac{3}{2} + \frac{6}{2} = \frac{(7 + 3 + 6)}{2}$$

$$\frac{a}{x} + \frac{b}{x} + \frac{c}{x} = \frac{a + b + c}{x}$$

3.3 Reciprocals

The value obtained by dividing 1 by a number is called the **reciprocal** of the number (reciprocals can be found using the $1/x$ key of a calculator). Finding the reciprocal of a reciprocal gets you back to the original number. For example:

$$\frac{1}{2} = 0.5, \qquad \frac{1}{0.5} = 2.$$

For a wave or oscillation:

$$\text{period } T = \frac{1}{f}, \qquad \text{frequency } f = \frac{1}{T}.$$

Reciprocals are sometimes written using a negative index:

$$x^{-1} = \frac{1}{x}.$$

To find the reciprocal of a fraction, simply turn it the other way up. For example:

$$\frac{1}{\frac{1}{2}} = \frac{2}{1} = 2$$

$$\frac{1}{\frac{2}{3}} = \frac{3}{2} = 1\tfrac{1}{2}$$

$$\left(\frac{3}{7}\right)^{-1} = \frac{7}{3}$$

This is not just an arbitrary rule. It makes sense if you think in terms of division sums. Consider the second example above. Question: 'How many times does $\frac{2}{3}$ go into 1?' Answer: 'one-and-a-half times.'

Adding and subtracting

One place where you need to add and subtract reciprocals is in calculations of resistors in parallel. To find the net resistance R of several resistors connected in parallel, you must first find the reciprocal of each resistor, then add the reciprocals together (to get $1/R$), then find the reciprocal of $1/R$ to get R.

For example, if $R_1 = 2.0\ \Omega$, $R_2 = 5.0\ \Omega$, $R_3 = 1.0\ \Omega$, then

$$\frac{1}{R_1} = \frac{1}{2}\ \Omega^{-1} = 0.50\ \Omega^{-1} \text{ (notice the unit of } 1/R)$$

$$\frac{1}{R_2} = 0.20\ \Omega^{-1},$$

$$\frac{1}{R_3} = 1.00\ \Omega^{-1}$$

(notice that $1/1 = 1$ – the number stays the same but the unit still changes). So

$$\frac{1}{R} = (0.50 + 0.20 + 1.00)\ \Omega^{-1} = 1.70\ \Omega^{-1}$$

$$R = \frac{1}{1.70}\ \Omega = 0.59\ \Omega$$

Notice that adding the reciprocals of two numbers is *not* the same as adding the two numbers and then finding the reciprocal of their sum.

Multiplying and dividing

Multiplying by the reciprocal of a number is the same as dividing by that number. For example

$$7 \times \frac{1}{2} = 7 \div 2 = 3.50$$

Dividing by the reciprocal of a number is the same as multiplying by that number. For example

$$4 \div \frac{1}{3} = 4 \times 3 = 12$$

$$9 \div \frac{3}{4} = 9 \times \frac{4}{3} = \frac{9 \times 4}{3} = 12$$

For a wave,

$$f = \frac{v}{\lambda}, \qquad \text{time period } T = \frac{1}{f} = \frac{1}{(v/\lambda)} = \frac{\lambda}{v}$$

We can simplify divisions involving fractions. For example:

$$\frac{3}{4} \div \frac{5}{4} = \frac{3}{4} \times \frac{4}{5} = \frac{3 \times 4}{4 \times 5} = \frac{3}{5} = 0.6.$$

3.4 Algebra and elimination

If we have two different relationships that both involve some of the same things, we can combine them to produce a new equation. This allows us to avoid measuring, or calculating, something that is not already known – we can eliminate it (remove it) from the equations. For example, we can take an expression for electrical power

$$P = IV$$

and use the resistance equation

$$V = IR$$

to write IR instead of V:

$$P = I \times IR = I^2R$$

This enables us to relate P directly to I and R without needing to know or calculate V. Similarly, if we want to eliminate I:

$$P = \frac{V}{R} \times V = \frac{V^2}{R}$$

3.5 Adding and subtracting fractions

You can of course add and subtract fractions on a calculator – you carry out several division sums and add or subtract the results. But for simple fractions it can often be quicker to do the sums 'by hand'.

The trick is to write the fractions so that they have the same **denominator** (the number underneath the fraction). Sometimes it is quite easy to spot how to do this. For example:

$$\frac{3}{4} + \frac{5}{6} = \frac{3 \times 3}{3 \times 4} + \frac{2 \times 5}{2 \times 6}$$

$$= \frac{9}{12} + \frac{10}{12} = \frac{9 + 10}{12} = \frac{19}{12}$$

Otherwise, make a common denominator by multiplying the original denominators together:

$$\frac{2}{17} + \frac{4}{3} = \frac{2 \times 3}{17 \times 3} + \frac{4 \times 17}{3 \times 17}$$

$$= \frac{6}{51} + \frac{68}{51} = \frac{6 + 68}{51} = \frac{74}{51}$$

Another example:

$$\frac{1}{2} + \frac{1}{3} = \frac{3}{6} + \frac{2}{6} = \frac{5}{6}$$

4 Solving equations

It may sound obvious, but the main thing to understand about equations is that the '=' sign means that the two things on either side are *equal* to one another. So whatever you do to one side, you must also do to the other, otherwise they would no longer be equal. (Beware of getting into the bad habit of writing '=' when you really mean 'and so the next step is…'.)

One way to think of an equation is as a 'recipe' for calculating. For example, $F = ma$ tells you how to calculate the net force F if you know the acceleration a that it gives to a mass m. In this example, F is the **subject** of the equation – it is written on its own (usually on the left).

4.1 Rearranging an equation

Quite often, the quantity you want to calculate is wrapped up in the right-hand side of an equation, and you need to make it the subject. When doing this, it helps if you try to understand what you are doing rather than blindly trying to apply a set of rules. It is also wise to write down each step, justifying each one to yourself as you do so. This might sound time-consuming, but it isn't really because it helps you to keep track of what you are doing and, if you do make a slip, it is quite easy to go back and check.

Look at the part of the equation that contains the quantity that you want to know. Think what you need to do to get that quantity on its own, and do the same thing(s) to both sides.

For example, suppose you want to know the acceleration that a force F gives to a mass m:

$$F = ma$$

To get a on its own, you need to divide the right-hand side by m ($ma \div m = a$), so do the same to the left-hand side:

$$\frac{F}{m} = a, \qquad \text{or} \qquad a = \frac{F}{m}.$$

Another example: suppose you want to calculate internal resistance r from

$$V = \mathcal{E} - Ir$$

It is a good idea first to arrange that the thing you are interested in has a positive sign. You can do this by adding Ir to both sides:

$$V + Ir = \mathcal{E}$$

then to get r on its own you subtract V from both sides:

$$Ir = \mathcal{E} - V$$

and then divide by I

$$r = \frac{\mathcal{E} - V}{I} \qquad \text{or} \qquad r = (\mathcal{E} - V)/I$$

(Notice that you have to divide the *whole* of the right-hand side by I – hence the brackets.)

4.2 Simultaneous equations

Simultaneous equations arise if we have two (or more) different ways of writing a relationship between quantities. If we have two unknown quantities, then they can both be found if we have two simultaneous equations. For three unknown quantities, we'd need three separate equations, and so on.

The trick in solving simultaneous equations is to carry out some algebra and arithmetic to get an expression that involves just *one* of the unknown things, and then use that value to calculate the other one.

For example, the equation $\mathcal{E} = V + Ir$ involves two things that can be measured (V and I). If neither $\mathcal{E}$ nor r is known, then they cannot be found from a single pair of values of V and I. However, if you obtain two *different* pairs of readings (V_1 and I_1, and V_2 and I_2) for the same power supply (using two different external loads), then you can write down two simultaneous equations – two different equations that both describe a relationship between the two unknown things $\mathcal{E}$ and r. These equations let you find both $\mathcal{E}$ and r. So

$$\mathcal{E} = V_1 + I_1 r$$
$$\mathcal{E} = V_2 + I_2 r$$

Since the right-hand side of each equation is equal to $\mathcal{E}$, then they must also be equal to each other:

$$V_1 + I_1 r = V_2 + I_2 r$$

Subtracting V_1 from each side

$$I_1 r = V_2 - V_1 + I_2 r$$

Subtracting $I_2 r$ from both sides (and being careful with signs and with the subscripts 1 and 2)

$$I_1 r - I_2 r = V_2 - V_1$$

Now r is a common factor on the left-hand side, so

$$r(I_1 - I_2) = V_2 - V_1$$

> **Maths reference**
>
> Brackets and common factors
> See Maths note 3.2

Dividing both sides by $(I_1 - I_2)$ (and using brackets to keep the subtracted things together)

$$r = \frac{(V_2 - V_1)}{(I_2 - I_1)} \qquad \text{or} \qquad r = (V_2 - V_1)/(I_1 - I_2)$$

This value of r can then be used in one of the original equations to find $\mathcal{E}$.

For example: a power supply gives readings of $V_1 = 3\,\text{V}$, $I_1 = 7\,\text{A}$, and $V_2 = 8\,\text{V}$, $I_2 = 2\,\text{A}$. So

$$r = \frac{8\,\text{V} - 3\,\text{V}}{7\,\text{A} - 2\,\text{A}} = \frac{5\,\text{V}}{5\,\text{A}} = 1\,\Omega$$

and

$$\mathcal{E} = V_1 + I_1 r = 3\,\text{V} + 7\,\text{A} \times 1\,\Omega = 3\,\text{V} + 7\,\text{V} = 10\,\text{V}$$

(you would find the same value using V_2 and I_2).

5 Relationships and graphs

Graphs are extremely useful in physics for giving us a pictorial representation of how one quantity is related to another. Trends in data are not always clear from a table of results, but become immediately evident when viewing a plot of the two quantities involved.

5.1 Graphs and proportionality

Many important relationships in physics involve the idea of direct proportion.

For example, if a conductor obeys Ohm's law, doubling the potential difference produces double the current, tripling the pd triples the current ... and so on. Mathematically, we say that the potential difference is **directly proportional** to the current. In symbols

$$V \propto I \qquad \text{or} \qquad V = kI$$

The symbol $\propto$ means 'is directly proportional to' and k is called a **constant of proportionality** and has a fixed value for a particular set of values of V and I. (The constant k in this example is the same thing as the electrical resistance R.)

If one quantity is directly proportional to another, then a graph of one plotted against the other is a straight line through the origin.

5.2 Linear relationships

The equation $V = \mathcal{E} - Ir$ is an example of a **linear relationship** between two variables, V and I in this case. A graph of V (on the vertical axis, the y-axis) against I (on the horizontal axis, the x-axis) gives a straight line. Linear relationships and graphs are often said to be of the type $y = mx + c$, where y stands for whatever is plotted on the y-axis and x for whatever is plotted on the x-axis, and m and c are constants (they remain fixed when x and y change). This type of graph has two properties that are often useful for doing calculations using experimental results. We can illustrate these with a graph of $y = 2x + 1$, i.e. $m = 2$, $c = 1$ (Figure 2).

On Figure 2, the line cuts the y-axis at $y = 1$ (using the equation, when $x = 0$, $y = c$). The line of such a graph always cuts the y-axis where $y = c$.

If y is directly proportional to x, then the line goes through the origin and $c = 0$.

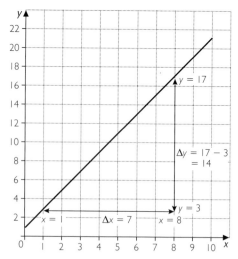

Figure 2 A graph of $y = 2x + 1$

5.3 Gradient of a linear graph

Figure 2 is a graph of the linear relationship $y = 2x + 1$.

The **gradient** (or slope) of the graph is defined as the rise of the graph (the increase in y, Δy) divided by the run (the corresponding increase in x, Δx) found by drawing a right angled triangle as shown in Figure 2. On Figure 2,

$$\Delta y = 14, \qquad \Delta x = 7,$$

$$\text{gradient} = \frac{\Delta y}{\Delta x} = 2.$$

Notice that Δy and Δx are numbers read from the graph scales, (*not* lengths measured with a ruler) and that any similar triangle drawn on the graph will give the same value of the gradient.

The gradient of a linear graph of y against x is always equal to the value m in the relationship $y = mx + c$.

The graph in Figure 2 has a positive gradient. If m is negative, then the graph slopes down from left to right.

If two variables measured in an experiment are related by a linear equation, then plotting them on a graph enables you to find the values of the constants relating them. It is helpful if you arrange the relationship so that it looks as much like $y = mx + c$ as possible. For example, by subtracting Ir from both sides you can write $\mathscr{E} = V + Ir$ as

$$V = (-r)I + \mathscr{E}$$

which can be compared directly with

$$y = mx + c$$

If you plot measured values of V on the y-axis against corresponding values of I on the x-axis, the graph will be a straight line that cuts the y-axis at $\mathscr{E}$, and with a gradient $m = -r$.

5.4 Inverse proportionality

If one quantity is **inversely proportional** to another, then as one increases, the other will decrease. For example, the acceleration produced by a given net force is inversely proportional to the mass on which it acts: doubling the mass halves the acceleration, tripling the mass divides the acceleration by three and so on – and vice versa.

Such a relationship is written using reciprocals and the symbol for direct proportion:

$$a \propto \frac{1}{m} \qquad a = \frac{k}{m}$$

Maths reference

Error bars and error boxes

See Maths note 7.5

Maths reference

Reciprocals

See Maths note 3.3

or

$$m \propto \frac{1}{a} \qquad m = \frac{k}{a}$$

(In this case, the constant of proportionality is the same as the net force F.)

If one quantity is inversely proportional to the other (Table 6), the graph of one plotted against the other is curved as in Figure 3.

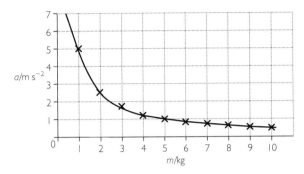

Figure 3 A graph showing how the accelertion a produced by a constant force F (= 5N) depends on mass m (data from Table 6)

m/kg	(1/m)/kg⁻¹	a/ms⁻²
1	1.000	5.00
2	0.500	2.50
3	0.333	1.67
4	0.250	1.25
5	0.200	1.00
6	0.167	0.83
7	0.143	0.71
8	0.125	0.63
9	0.111	0.55
10	0.100	0.50

Table 6 Data for Figures 3 and 4

But if one quantity is plotted against the *reciprocal* of the other, then the graph is a straight line through the origin, as shown in Figure 4.

5.5 Testing mathematical relationships

Sometimes we are interested in finding a mathematical relationship between two measured quantities. This usually involves some educated guesswork, based on ideas about the underlying physics and/or from looking at the numbers. Plotting graphs provides a way of testing the guesses.

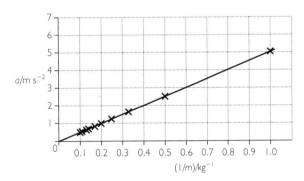

Figure 4 The data from Figure 3 plotted as a against $1/m$

Direct proportion

For example, if both quantities increase together, you might guess that one is directly proportional to the other. Plot a graph of one against the other and see whether you can draw a straight line through all the error boxes.

Examples that give straight-line graphs include:

$s \propto t$ for motion at constant speed

$I \propto V$ for an ohmic conductor.

If the plot does not give a straight line, try something else. For example, motion from rest at constant acceleration is described by the equation

$$s = \tfrac{1}{2}at^2$$
$$s \propto t^2$$

Maths reference
Experimental uncertainty
See Maths note 7.1
Error bars and error boxes
See Maths note 7.5

A graph of distance s against time t is a curve, but a graph of s against t^2 is a straight line with gradient $a/2$ or $\frac{1}{2}a$.

Sometimes you need to use the square root of a quantity to get a straight line. For example, for a simple pendulum a plot of its period T against the square root of its length l gives a straight line:

$$T \propto \sqrt{\ell}$$

Inverse proportion

If one quantity increases as the other decreases, you might guess that you are looking at inverse proportionality, so try plotting a graph using the reciprocal of one quantity.

If this does not give a straight line, try plotting the square, or the square root, of the reciprocal.

For example, suppose you measure the frequency f of the note from a plucked string of mass per unit length μ. Frequency f decreases as you increase μ, but suppose you find that a graph of f against $1/\mu$ is not a straight line.

If a graph of f against $\dfrac{1}{\mu^2}$ is a straight line, then $f \propto \dfrac{1}{\mu^2}$

If you need to plot f against $\dfrac{1}{\sqrt{\mu}}$ to get a straight line, then $f \propto \dfrac{1}{\sqrt{\mu}}$

6 Trigonometry and angular measurements

6.1 Degrees and radians

A **radian**, or **rad** for short, is a unit for measuring angles commonly used in physics instead of degrees. Figure 5 shows how the size of an angle, in radians, is defined.

For a full circle, length of arc = length of circumference = $2\pi r$.

Size of angle = $\dfrac{2\pi r}{r}$ = 2π radians, i.e. approximately 6.28 rad.

Table 7 lists some useful conversions between radians and degrees.

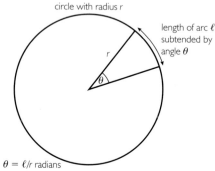

Figure 5 The size of an angle measured in radians

Angle	Size in degrees	Size in radians
full circle	360°	2π rad = 6.28 rad
half circle	180°	π rad = 3.14 rad
	114.6°	2.0 rad
quarter circle	90°	$\pi/2$ rad = 1.57 rad
	60°	$\pi/3$ rad = 1.05 rad
	57.3°	1.0 rad
	45°	$\pi/4$ rad = 0.79 rad
	30°	$\pi/6$ rad = 0.52 rad
	28.6°	0.5 rad

Table 7 Some conversions between radians and degrees

Note that π is a *number* (approximately 3.14) that frequently, but not always, appears in angles measured in radians.

6.2 Sine, cosine and tangent of an angle

Figure 6 shows a right angled triangle. The sides of the triangle are related by Pythagoras's theorem:

$$c^2 = b^2 + a^2$$
$$c = \sqrt{(a^2 + b^2)}$$

(Care! You can't 'cancel' the squares inside the bracket.)

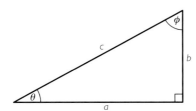

Figure 6 A right angled triangle

All similar triangles, i.e. those with the same angle θ, will have sides in the same proportion to one another. The ratios of the sides depend only on the angle θ.

The sine, cosine and tangent of the angle θ are known as **trigonometric ratios**.

- Sine of angle θ, $\sin \theta = \dfrac{\text{opposite side}}{\text{hypotenuse}} = \dfrac{b}{c}$

- Cosine of θ, $\cos \theta = \dfrac{\text{adjacent side}}{\text{hypotenuse}} = \dfrac{a}{c}$

- Tangent of θ, $\tan \theta = \dfrac{\text{opposite side}}{\text{adjacent side}} = \dfrac{b}{a}$

We can combine these to give another useful relationship. Since

$$\frac{b}{a} = \frac{b}{c} \div \frac{a}{c} \quad (c \text{ cancels}),$$

we can write

$$\tan \theta = \frac{\sin \theta}{\cos \theta}$$

Also

$$\sin \phi = \frac{a}{c} = \cos \theta \quad \text{and} \quad \cos \phi = \frac{b}{c} = \sin \theta$$

i.e. if two angles add up to 90°, then the cosine of one is equal to the sine of the other.

Using Pythagoras's theorem leads to another useful result. Dividing $c^2 = a^2 + b^2$ by c^2:

$$1 = \frac{a^2}{c^2} + \frac{b^2}{c^2} = \left(\frac{a}{c}\right)^2 + \left(\frac{b}{c}\right)^2$$

$$1 = (\cos \theta)^2 + (\sin \theta)^2,$$

which is true for any angle and is usually written as

$$\cos^2 \theta + \sin^2 \theta = 1$$

6.3 Graphs of trigonometric functions

For angles greater than 90°, Figure 7 shows how sin, cos and tan are defined. For some angles, negative numbers are involved. Figure 8 shows how the sin, cos and tan vary with angle θ. Note that we have labelled the axis in degrees and in radians.

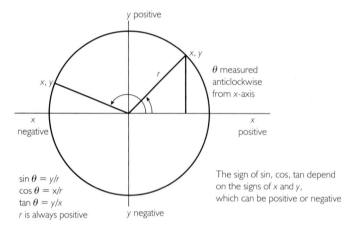

y positive

x, y

θ measured anticlockwise from *x*-axis

r

x, y

x negative

x positive

$\sin \theta = y/r$
$\cos \theta = x/r$
$\tan \theta = y/x$
r is always positive

y negative

The sign of sin, cos, tan depend on the signs of *x* and *y*, which can be positive or negative

Figure 7 Defining sin, cos and tan for angles greater than 90°

Maths reference

Degrees and radians
See Maths note 6.1

Notice that $\sin \theta$ and $\cos \theta$ are always between +1 and −1, but $\tan \theta$ is infinite for some angles (notice the different scale in Figure 8(c)).

Also notice some useful values, e.g. $\sin 30° = \cos 60° = 0.5$. Look at the values of $\sin \theta$ and $\cos \theta$ when θ is a multiple of 90°.

6.4 Inverse sin, etc.

The angle whose sin is x is written $\sin^{-1} x$. We can write the relationships from Figure 6 as

$$\theta = \sin^{-1}\frac{b}{c} \qquad \theta = \cos^{-1}\frac{a}{c} \qquad \theta = \tan^{-1}\frac{b}{a} \qquad \phi = \sin^{-1}\frac{a}{c}$$

Beware! The index −1 here does *not* indicate a reciprocal:

$$\sin^{-1} x \text{ is } not \text{ the same as } \frac{1}{\sin x}$$

6.5 Trigonometry on a calculator

You can find the sine, cosine and tangent of an angle on a calculator. For example, to find sin 30°, type 30 and press sin.

Many scientific calculators can be switched between 'degree' and 'radian' modes. The display will indicate which one you are in.

If you switch your calculator to 'radian' mode, you can find sin, etc., of angles in radians without having to convert to degrees. Check that you know how to do this.

With your calculator in radian mode, type π, ÷, 2 (you may need to press = as well) and then press sin or cos. You should get $\sin (\pi/2) = 1$, $\cos (\pi/2) = 0$. If you have your calculator in degree mode by mistake, you will find the sin or cos of 1.57° (3.14° ÷ 2).

Try finding the sin, cos and tan of some angles in degrees and in radians. Check that you get the same values as shown in Figure 8.

(a) sin θ

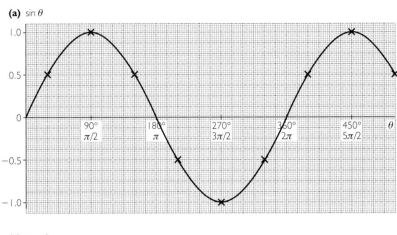

(b) cos θ

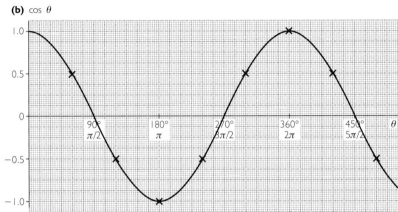

(c) tan θ

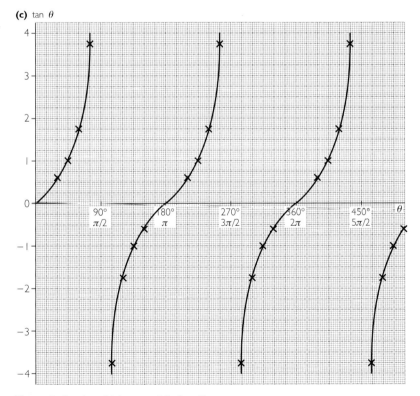

Figure 8 Graphs of trigonometric functions

If you know the sin, cos or tan of an angle and wish to determine the size of the angle, use the 'inv' key.

For example, to find the angle whose sin is 0.5, type 0.5, press inv and then press sin. You should get 30 if you have your calculator in degree mode. If you do this with your calculator in radian mode, you will get 0.5236 ($\approx \pi/6$).

6.6 The small angle approximations

There are some useful approximations involving the trigonometric ratios of small angles. These become evident when we express the sine and tangent of an angle θ in terms of the right angled triangles shown in Figure 9.

From the triangle OAC

$$\sin \theta = \frac{AC}{OC} = \frac{AC}{r}$$

and

$$\cos \theta = \frac{OA}{OC} = \frac{OA}{r}$$

From the triangle OBD

$$\tan \theta = \frac{BD}{OB} = \frac{BD}{r}$$

and

$$\cos \theta = \frac{OB}{OD} = \frac{r}{OB}$$

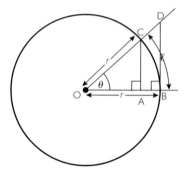

Figure 9 The sine and tangent of an angle

Figure 9 shows that $\tan \theta$ is always greater than $\sin \theta$ because BD is greater than AC.

As θ is made smaller, the lines AC and BD become closer together and more equal in length, and the lines OA and OD become closer to r, so *for small angles*:

$$\sin \theta \approx \tan \theta$$

and

$$\cos \theta \approx 1$$

With your calculator in degree mode, try finding the sin, cos and tan of the angles listed in Table 7, and some smaller angles. Notice that the approximations get better as the angles get smaller.

Small angles in radians

Comparison with Figure 5 shows that the size of θ *measured in radians* lies between $\sin \theta$ and $\tan \theta$ (the arc length ℓ is longer than AC and shorter than OD):

$$\sin \theta < \theta < \tan \theta$$

When θ is small,

$$AC \approx \ell \approx BD$$

and so *for small angles measured in radians* we have some additional approximations:

$$\sin \theta \approx \theta$$

and

$$\tan \theta \approx \theta$$

Switch your calculator into radian mode, and again try finding the sin, cos and tan of various angles. Notice that the approximation gets better at small angles.

7 Size and precision

7.1 Precision in measurements; experimental uncertainty

In any measurement, there is a limit to the precision of your result. Sometimes this **experimental uncertainty** arises because you get different answers when you repeat the measurement. For example, if you time an athlete running 100 metres, the same athlete will probably record different times on different occasions. The uncertainty in the measurements is indicated by the 'scatter' in the results.

For example, suppose a certain athlete records times of 12.5 s, 12.1 s, 12.6 s, 12.5 s and 12.3 s. The average time is $t = 12.4$ s. The difference between the average and the biggest or smallest value indicates the uncertainty Δt – in this case, $\Delta t \approx 0.3$ s.

Sometimes the uncertainty arises because it is difficult to judge exactly what to measure. For example, if you are using a signal generator to produce a sound from a speaker and ajusting it to give the same pitch as a note from a guitar, it might be hard to judge the frequency exactly. If you think the best frequency f is 260 Hz, but are unsure by about 10 Hz either way, then the uncertaintly would be $\Delta f \approx 10$ Hz.

Even if there is no problem deciding exactly what to measure, and you get the same answer each time you repeat the measurement, there is still an uncertainty because the measurement is limited by the instrument you are using. For example, if you use a digital ammeter to measure a current I, and you get 0.357 A each time, you can only be sure that the current is closer to 0.357 A than it is to either 0.356 A or 0.358 A – it could lie anywhere between 0.3565 A and 0.3575 A. So the uncertainty is $\Delta I \approx 0.0005$ A.

Some books refer to **experimental error** rather than uncertainty. Don't be misled into thinking that they mean a mistake. However carefully and correctly you carry out a measurement, there will always be an uncertainty.

Experimental uncertainties apply to *all* measured quantities – including those you look up in a data book, though these values have usually been measured with much greater precision than you can achieve in a school or college laboratory.

7.2 Calculations with uncertainties; significant figures

If you carry out a calculation using a measured value, there will always be an uncertainty in your answer. You can use the uncertainties in the measurements to work out the uncertainty in the calculated value.

For example, suppose you measure a current of $I = 0.24$ A $\pm$ 0.01 A and a corresponding pd of $V = 0.67$ V $\pm$ 0.02 V.

On a calculator, the resistance found using the 'best' values is

$$R_{best} = \frac{V}{I} = \frac{0.67V}{0.24A} = 2.791\ 6667\ \Omega$$

But, using the largest possible V (0.39 V) and the smallest possible I (0.23 A), the calculated resistance could be as large as

$$R_{max} = \frac{0.69V}{0.23A} = 3\ \Omega$$

Or, using the smallest V and the largest I, it could be as small as

$$R_{min} = \frac{0.65V}{0.25A} = 2.6\ \Omega$$

There are several things to notice! First, there are quite large differences between the three values. Second, the first value extends to the full length of the calculator display, whereas the others do not.

The large differences show that you cannot *possibly* say that the resistance is precisely 2.791 6667 Ω. This value is close to 2.8 Ω, and the other two differ by 0.2 Ω in either direction, i.e. the uncertainty in R is $\Delta R \approx 0.2\ \Omega$. The resistance can therefore be written as

$$R = 2.8\ \Omega \pm 0.2\ \Omega.$$

The second figure in this answer (the 8 after the decimal point) is uncertain, and so any further figures are meaningless.

Another way of putting this is to say that the answer has (only) two **significant figures** – the one before the decimal point and the first one after it. The rest of the figures in the original 'best' answer are meaningless. They are *not* significant.

7.3 A useful rule of thumb

In a calculation, the answer cannot be known any more precisely than the values used to calculate it. As a useful rule of thumb, the final answer has no more **significant figures** than the *least* precise value used in the calculation. (The example in Maths note 7.2 illustrates this.)

Suppose you did a calculation to find the frequency f of light whose wavelength is 468 nm (4.68×10^{-7} m). The speed of light is known very precisely: $2.997\ 925 \times 10^8$ m s^{-1}.

Using speed $\div$ wavelength

$$f = \frac{2.997\ 925 \times 10^8 \text{ m s}^{-1}}{4.68 \times 10^{-7} \text{ m}}$$

$$= 6.4058 \times 10^{14} \text{ Hz}$$

However, we only knew the wavelength to three significant figures, so we cannot quote the frequency of this precisely. We must stick to the three significant figures and write

$$f = 6.41 \times 10^{14} \text{ Hz}$$

There was in fact no point in using the very precise value for the speed of light. Values listed in data books are often rounded to, say, three significant figures if they are likely to be used only in calculations requiring this precision or less.

7.4 Significant figures and orders of magnitude

The speed of light to seven significant figures is $2.997\ 925 \times 10^8$ m s^{-1}; the significant figures are 2997925.

Zeros in front of a number are not significant. The speed of light could be written (rather oddly) as $002.997\ 925 \times 10^8$ m s^{-1} or $0.000\ 299\ 7925 \times 10^{12}$ m s^{-1} without making any difference to its value.

However, zeros at the end of a number are (or at least can be!) significant. If you wrote the speed of light as 299 792 500 m s⁻¹, that would imply that you knew that the last two figures were definitely zeros and not some other numbers. If they are, in fact, not known, it is better to use standard form so that the meaningless zeros can be dropped.

To five significant figures, the speed of light would be 2.9979×10^8 m s⁻¹. To three significant figures, it would be 3.00×10^8 m s⁻¹. Here the zeros *are* significant and should be written down, because 2.997... rounds to 3.00.

To one significant figure the speed of light would be 3×10^8 m s⁻¹.

If a value is rounded to just the nearest power of 10, then we say we are giving just the **order of magnitude**. Two values are said to have the same order of magnitude if one is between 1 and 10 times the other. For example, the wavelengths of red and blue light (about 400 nm and 700 nm) are within the same order of magnitude. But the wavelengths of infrared radiation range from about 10^{-6} m to about 10^{-3} m – they cover three orders of magnitude.

7.5 Error bars and error boxes

When plotting a graph of experimental data, you should take account of the uncertainties. Rather than representing each measurement by a point, you should draw an **error bar** to represent the range of possible values. Then use the vertical and horizontal error bars to draw an **error box** around each plotted point. Once you have plotted the error boxes, you can then draw a trend line on your graph. It might be possible to draw a straight line passing through all the boxes, even if you could not draw one through all the points.

8 Logarithms

8.1 Logs and powers of 10

If a number can be written as *just* a 'power of 10', then the power is the **logarithm** of that number; strictly speaking, it is the **logarithm to base 10**, or **common logarithm**, of the number, but it is often simply called the **log**.

Table 8 lists some examples using whole-number powers.

Number x	$\log_{10}(x)$
$100\,000 = 10^5$	5
$10\,000 = 10^4$	4
$1000 = 10^3$	3
$100 = 10^2$	2
$10 = 10^1$	1
$1 = 10^0$	0
$0.1 = 10^{-1}$	−1
$0.001 = 10^{-2}$	−2

Table 8 Some numbers and their common logarithms

In fact *any positive number* can be expressed as a power of 10, using powers that are not whole numbers. Most whole numbers have logs that are not themselves whole numbers or simple fractions. For example:

$$10^{0.6021} = 4.000$$

so

$$\log_{10}(4.000) = 0.6021$$

All numbers between 1 and 10 have logs that lie between 0 and 1. For example:

$$10^{0.333} = 2.153$$

so

$$\log_{10}(2.513) = 0.333$$

Similarly, all numbers between 10 and 100 have logs that lie between 1 and 2; all numbers between 100 and 1000 have logs between 2 and 3, and so on.

All numbers less than 1 have negative logs. For example:

$$\log_{10}(0.5) = -0.3010$$
$$\log_{10}(0.1) = -1.000$$

8.2 Logs on a calculator

To find the common log of a number using a calculator, type in the number and then press the key marked log or lg.

This process can be reversed to find the **antilog** of a number. Type in the log whose number you want to find, then press the keys marked INV and log (or lg). By doing this, you can show that 4.000 is the antilog of 0.6021, and 2.513 is the antilog of 0.333.

Notice that using the INV and log keys to find the antilog of a number x gives exactly the same result as using the y^x key to find 10^x.

> **Maths reference**
>
> Powers that are not whole numbers.
>
> See Maths note 1.5

Index